Mechanical Engineering Series

Frederick F. Ling
Series Editor

Springer
New York
Berlin
Heidelberg
Hong Kong
London
Milan
Paris
Tokyo

Mechanical Engineering Series

J. Angeles, **Fundamentals of Robotic Mechanical Systems: Theory, Methods, and Algorithms, 2nd ed.**

P. Basu, C. Kefa, and L. Jestin, **Boilers and Burners: Design and Theory**

J.M. Berthelot, **Composite Materials: Mechanical Behavior and Structural Analysis**

I.J. Busch-Vishniac, **Electromechanical Sensors and Actuators**

J. Chakrabarty, **Applied Plasticity**

G. Chryssolouris, **Laser Machining: Theory and Practice**

V.N. Constantinescu, **Laminar Viscous Flow**

G.A. Costello, **Theory of Wire Rope, 2nd ed.**

K. Czolczynski, **Rotordynamics of Gas-Lubricated Journal Bearing Systems**

M.S. Darlow, **Balancing of High-Speed Machinery**

J.F. Doyle, **Nonlinear Analysis of Thin-Walled Structures: Statics, Dynamics, and Stability**

J.F. Doyle, **Wave Propagation in Structures: Spectral Analysis Using Fast Discrete Fourier Transforms, 2nd ed.**

P.A. Engel, **Structural Analysis of Printed Circuit Board Systems**

A.C. Fischer-Cripps, **Introduction to Contact Mechanics**

A.C. Fischer-Cripps, **Nanoindentation**

J. García de Jalón and E. Bayo, **Kinematic and Dynamic Simulation of Multibody Systems: The Real-Time Challenge**

W.K. Gawronski, **Advanced Structural Dynamics and Active Control of Structures**

W.K. Gawronski, **Dynamics and Control of Structures: A Modal Approach**

K.C. Gupta, **Mechanics and Control of Robots**

J. Ida and J.P.A. Bastos, **Electromagnetics and Calculations of Fields**

M. Kaviany, **Principles of Convective Heat Transfer, 2nd ed.**

M. Kaviany, **Principles of Heat Transfer in Porous Media, 2nd ed.**

(continued after index)

Wodek K. Gawronski

Advanced Structural Dynamics and Active Control of Structures

With 157 Figures

Springer

Wodek K. Gawronski
Jet Propulsion Laboratory
California Institute of Technology
Pasadena, CA 91109, USA
wodek.k.gawronski@jpl.nasa.gov

Series Editor
Frederick F. Ling
Ernest F. Gloyna Regents Chair in Engineering, Emeritus
Department of Mechanical Engineering
The University of Texas at Austin
Austin, TX 78712-1063, USA
and
William Howard Hart Professor Emeritus
Department of Mechanical Engineering,
Aeronautical Engineering and Mechanics
Rensselaer Polytechnic Institute
Troy, NY 12180-3590, USA

Library of Congress Cataloging-in-Publication Data
Gawronski, Wodek, 1944–
Advanced structural dynamics and active control of structures/Wodek Gawronski.
p. cm. — (Mechanical engineering series)
ISBN 0-387-40649-2 (alk. paper)
1. Structural dynamics. 2. Structural control (Engineering) I. Title. II. Mechanical engineering series (Berlin, Germany)
TA654.G36 2004
624.1′71—dc22 2003058443

Based on *Dynamics and Control of Structures: A Modal Approach*, by Wodek K. Gawronski, © 1998 Springer-Verlag New York, Inc.

ISBN 0-387-40649-2 Printed on acid-free paper.

Printed in the United States of America.

9 8 7 6 5 4 3 2 1 SPIN 10943243

www.springer-ny.com

Springer-Verlag New York Berlin Heidelberg
A member of BertelsmannSpringer Science+Business Media GmbH

To my friends—

Jan Kruszewski and
Hans Günther Natke

—scholars of dedication and imagination

Although this may seem a paradox,
all exact science is dominated by the idea of approximation.
—*Bertrand Russell*

Mechanical Engineering Series

Frederick F. Ling
Series Editor

The Mechanical Engineering Series features graduate texts and research monographs to address the need for information in contemporary mechanical engineering, including areas of concentration of applied mechanics, biomechanics, computational mechanics, dynamical systems and control, energetics, mechanics of materials, processing, production systems, thermal science, and tribology.

Series Preface

Mechanical engineering, an engineering discipline forged and shaped by the needs of the industrial revolution, is once again asked to do its substantial share in the call for industrial renewal. The general call is urgent as we face profound issues of productivity and competitiveness that require engineering solutions. The Mechanical Engineering Series features graduate texts and research monographs intended to address the need for information in contemporary areas of mechanical engineering.

The series is conceived as a comprehensive one that covers a broad range of concentrations important to mechanical engineering graduate education and research. We are fortunate to have a distinguished roster of consulting editors on the advisory board, each an expert in one of the areas of concentration. The names of the consulting editors are listed on the facing page of this volume. The areas of concentration are applied mechanics, biomechanics, computational mechanics, dynamical systems and control, energetics, mechanics of materials, processing, production systems, thermal science, and tribology.

Frederick F. Ling
New York, New York

Preface

Science is for those who learn;
poetry for those who know.
—Joseph Roux

This book is a continuation of my previous book, *Dynamics and Control of Structures* [44]. The expanded book includes three additional chapters and an additional appendix: Chapter 3, "Special Models"; Chapter 8, "Modal Actuators and Sensors"; and Chapter 9, "System Identification." Other chapters have been significantly revised and supplemented with new topics, including discrete-time models of structures, limited-time and -frequency grammians and reduction, almost-balanced modal models, simultaneous placement of sensors and actuators, and structural damage detection. The appendices have also been updated and expanded. Appendix A consists of thirteen new Matlab programs. Appendix B is a new addition and includes eleven Matlab programs that solve examples from each chapter. In Appendix C model data are given.

Several books on structural dynamics and control have been published. Meirovitch's textbook [108] covers methods of structural dynamics (virtual work, d'Alambert's principle, Hamilton's principle, Lagrange's and Hamilton's equations, and modal analysis of structures) and control (pole placement methods, LQG design, and modal control). Ewins's book [33] presents methods of modal testing of structures. Natke's book [111] on structural identification also contains excellent material on structural dynamics. Fuller, Elliot, and Nelson [40] cover problems of structural active control and structural acoustic control. Inman's book [79] introduces the basic concepts of vibration control, while Preumont in [120] presents modern approaches to structural control, including LQG controllers, sensors, and actuator placement, and piezoelectric materials with numerous applications in aerospace and civil engineering. The Junkins and Kim book [87] is a graduate-level textbook, while the Porter and Crossley book [119] is one of the first books on modal control. Skelton's work [125] (although on control of general linear systems) introduces methods designed specifically for the control of flexible structures. For example, the component cost approach to model or controller reduction is a tool frequently used in this field. The monograph by Joshi [83] presents developments on dissipative and LQG controllers supported by numerous applications. Genta's book [65] includes rotor dynamics; the book by Kwon and Bang [92] is dedicated mainly to structural finite-element models, but a part of it is dedicated to structural dynamics and control. The work by Hatch [70] explains vibrations and dynamics problems in practical ways, is illustrated with numerous examples, and supplies Matlab programs to solve vibration problems. The Maia and Silva book [107] is a study on modal analysis and testing, while the Heylen, Lammens, and Sas book [71] is an up-to-date and attractive presentation of modal analysis. The De Silva book [26] is a comprehensive source on vibration analysis and testing. Clark, Saunders, and Gibbs [17] present recent developments in dynamics and control of structures; and Elliott [31] applies structural dynamics and control problems to acoustics. My book [47] deals with structural dynamics and control problems in balanced coordinates. The recent advances in structural dynamics and control can be found in [121].

This book describes comparatively new areas of structural dynamics and control that emerged from recent developments. Thus:

- State-space models and modal methods are used in structural dynamics as well as in control analysis. Typically, structural dynamics problems are solved using second-order differential equations.
- Control system methods (such as the state-space approach, controllability and observability, system norms, Markov parameters, and grammians) are applied to solve structural dynamics problems (such as sensor and actuator placement, identification, or damage detection).
- Structural methods (such as modal models and modal independence) are used to solve control problems (e.g., the design of LQG and H_∞ controllers), providing new insight into well-known control laws.
- The methods described are based on practical applications. They originated from developing, testing, and applying techniques of structural dynamics, identification, and control to antennas and radiotelescopes. More on the dynamics and control problems of the NASA Deep Space Network antennas can be found at http://tmo.jpl.nasa.gov/tmo/progress_report/.
- This book uses approximate analysis, which is helpful in two ways. First, it simplifies analysis of large structural models (e.g., obtaining Hankel singular values for a structure with thousands of degrees of freedom). Second, approximate values (as opposed to exact ones) are given in closed form, giving an opportunity to conduct a parametric study of structural properties.

This book requires introductory knowledge of structural dynamics and of linear control; thus it is addressed to the more advanced student. It can be used in graduate courses on vibration and structural dynamics, and in control system courses with application to structural control. It is also useful for engineers who deal with structural dynamics and control.

Readers who would like to contact me with comments and questions are invited to do so. My e-mail address is Wodek.K.Gawronski@jpl.nasa.gov. Electronic versions of Matlab programs from Appendix A, examples from Appendix B, and data from Appendix C can also be obtained from this address.

I would like to acknowledge the contributions of my colleagues who have had an influence on this work: Kyong Lim, NASA Langley Research Center (sensor/actuator placement, filter design, discrete-time grammians, and H_∞ controller analysis); Hagop Panossian, Boeing North American, Inc., Rocketdyne (sensor/actuator placement of the International Space Station structure); Jer-Nan Juang, NASA Langley Research Center (model identification of the Deep Space Network antenna); Lucas Horta, NASA Langley Research Center (frequency-dependent grammians for discrete-time systems); Jerzy Sawicki, Cleveland State University (modal error estimation of nonproportional damping); Abner Bernardo, Jet Propulsion Laboratory, California Institute of Technology (antenna data collection); and Angel Martin, the antenna control system supervisor at the NASA Madrid Deep Space Communication Complex (Spain) for his interest and encouragement. I thank Mark Gatti, Scott Morgan, Daniel Rascoe, and Christopher Yung, managers at the Communications Ground Systems Section, Jet Propulsion Laboratory, for their support of the Deep Space Network antenna study, some of which is included in this book. A portion of the research described in this book was carried out at the Jet Propulsion Laboratory, California Institute of Technology, under contract with the National Aeronautics and Space Administration.

Wodek K. Gawronski
Pasadena, California
January 2004

Contents

List of Symbols

Each equation in the book
... would halve the sales.
—*Stephen Hawking*

General

A^T	transpose of matrix A
A^*	complex-conjugate transpose of matrix A
A^{-1}	inverse of square nonsingular matrix A
$\mathrm{tr}(A)$	trace of a matrix A, $\mathrm{tr}(A) = \sum_i a_{ii}$
$\|A\|_2$	Euclidean (Frobenius) norm of a real-valued matrix A: $\|A\|_2 = \sqrt{\sum_{i,j} a_{ij}^2} = \sqrt{\mathrm{tr}(A^T A)}$
$\mathrm{diag}(a_i)$	diagonal matrix with elements a_i along the diagonal
$\mathrm{eig}(A)$	eigenvalue of a square matrix A
$\lambda_i(A)$	ith eigenvalue of a square matrix A
$\lambda_{\max}(A)$	maximal eigenvalue of a square matrix A
$\sigma_i(A)$	ith singular value of a matrix A
$\sigma_{\max}(A)$	maximal singular value of a matrix A
I_n	identity matrix, $n \times n$
$0_{n\times m}$	zero matrix, $n \times m$

Linear Systems

(A,B,C,D)	quadruple of the system state-space representation
(A,B,C)	triple of the system state-space representation
(A_d,B_d,C_d)	discrete-time state-space representation
$(A_{lqg},B_{lqg},C_{lqg})$	LQG controller state-space representation
$(A_\infty,B_\infty,C_\infty)$	H_∞ controller state-space representation
(A_o,B_o,C_o)	closed-loop state-space representation
G	transfer function
G_d	discrete-time transfer function
H_1	Hankel matrix
H_2	shifted Hankel matrix
h_k	kth Markov parameter
U	input measurement matrix
Y	output measurement matrix
x	system state
x_e	system estimated state
u	system (control) input
y	system (measured) output
z	performance output
w	disturbance input
B_1	matrix of disturbance inputs
B_2	matrix of control inputs
C_1	matrix of performance outputs
C_2	matrix of measured outputs
$\|G\|_2$	continuous-time system H_2 norm
$\|G\|_\infty$	continuous-time system H_∞ norm
$\|G\|_h$	continuous-time system Hankel norm
$\|G_d\|_2$	discrete-time system H_2 norm
$\|G_d\|_\infty$	discrete-time system H_∞ norm
$\|G_d\|_h$	discrete-time system Hankel norm
$\mathcal{C}$	controllability matrix
$\mathcal{O}$	observability matrix
W_c	controllability grammian
W_o	observability grammian
γ_i	ith Hankel singular value
$\gamma_{\max}$	the largest Hankel singular value of a system
Γ	matrix of Hankel singular values
CARE	controller algebraic Riccati equation
FARE	filter (or estimator) algebraic Riccati equation
HCARE	H_∞ controller algebraic Riccati equation

HFARE	H_∞ filter (or estimator) algebraic Riccati equation
S_c	solution of CARE
S_e	solution of FARE
$S_{\infty c}$	solution of HCARE
$S_{\infty e}$	solution of HFARE
μ_i	ith LQG singular value
$\mu_{\infty i}$	ith H_∞ singular value
M	matrix of the LQG singular values, $M = \mathrm{diag}(\mu_i)$
M_∞	matrix of the H_∞ singular values, $M_\infty = \mathrm{diag}(\mu_{\infty i})$
ρ	parameter of the H_∞ controller
K_c	controller gain
K_e	estimator gain
ε	tracking error
t	time sequence
Δt	sampling time
N	number of states
s	number of inputs
r	number of outputs

Structures

D	damping matrix
K	stiffness matrix
M	mass matrix
D_m	modal damping matrix
K_m	modal stiffness matrix
M_m	modal mass matrix
q	structural displacement (nodal)
q_m	structural displacement (modal)
q_{ab}	structural displacement (almost-balanced)
q_i	displacement of the ith degree of freedom
q_{mi}	displacement of the ith mode
q_{abi}	displacement of the ith almost-balanced mode
ϕ_i	ith structural mode
ϕ_{abi}	almost-balanced ith structural mode
Φ	modal matrix
Φ_{ab}	almost-balanced modal matrix
ω_i	ith natural frequency
Ω	matrix of natural frequencies

ζ_i	ith modal damping
Z	matrix of modal damping coefficients
B_o	nodal input matrix
C_{oq}	nodal displacement output matrix
C_{ov}	nodal velocity output matrix
B_m	modal input matrix
C_{mq}	modal displacement output matrix
C_{mv}	modal velocity output matrix
C_m	modal output matrix, $C_m = C_{mq}\Omega^{-1} + C_{mv}$
b_{mi}	input matrix of the ith mode, ith row of B_m
c_{mi}	output matrix of the ith mode, ith column of C_m
$\Vert B_m \Vert_2$	modal input gain
$\Vert C_m \Vert_2$	modal output gain, $\Vert C_m \Vert_2^2 = \Vert C_{mq}\Omega^{-1} \Vert_2^2 + \Vert C_{mv} \Vert_2^2$
$\Vert b_{mi} \Vert_2$	input gain of the ith mode
$\Vert c_{mi} \Vert_2$	output gain of the ith mode
$\Delta\omega_i$	ith half-power frequency, $\Delta\omega_i = 2\zeta_i\omega_i$
σ_{2ij}	H_2 placement index for the ith actuator (sensor) and the kth mode
$\sigma_{\infty ij}$	H_∞ placement index for the ith actuator (sensor) and the kth mode
Σ_2	H_2 placement matrix
Σ_∞	H_∞ placement matrix
$I(k)$	membership index of the kth sensor
β_i	pole shift factor
n_d	number of degrees of freedom
n	number of modes
N	number of states
s	number of inputs
r	number of outputs
S	number of candidate actuator locations
R	number of candidate sensor locations

1
Introduction to Structures

↳ *examples, definition, and properties*

A vibration is a motion
that can't make up its mind
which way it wants to go.
—*From Science Exam*

Flexible structures in motion have specific features that are not a secret to a structural engineer. One of them is resonance—strong amplification of the motion at a specific frequency, called natural frequency. There are several frequencies that structures resonate at. A structure movement at these frequencies is harmonic, or sinusoidal, and remains at the same pattern of deformation. This pattern is called a mode shape, or mode. The modes are not coupled, and being independent they can be excited separately. More interesting, the total structural response is a sum of responses of individual modes. Another feature—structural poles—are complex conjugate. Their real parts (representing modal damping) are typically small, and their distance from the origin is the natural frequency of a structure.

1.1 Examples

In this book we investigate several examples of flexible structures. This includes a simple structure, composed of three lumped masses, a two-dimensional (2D) truss and a three-dimensional (3D) truss, a beam, the Deep Space Network antenna, and the International Space Station structure. They represent different levels of complexity.

1.1.1 A Simple Structure

A three-mass system—a simple structure—is used mainly for illustration purposes, and to make examples easy to follow. Its simplicity allows for easy analysis, and for straightforward interpretation. Also, solution properties and numerical data can be displayed in a compact form.

The system is shown in Fig. 1.1. In this figure m_1, m_2, and m_3 represent system masses, k_1, k_2, k_3, and k_4, are stiffness coefficients, while d_1, d_2, d_3, and d_4, are damping coefficients. This structure has six states, or three degrees of freedom.

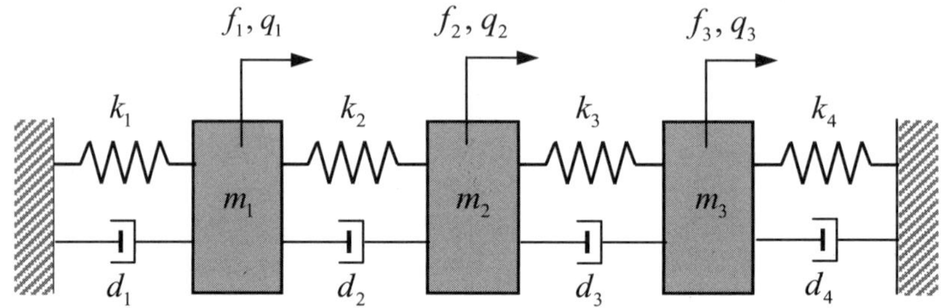

Figure 1.1. A simple structure.

1.1.2 A 2D Truss

The truss structure in Fig. 1.2 is a more complex example of a structure, which can still easily be simulated by the reader, if necessary. For this structure, l_1=15 cm, l_2=20 cm are dimensions of truss components. Each truss has a cross-sectional area of 1 cm^2, elastic modulus of 2.0×10^7 N/cm^2, and mass density of 0.00786 kg/cm^3. This structure has 32 states (or 16 degrees of freedom). Its stiffness and mass matrices are given in Appendix C.1.

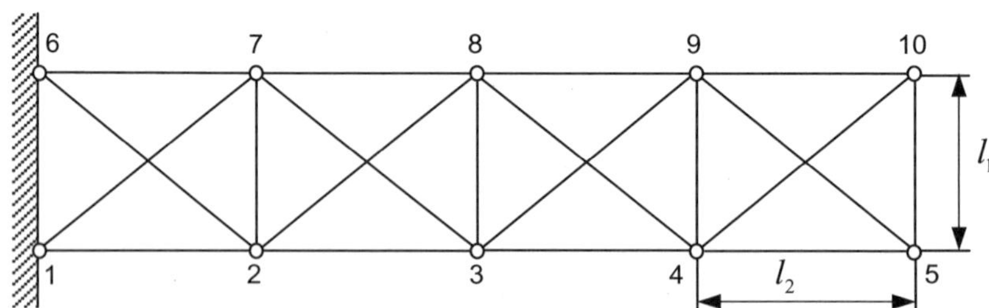

Figure 1.2. A 2D truss structure.

1.1.3 A 3D Truss

A 3D truss is shown in Fig. 1.3. For this truss, the length is 60 cm, the width 8 cm, the height 10 cm, the elastic modulus is 2.1×10^7 N/cm^2, and the mass density is 0.00392 kg/cm^3. The truss has 72 degrees of freedom (or 144 states).

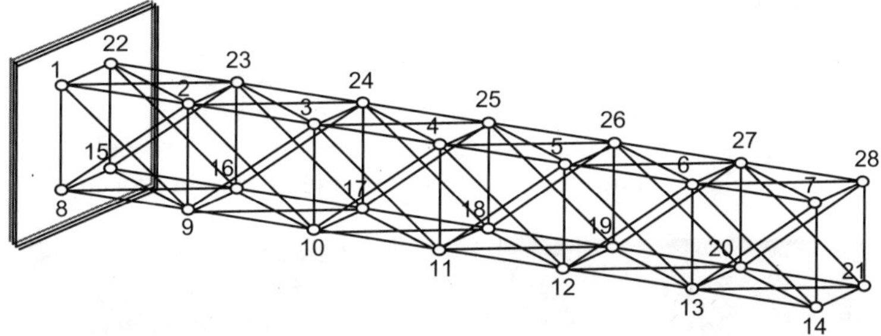

Figure 1.3. A 3D truss structure.

1.1.4 A Beam

A clamped beam is shown in Fig. 1.4. It is divided into n elements, with n–1 nodes, and two fixed nodes. In some cases later in this book we use n=15 elements for simple illustration, and sometimes n=60 or n=100 elements for more sophisticated examples of beam dynamics. Each node has three degrees of freedom: horizontal displacement, x, vertical displacement, y, and in plane rotation, θ. In total it has 3(n–1) degrees of freedom. The beam is 150 cm long, with a cross-section of 1 cm^2. The external (filled) nodes are clamped. The beam mass and stiffness matrices for n=15 are given in Appendix C.2.

Figure 1.4. A beam divided into n finite elements.

1.1.5 The Deep Space Network Antenna

The NASA Deep Space Network antenna structure illustrates a real-world flexible structure. The Deep Space Network antennas, operated by the Jet Propulsion Laboratory, consist of several antenna types and are located at Goldstone (California), Madrid (Spain), and Canberra (Australia). The Deep Space Network serves as a communication tool for space exploration. A new generation of Deep Space Network antenna with a 34-m dish is shown in Fig. 1.5. This antenna is an articulated large flexible structure, which can rotate around azimuth (vertical) and elevation (horizontal) axes. The rotation is controlled by azimuth and elevation servos, as shown in Fig. 1.6. The combination of the antenna structure and its azimuth and elevation drives is the open-loop model of the antenna. The open-loop plant has two inputs (azimuth and elevation rates) and two outputs (azimuth and elevation position), and the position loop is closed between the encoder outputs and the rate inputs. The drives consist of gearboxes, electric motors, amplifiers, and tachometers. For more details about the antenna and its control systems, see [59] and [42], or visit the web page http://ipnpr.jpl.nasa.gov/. The finite-element model of the antenna structure consists of about 5000 degrees of freedom, with some nonlinear properties (dry friction, backlash, and limits imposed on its rates, and accelerations). However, the model of the structure and the drives used in this book are linear, and are obtained from the field test data using system identification procedures.

Figure 1.5. The Deep Space Network antenna at Goldstone, California (courtesy of NASA/JPL/Caltech, Pasadena, California). It can rotate with respect to azimuth (vertical) axis, and the dish with respect to elevation (horizontal axis).

In the following we briefly describe the field test. We tested the antenna using a white noise input signal of sampling frequency 30.6 Hz, as shown in Fig. 1.7(a). The antenna elevation encoder output record is shown in Fig. 1.7(b). From these records we determined the transfer function, from the antenna rate input to the encoder output, see Fig. 1.8(a),(b), dashed line. Next, we used the Eigensystem Realization Algorithm (ERA) identification algorithm (see [84], and Chapter 9 of this book) to determine the antenna state-space representation. For this representation we obtained the plot of the transfer function plot as shown in Fig. 1.8(a),(b), solid line. The plot displays good coincidence between the measured and identified transfer function.

The flexible properties are clearly visible in the identified model. The identified state-space representation of the antenna model is given in Appendix C.3.

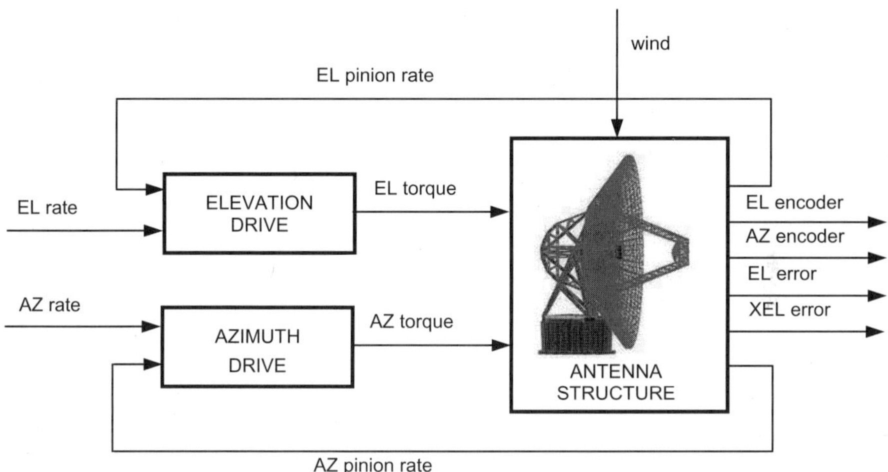

Figure 1.6. The open-loop model of the Deep Space Network antenna (AZ = azimuth, EL = elevation, XEL = cross-elevation): The AZ and EL positions are measured with encoders, EL and XEL errors are RF beam pointing errors.

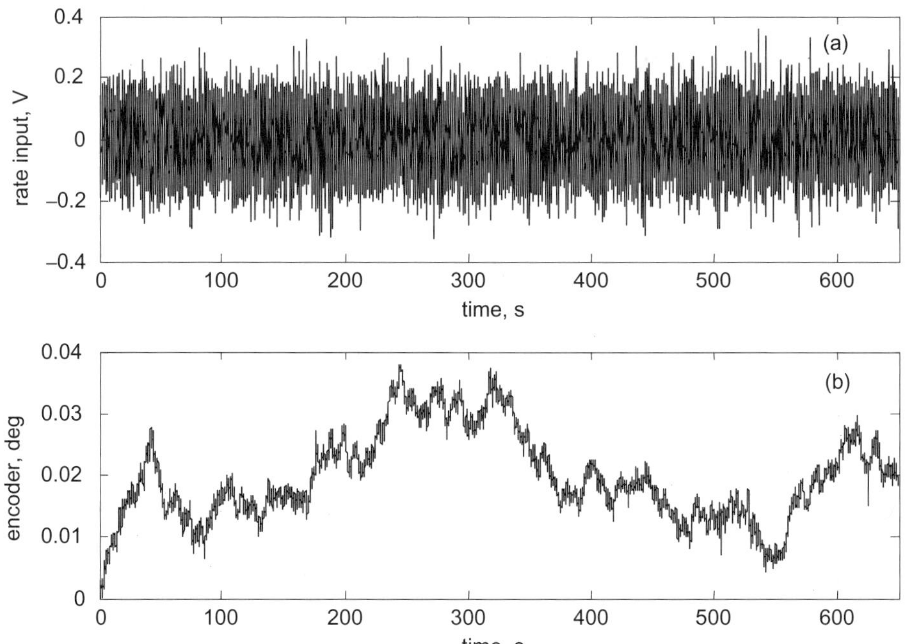

Figure 1.7. Signals in the identification of the antenna model: (a) Input white noise (voltage); and (b) output–antenna position measured by the encoder.

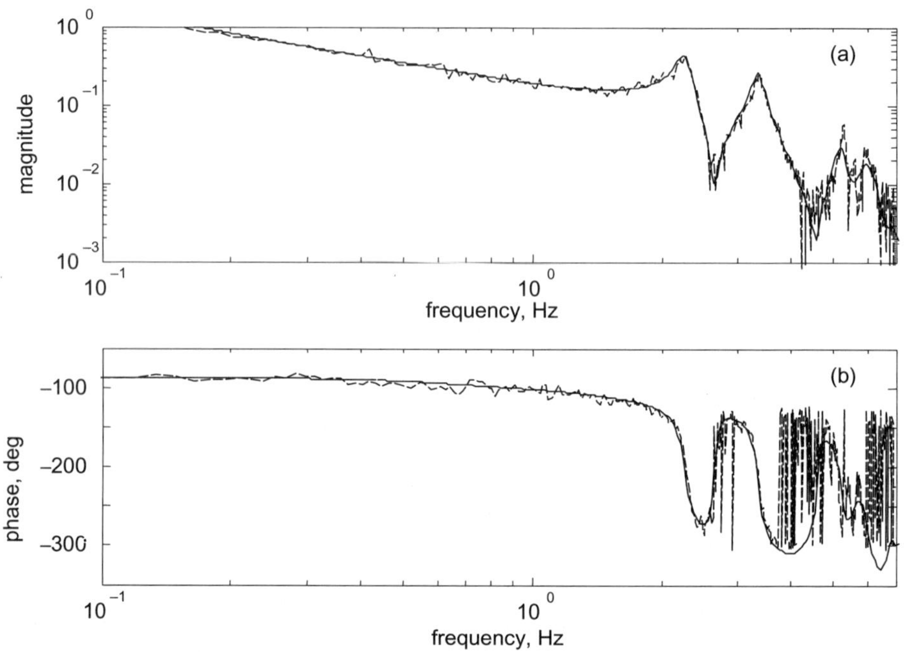

Figure 1.8. The antenna transfer functions obtained from the data (dashed line), and obtained from the identified model (solid line): (a) Magnitude; and (b) phase.

1.1.6 The International Space Station Structure

The Z1 module of the International Space Station structure is a large structure of a cubical shape with a basic truss frame, and with numerous appendages and attachments such as control moment gyros and a cable tray. Its finite-element model is shown in Fig. 1.9. The total mass of the structure is around 14,000 kg. The finite-element model of the structure consists of 11,804 degrees of freedom with 56 modes, of natural frequencies below 70 Hz. This structure was analyzed for the preparation of the modal tests. The determination of the optimal locations of shakers and accelerometers is presented in Chapter 7.

1.2 Definition

The term *flexible structure* or, briefly, *structure* has different interpretations and definitions, depending on source and on application. For the purposes of this book we define a structure as a linear system, which is

- finite-dimensional;
- controllable and observable;
- its poles are complex with small real parts; and
- its poles are nonclustered.

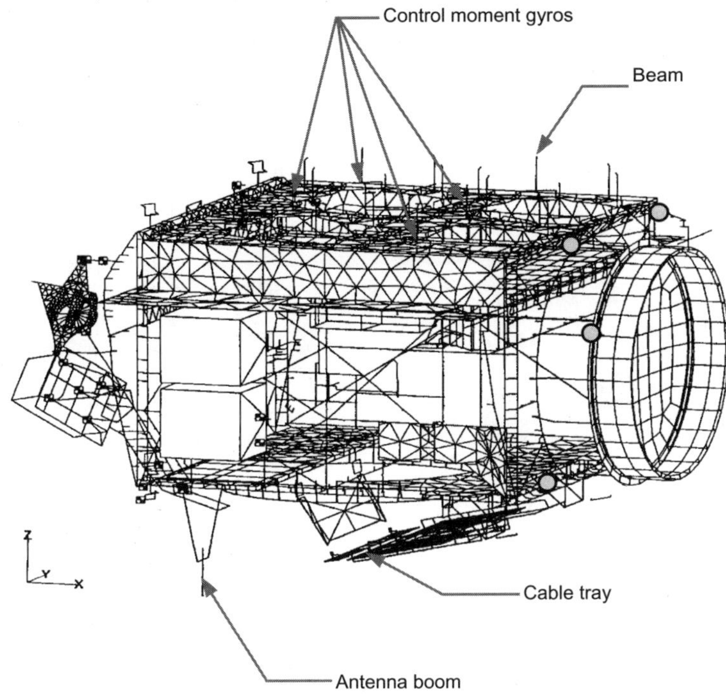

Figure 1.9. The finite-element model of the International Space Station structure.

Based on this definition, we derive many interesting properties of structures and their controllers later in this book.

The above conditions are somehow restrictive, and introduced to justify the mathematical approach used in this book. However, our experience shows that even if these conditions are violated or extended the derived properties still hold. For example, for structures with heavy damping (with larger real parts of complex poles), or with some of the poles close to each other, the analysis results in many cases still apply.

1.3 Properties

In this section we briefly describe the properties of flexible structures. The properties of a typical structure are illustrated in Fig. 1.10.

- Motion of a flexible structure can be described in independent coordinates, called modes. One can excite a single mode without excitation of the remaining ones. Displacement of each point of structure is sinusoidal of fixed frequency. The shape of modal deformation is called a modal shape, or mode. The frequency of modal motion is called natural frequency.
- Poles of a flexible structure are complex conjugate, with small real parts; their locations are shown in Fig. 1.10(a).

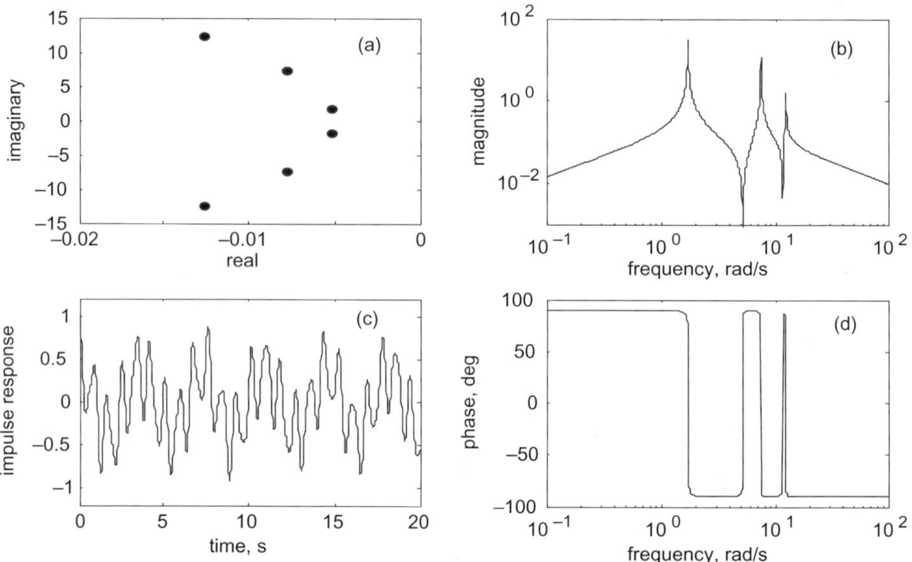

Figure 1.10. Properties of a typical flexible structure: (a) Poles are complex with small real parts; (b) magnitude of a transfer function shows resonant peaks; (c) impulse response is composed of harmonic components; and (d) phase of a transfer function displays 180 deg shifts at resonant frequencies.

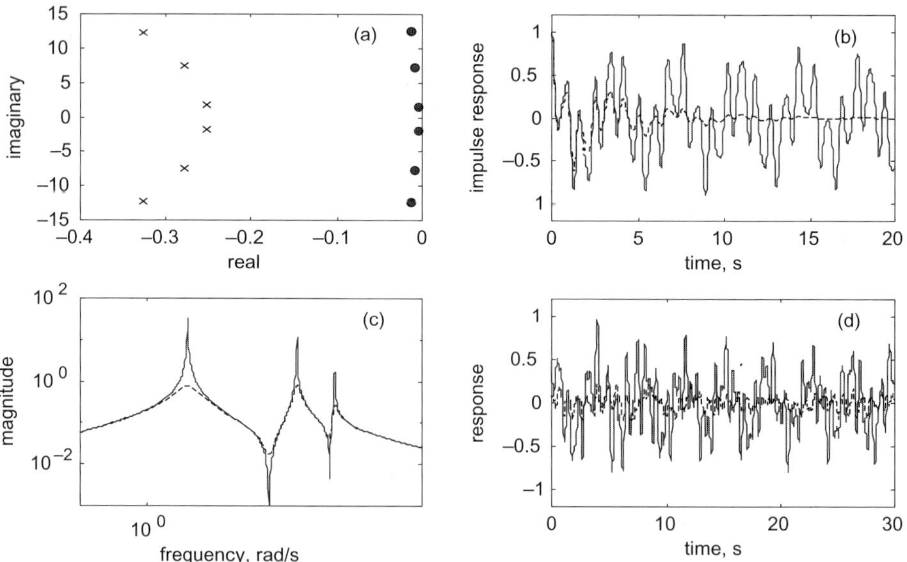

Figure 1.11. Structure response depends strongly on damping: (a) Poles of a structure with small (•) and larger (×) damping – damping impacts the real parts; (b) impulse response for small (solid line) and larger (dashed line) damping – damping impacts the transient time; (c) magnitude of the transfer function for small (solid line) and larger (dashed line) damping – damping impacts the resonance peaks; and (d) response to the white noise input for small (solid line) and larger damping (dashed line) – damping impacts the rms of the response.

- The magnitude of a flexible structure transfer function is characterized by the presence of resonance peaks; see Fig. 1.10(b).
- The impulse response of a flexible structure consists of harmonic components, related to complex poles, or to resonance peaks; this is shown in Fig. 1.10(c).
- The phase of a transfer function of a flexible structure shows 180 degree shifts at natural frequencies, see Fig. 1.10(d).

Poles of a flexible structure are complex conjugate. Each complex conjugate pair represents a structural mode. The real part of a pole represents damping of the mode. The absolute value of the pole represents the natural frequency of the mode.

Consider two different structures, as in Fig. 1.11(a). The first one has poles denoted with black circles (•), the second one with crosses (×). The locations of the poles indicate that they have the same natural frequencies, but different damping. The structure with poles marked with black circles has larger damping than the one with poles marked with the crosses. The figure illustrates that structural response depends greatly on the structural damping. For small damping the impulse response of a structure decays slower than the response for larger damping, see Fig. 1.11(b). Also, the magnitude of the response is visible in the plots of the magnitude of the transfer function in Fig. 1.11(c). For small damping the resonance peak is larger than that for larger damping. Finally, the damping impacts the root-mean-square (rms) of the response to white noise. For example, Fig. 1.11(d) shows that for small damping the rms response of a structure is larger than the response for larger damping.

When a structure is excited by a harmonic force, its response shows maximal amplitude at natural frequencies. This is a resonance phenomenon – a strong amplification of the motion at natural frequency. There are several frequencies that structures resonate at. A structure movement at these frequencies is harmonic, or sinusoidal, and remains at the same pattern of deformation. This pattern is called a mode shape, or mode. The resonance phenomenon leads to an additional property – the independence of each mode. Each mode is excited almost independently, and the total structural response is the sum of modal responses. For example, let a structure be excited by a white noise. Its response is shown in Fig. 1.12(a). Also, let each mode be excited by the same noise. Their responses are shown in Fig. 1.12(b),(c),(d). The spectrum of the structural response is shown in Fig. 1.13(a), and the spectra of responses of each individual mode are shown in Fig. 1.13(b),(c),(d). Comparing Fig1.13a with Fig.1.13b,c,d we see that the resonance peak for each natural frequency is the same, either it was total structure excited, or individual mode excited. This shows that the impact of each mode on each other is negligible.

The independence of the modes also manifests itself in a possibility of exciting each individual mode. One can find a special input configuration that excites a selected mode. For example, for the simple structure presented above we found an excitation that the impulse response has only one harmonic, see Fig. 1.14(a), and the magnitude of the transfer function of the structure shows a single resonance peak, see Fig. 1.14(b). However, there is no such input configuration that is able to excite a single node (or selected point) of a structure. Thus structural modes are independent, while structural nodes are not.

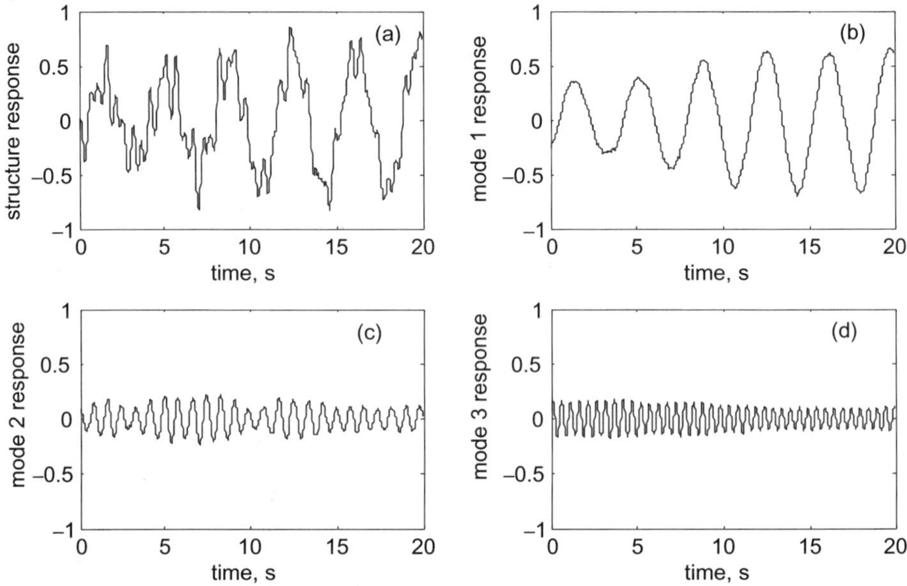

Figure 1.12. Response to the white noise input: (a) Total structure response is composed of three modal responses; (b) mode 1 response of the first natural frequency; (c) mode 2 response of the second natural frequency; and (d) mode 3 response of the third natural frequency.

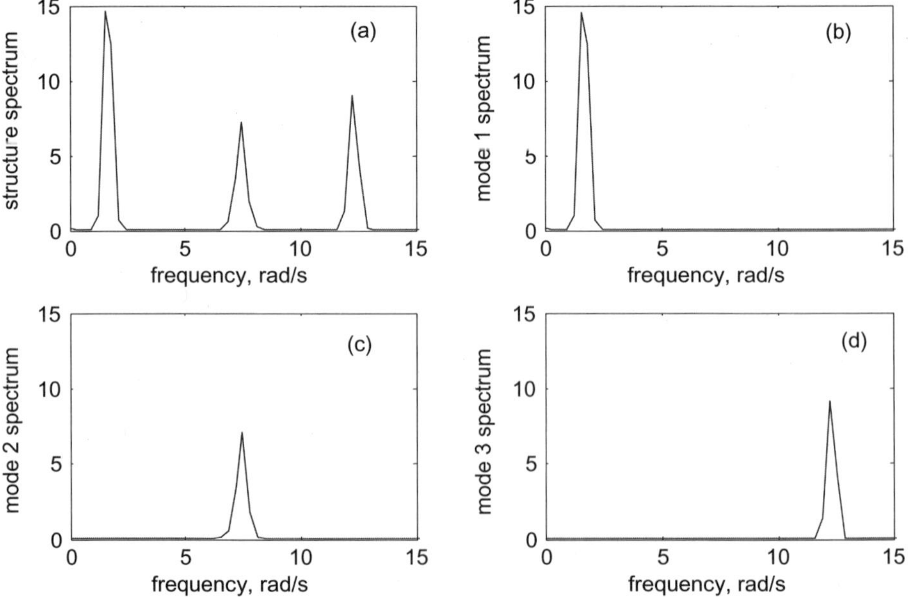

Figure 1.13. Spectra of the response to the white noise input: (a) Total structure spectrum consists of three modal spectra; (b) mode 1 spectrum of the first natural frequency; (c) mode 2 spectrum of the second natural frequency; and (d) mode 3 spectrum of the third natural frequency.

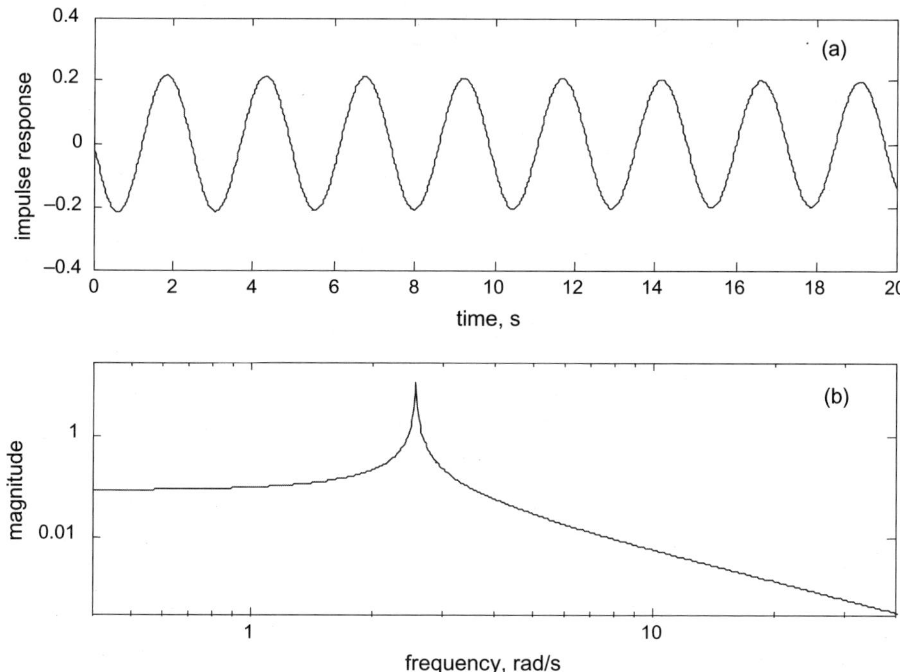

Figure 1.14. An input configuration that excites a single mode: (a) Impulse response; and (b) magnitude of the transfer function.

2
Standard Models

how to describe typical structures

The best model of a cat is another, or
preferably the same, cat.
—*Arturo Rosenblueth with Norbert Wiener*

In this and the following chapter we explain structural models that describe standard —or more common—structures. The standard models include structures that are stable, linear, continuous-time, and with proportional damping.

We derive the structural analytical models either from physical laws, such as Newton's motion laws, Lagrange's equations of motion, or D'Alembert's principle [108], [111]; or from finite-element models; or from test data using system identification methods. The models are represented either in time domain (differential equations), or in frequency domain (transfer functions).

We use linear differential equations to represent linear structural models in time domain, either in the form of second-order differential equations or in the form of first-order differential equations (as a state-space representation). In the first case, we use the degrees of freedom of a structure to describe structural dynamics. In the second case we use the system states to describe the dynamics. Structural engineers prefer degrees of freedom and the second-order differential equations of structural dynamics; this is not a surprise, since they have a series of useful mathematical and physical properties. This representation has a long tradition and using it many important results have been derived. The state-space model, on the other hand, is a standard model used by control engineers. Most linear control system analyses and design methods are given in the state-space form. The state-space standardization of structural models allows for the extension of known control system properties into structural dynamics. In this chapter we use both second-order and state-space models, and show their interrelations.

Besides the choice of form of equations, we represent the analytical model in different coordinates. The choice of coordinates in which the system model is represented is rather arbitrary. However, two coordinate systems, nodal and modal, are commonly used. Nodal coordinates are defined through displacements and velocities of selected structural locations, called nodes; and modal coordinates are defined through the displacements and velocities of structural (or natural) modes. In this book we use both coordinate systems; however, we put more weight on the modal coordinate system.

At the beginning of this chapter we present a generic state-space system model and its transfer function; next, structural state-space models follow the second-order models of flexible structures.

2.1 Models of a Linear System

Models of a linear system are described by linear differential equations. The equations can be organized in a standard form called state-space representation. This form is a set of first-order differential equations with unit coefficient at the first derivative. The models can also be represented in the form of a transfer function, after applying the Laplace or Fourier transformation. The state-space representation carries information about the internal structure (represented by states) of the model, while the transfer function describes the model in terms of its input–output properties (although its internal state can be somehow recovered). Also, the state-space models are more convenient and create less numerical difficulties than transfer functions when one deals with high-order models.

2.1.1 State-Space Representation

A linear time-invariant system of finite dimensions is described by the following linear constant coefficient differential equations:

$$\begin{aligned} \dot{x} &= Ax + Bu, \\ y &= Cx, \end{aligned} \tag{2.1}$$

with the initial state $x(0) = x_o$. In the above equations the N-dimensional vector x is called the state vector, x_o is the initial condition of the state, the s-dimensional vector u is the system input, and the r-dimensional vector y is the system output. The A, B, and C matrices are real constant matrices of appropriate dimensions (A is $N \times N$, B is $N \times s$, and C is $r \times N$). We call the triple (A, B, C) the system state-space representation.

Every linear system, or system of linear-time invariant differential equations can be presented in the above form (with some exceptions discussed in Chapter 3). It is important to have a unique form as a standard form in order to develop interchangeable software and compatible methods of analysis. However, for the same system presented by the state equations (2.1) the matrices A, B, C and the state vector are not unique: different representations (A, B, C) can give an identical input–output relationship. Indeed, one can introduce a new state variable, x_n, such that

$$x = Rx_n, \tag{2.2}$$

where R is a nonsingular transformation matrix. Introducing x from (2.2) to (2.1) we obtain the new state equations

$$\begin{aligned} \dot{x}_n &= A_n x_n + B_n u, \\ y &= C_n x_n, \end{aligned} \tag{2.3}$$

where

$$A_n = R^{-1}AR, \qquad B_n = R^{-1}B, \qquad C_n = CR. \tag{2.4}$$

Note that u and y are identical in (2.1) and (2.3); i.e., the input–output relationship is identical in the new representation (A_n, B_n, C_n) and in the original representation (A,B,C). This might suggest that there is no difference as to what coordinates we use for a system analysis. But this is not necessarily true. Although input–output relations remain invariant, it makes a difference for system analysis or controller design what state or representation is chosen. For example, some representations have useful physical interpretations; others are more convenient for analysis and design.

2.1.2 Transfer Function

Besides the state-space representation a linear system can be alternatively represented by its transfer function. The transfer function $G(s)$ is defined as a complex gain between $y(s)$ and $u(s)$,

$$y(s) = G(s)u(s), \tag{2.5}$$

where $y(s)$ and $u(s)$ are the Laplace transforms of the output $y(t)$ and input $u(t)$, respectively. Using the Laplace transformation of (2.1) for the zero initial condition, $x(0) = 0$, we express the transfer function in terms of the state parameters (A,B,C),

$$G(s) = C(sI - A)^{-1}B. \tag{2.6}$$

The transfer function is invariant under the coordinate transformation (i.e., $C(sI - A)^{-1}B = C_n(sI - A_n)^{-1}B_n$, which can be checked by introducing (2.4) into the above equation.

2.2 Second-Order Structural Models

In this and the following sections we will discuss the structural models. One of them is the second-order structural model. It is represented by the second-order linear differential equations, and is commonly used in the analysis of structural dynamics. Similarly to the state-space models the second-order models also depend on the choice of coordinates. Typically, the second-order models are represented either in the nodal coordinates, and are called nodal models, or in the modal coordinates, and are called modal models.

2.2.1 Nodal Models

The nodal models are derived in nodal coordinates, in terms of nodal displacements, velocities, and accelerations. The model is characterized by the mass, stiffness, and damping matrices, and by the sensors and actuators locations. These models are typically obtained from the finite-element codes or from other Computer-Aided-Design-type software.

As a convention, we denote a dot as a first derivative with respect to time (i.e., $\dot{x} = dx/dt$), and a double dot as a second derivative with respect to time (i.e., $\ddot{x} = d^2x/dt^2$). Let n_d be a number of degrees of freedom of the system (linearly independent coordinates describing the finite-dimensional structure), let r be a number of outputs, and let s be a number of inputs. A flexible structure in nodal coordinates is represented by the following second-order matrix differential equation:

$$
\begin{aligned}
M\ddot{q} + D\dot{q} + Kq &= B_o u, \\
y &= C_{oq}q + C_{ov}\dot{q}.
\end{aligned}
\tag{2.7}
$$

In this equation q is the $n_d \times 1$ nodal displacement vector; $\dot{q}$ is the $n_d \times 1$ nodal velocity vector; $\ddot{q}$ is the $n_d \times 1$ nodal acceleration vector; u is the $s \times 1$ input vector; y is the output vector, $r \times 1$; M is the mass matrix, $n_d \times n_d$; D is the damping matrix, $n_d \times n_d$; and K is the stiffness matrix, $n_d \times n_d$. The input matrix B_o is $n_d \times s$, the output displacement matrix C_{oq} is $r \times n_d$, and the output velocity matrix C_{ov} is $r \times n_d$. The mass matrix is positive definite (all its eigenvalues are positive), and the stiffness and damping matrices are positive semidefinite (all their eigenvalues are nonnegative).

Example 2.1. Determine the nodal model for a simple system from Fig. 1.1. For this system we selected masses $m_1 = m_2 = m_3 = 1$, stiffness $k_1 = k_2 = k_3 = 3$, $k_4 = 0$, and a damping matrix proportional to the stiffness matrix, D = $0.01K$, or

$d_i = 0.01k_i$, i = 1, 2, 3, 4. There is a single input force at mass 3, and three outputs: displacement and velocity of mass 1 and velocity of mass 3.

For this system the mass matrix is $M = diag(m_1, m_2, m_3)$, thus $M = I_3$. The stiffness and damping matrices are

$$K = \begin{bmatrix} k_1 + k_2 & -k_2 & 0 \\ -k_2 & k_2 + k_3 & -k_3 \\ 0 & -k_3 & k_3 + k_4 \end{bmatrix}, \quad D = \begin{bmatrix} d_1 + d_2 & -d_2 & 0 \\ -d_2 & d_2 + d_3 & -d_3 \\ 0 & -d_3 & d_3 + d_4 \end{bmatrix},$$

therefore,

$$K = \begin{bmatrix} 6 & -3 & 0 \\ -3 & 6 & -3 \\ 0 & -3 & 3 \end{bmatrix}, \quad \text{and} \quad D = \begin{bmatrix} 0.06 & -0.03 & 0.00 \\ -0.03 & 0.06 & -0.03 \\ 0.00 & -0.03 & 0.03 \end{bmatrix}.$$

The input and output matrices are

$$B_o = \begin{bmatrix} 0 \\ 0 \\ 1 \end{bmatrix}, \quad C_{oq} = \begin{bmatrix} 1 & 0 & 0 \\ 0 & 0 & 0 \\ 0 & 0 & 0 \end{bmatrix}, \quad \text{and} \quad C_{ov} = \begin{bmatrix} 0 & 0 & 0 \\ 1 & 0 & 0 \\ 0 & 0 & 1 \end{bmatrix}.$$

On details of the derivation of this type of equation, see [70], [120].

2.2.2 Modal Models

The second-order models are defined in modal coordinates. These coordinates are often used in the dynamics analysis of complex structures modeled by the finite elements to reduce the order of a system. It is also used in the system identification procedures, where modal representation is a natural outcome of the test.

Modal models of structures are the models expressed in modal coordinates. Since these coordinates are independent, it leads to a series of useful properties that simplify the analysis (as will be shown later in this book). The modal coordinate representation can be obtained by the transformation of the nodal models. This transformation is derived using a modal matrix, which is determined as follows.

Consider free vibrations of a structure without damping, i.e., a structure without external excitation ($u \equiv 0$) and with the damping matrix D = 0. The equation of motion (2.7) in this case turns into the following equation:

$$M\ddot{q} + Kq = 0. \tag{2.8}$$

The solution of the above equation is $q = \phi e^{j\omega t}$. Hence, the second derivative of the solution is $\ddot{q} = -\omega^2 \phi e^{j\omega t}$. Introducing the latter q and $\ddot{q}$ into (2.8) gives

$$(K - \omega^2 M)\phi e^{j\omega t} = 0. \tag{2.9}$$

This is a set of homogeneous equations, for which a nontrivial solution exists if the determinant of $K - \omega^2 M$ is zero,

$$\det(K - \omega^2 M) = 0. \tag{2.10}$$

The above determinant equation is satisfied for a set of n values of frequency ω. These frequencies are denoted $\omega_1, \omega_2, ..., \omega_n$, and their number n does not exceed the number of degrees of freedom, i.e., $n \le n_d$. The frequency ω_i is called the ith natural frequency.

Substituting ω_i into (2.9) yields the corresponding set of vectors $\{\phi_1, \phi_2, ..., \phi_n\}$ that satisfy this equation. The ith vector ϕ_i corresponding to the ith natural frequency is called the ith natural mode, or ith mode shape. The natural modes are not unique, since they can be arbitrarily scaled. Indeed, if ϕ_i satisfies (2.9), so does $\alpha\phi_i$, where α is an arbitrary scalar.

For a notational convenience define the matrix of natural frequencies

$$\Omega = \begin{bmatrix} \omega_1 & 0 & \cdots & 0 \\ 0 & \omega_2 & \cdots & 0 \\ \cdots & \cdots & \cdots & \cdots \\ 0 & 0 & \cdots & \omega_n \end{bmatrix} \tag{2.11}$$

and the matrix of mode shapes, or modal matrix Φ, of dimensions $n_d \times n$, which consists of n natural modes of a structure

$$\Phi = [\phi_1 \quad \phi_2 \quad \ldots \quad \phi_n] = \begin{bmatrix} \phi_{11} & \phi_{21} & \cdots & \phi_{n1} \\ \phi_{12} & \phi_{22} & \cdots & \phi_{n2} \\ \cdots & \cdots & \cdots & \cdots \\ \phi_{1n_d} & \phi_{2n_d} & \cdots & \phi_{nn_d} \end{bmatrix}, \tag{2.12}$$

where ϕ_{ij} is the jth displacement of the ith mode, that is,

$$\phi_i = \begin{Bmatrix} \phi_{i1} \\ \phi_{i2} \\ \vdots \\ \phi_{in} \end{Bmatrix}. \tag{2.13}$$

The modal matrix Φ has an interesting property: it diagonalizes mass and stiffness matrices M and K,

$$M_m = \Phi^T M \Phi, \tag{2.14}$$

$$K_m = \Phi^T K \Phi. \tag{2.15}$$

The obtained diagonal matrices are called modal mass matrix (M_m) and modal stiffness matrix (K_m). The same transformation, applied to the damping matrix

$$D_m = \Phi^T D \Phi, \tag{2.16}$$

gives the modal damping matrix D_m, which is not always obtained as a diagonal matrix. However, in some cases, it is possible to obtain D_m diagonal. In these cases the damping matrix is called a matrix of proportional damping. The proportionality of damping is commonly assumed for analytical convenience. This approach is justified by the fact that the nature of damping is not known exactly, that its values are rather roughly approximated, and that the off-diagonal terms in most cases—as will be shown later—have negligible impact on the structural dynamics. The damping proportionality is often achieved by assuming the damping matrix as a linear combination of the stiffness and mass matrices; see [18], [70],

$$D = \alpha_1 K + \alpha_2 M, \tag{2.17}$$

where α_1 and α_2 are nonnegative scalars.

Modal models of structures are the models expressed in modal coordinates. In order to do so we use a modal matrix to introduce a new variable, q_m, called modal displacement. This is a variable that satisfies the following equation:

$$q = \Phi q_m. \tag{2.18}$$

In order to obtain the equations of motion for this new variable, we introduce (2.18) to (2.7) and additionally left-multiply (2.7) by Φ^T, obtaining

$$\Phi^T M \Phi \ddot{q}_m + \Phi^T D \Phi \dot{q}_m + \Phi^T K \Phi q_m = \Phi^T B_o u,$$
$$y = C_{oq} \Phi q_m + C_{ov} \Phi \dot{q}_m.$$

Assuming a proportional damping, and using (2.14), (2.15), and (2.16) we obtain the above equation in the following form:

$$M_m \ddot{q}_m + D_m \dot{q}_m + K_m q_m = \Phi^T B_o u,$$
$$y = C_{oq} \Phi q_m + C_{ov} \Phi \dot{q}_m.$$

Next, we multiply (from the left) the latter equation by M_m^{-1}, which gives

$$\ddot{q}_m + M_m^{-1} D_m \dot{q}_m + M_m^{-1} K_m q_m = M_m^{-1} \Phi^T B_o u,$$
$$y = C_{oq} \Phi q_m + C_{ov} \Phi \dot{q}_m.$$

The obtained equations look quite messy, but the introduction of appropriate notations simplifies them,

$$\begin{aligned} \ddot{q}_m + 2Z\Omega \dot{q}_m + \Omega^2 q_m &= B_m u, \\ y = C_{mq} q_m + C_{mv} \dot{q}_m. \end{aligned} \tag{2.19}$$

In (2.19) Ω is a diagonal matrix of natural frequencies, as defined before. Note, however, that this is obtained from the modal mass and stiffness matrices as follows:

$$\Omega^2 = M_m^{-1} K_m. \tag{2.20}$$

In (2.19) Z is the modal damping matrix. It is a diagonal matrix of modal damping,

$$Z = \begin{bmatrix} \zeta_1 & 0 & \cdots & 0 \\ 0 & \zeta_2 & \cdots & 0 \\ \cdots & \cdots & \cdots & \cdots \\ 0 & 0 & \cdots & \zeta_n \end{bmatrix}, \tag{2.21}$$

where ζ_i is the damping of the ith mode. We obtain this matrix using the following relationship $M_m^{-1} D_m = 2Z\Omega$, thus,

$$Z = 0.5 M_m^{-1} D_m \Omega^{-1} = 0.5 M_m^{-\frac{1}{2}} K_m^{-\frac{1}{2}} D_m. \tag{2.22}$$

Next, we introduce the modal input matrix B_m in (2.19),

$$B_m = M_m^{-1} \Phi^T B_o. \tag{2.23}$$

Finally, in (2.19) we use the following notations for the modal displacement and rate matrices:

$$C_{mq} = C_{oq}\Phi, \tag{2.24}$$

$$C_{mv} = C_{ov}\Phi. \tag{2.25}$$

Note that (2.19) (a modal representation of a structure) is a set of uncoupled equations. Indeed, due to the diagonality of Ω and Z, this set of equations can be written, equivalently, as

$$\begin{aligned} &\ddot{q}_{mi} + 2\zeta_i\omega_i\dot{q}_{mi} + \omega_i^2 q_{mi} = b_{mi}u \\ &y_i = c_{mqi}q_{mi} + c_{mvi}\dot{q}_{mi}, \qquad i = 1,\ldots,n, \\ &y = \sum_{i=1}^{n} y_i, \end{aligned} \tag{2.26}$$

where b_{mi} is the ith row of B_m and c_{mqi}, c_{mvi} are the ith columns of C_{mq} and C_{mv}, respectively. The coefficient ζ_i is called a modal damping of the ith mode. In the above equations y_i is the system output due to the ith mode dynamics, and the quadruple $(\omega_i, \zeta_i, b_{mi}, c_{mi})$ represents the properties of the ith natural mode. Note that the structural response y is a sum of modal responses y_i, which is a key property used to derive structural properties in modal coordinates.

This completes the modal model description. In the following we introduce the transfer function obtained from the modal equations. The generic transfer function is obtained from the state-space representation using (2.6). For structures in modal coordinates it has a specific form.

Transfer Function of a Structure. The transfer function of a structure is derived from (2.19),

$$G(\omega) = (C_{mq} + j\omega C_{mv})(\Omega^2 - \omega^2 I_n + 2j\omega Z\Omega)^{-1} B_m. \tag{2.27}$$

However, this can be presented in a more useful form, since the matrices Ω and Z are diagonal, allowing for representation of each single mode.

Transfer Function of a Mode. The transfer function of the ith mode is obtained from (2.26),

$$G_{mi}(\omega) = \frac{(c_{mqi} + j\omega c_{mvi})b_{mi}}{\omega_i^2 - \omega^2 + 2j\zeta_i\omega_i\omega}. \tag{2.28}$$

The structural and modal transfer functions are related as follows:

Property 2.1. *Transfer Function in Modal Coordinates.* *The structural transfer function is a sum of modal transfer functions*

(a)
$$G(\omega)=\sum_{i=1}^{n} G_{mi}(\omega) \tag{2.29}$$

or, in other words,

$$G(\omega)=\sum_{i=1}^{n} \frac{(c_{mqi}+j\omega c_{mvi})b_{mi}}{\omega_i^2-\omega^2+2j\zeta_i\omega_i\omega}, \tag{2.30}$$

and the structural transfer function at the ith resonant frequency is approximately equal to the ith modal transfer function at this frequency

(b)
$$G(\omega_i)\cong G_{mi}(\omega_i)=\frac{(-jc_{mqi}+\omega_i c_{mvi})b_{mi}}{2\zeta_i\omega_i^2}, \qquad i=1,\ldots,n. \tag{2.31}$$

Proof. By inspection of (2.27) and (2.28). ▣

Structural Poles. The poles of a structure are the zeros of the characteristic equations (2.26). The equation $s^2+2\zeta_i\omega_i s+\omega_i^2=0$ is the characteristic equation of the ith mode. For small damping the poles are complex conjugate, and in the following form:

$$\begin{aligned} s_1 &= -\zeta_i\omega_i + j\omega_i\sqrt{1-\zeta_i^2}, \\ s_2 &= -\zeta_i\omega_i - j\omega_i\sqrt{1-\zeta_i^2}. \end{aligned} \tag{2.32}$$

The plot of the poles is shown in Fig. 2.1, which shows how the location of a pole relates to the natural frequency and modal damping.

Example 2.2. Determine the modal model of a simple structure from Example 2.1.

The natural frequency matrix is

$$\Omega=\begin{bmatrix} 3.1210 & 0 & 0 \\ 0 & 2.1598 & 0 \\ 0 & 0 & 0.7708 \end{bmatrix},$$

and the modal matrix is

(a)
$$\Phi = \begin{bmatrix} 0.5910 & 0.7370 & 0.3280 \\ -0.7370 & 0.3280 & 0.5910 \\ 0.3280 & -0.5910 & 0.7370 \end{bmatrix}.$$

The modes are shown in Fig. 2.2.

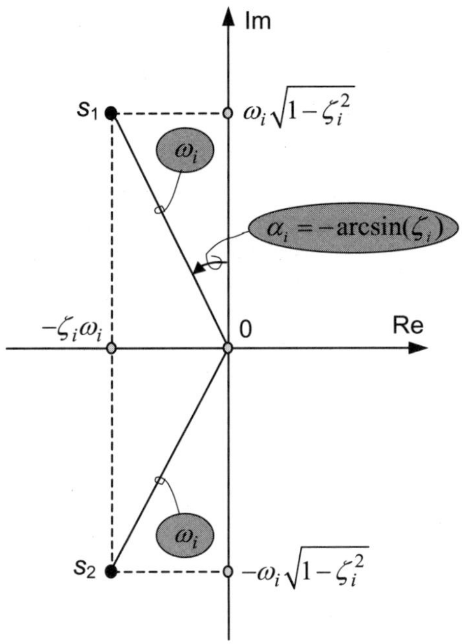

Figure 2.1. Pole location of the *i*th mode of a lightly damped structure: It is a complex pair with the real part proportional to the *i*th modal damping; the imaginary part approximately equal to the *i*th natural frequency; and the radius is the exact natural frequency.

The modal mass is $M_m = I_3$, the modal stiffness is $K_m = \Omega^2$, and the modal damping, from (2.22), is

$$Z = \begin{bmatrix} 0.0156 & 0 & 0 \\ 0 & 0.0108 & 0 \\ 0 & 0 & 0.0039 \end{bmatrix}.$$

We obtain the modal input and output matrices from (2.23), (2.24), and (2.25):

$$B_m = \begin{bmatrix} 0.3280 \\ -0.5910 \\ 0.7370 \end{bmatrix},$$

$$C_{mq} = \begin{bmatrix} 0.5910 & 0.7370 & 0.3280 \\ 0 & 0 & 0 \\ 0 & 0 & 0 \end{bmatrix},$$

and

$$C_{mv} = \begin{bmatrix} 0 & 0 & 0 \\ 0.5910 & 0.7370 & 0.3280 \\ 0.3280 & -0.5910 & 0.7370 \end{bmatrix}.$$

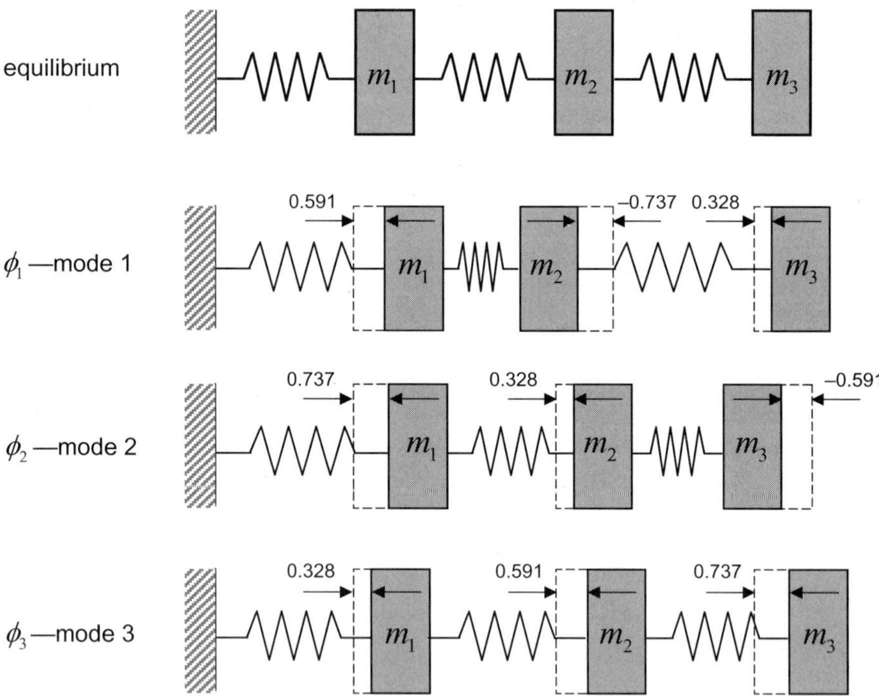

Figure 2.2. Modes of a simple system: For each mode the mass displacements are sinusoidal and have the same frequency, and the displacements are shown at their extreme values (see the equation (a)).

Example 2.3. Determine the first four natural modes and frequencies of the beam presented in Fig. 1.5.

Using the finite-element model we find the modes, which are shown in Fig. 2.3. For the first mode the natural frequency is $\omega_1 = 72.6$ rad/s, for the second mode the natural frequency is $\omega_2 = 198.8$ rad/s, for the third mode the natural frequency is $\omega_3 = 386.0$ rad/s, and for the fourth mode the natural frequency is $\omega_4 = 629.7$ rad/s.

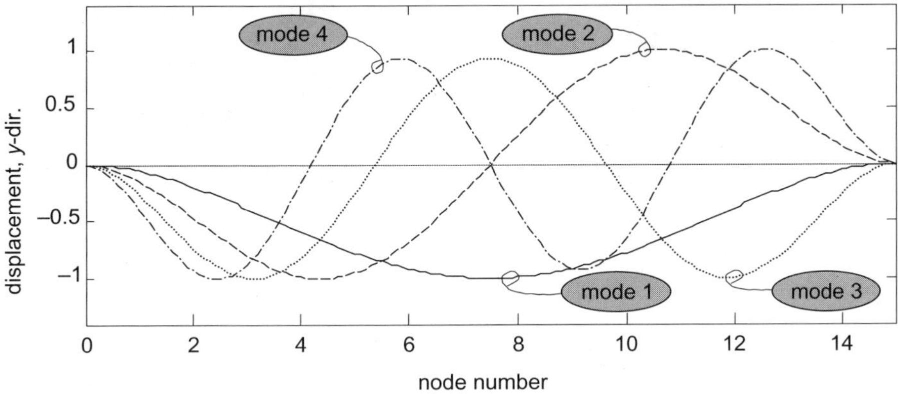

Figure 2.3. Beam modes: For each mode the beam displacements are sinusoidal and have the same frequency, and the displacements are shown at their extreme values.

Example 2.4. Determine the first four natural modes and frequencies of the antenna presented in Fig. 1.6.

We used the finite-element model of the antenna to solve this problem. The modes are shown in Fig. 2.4. For the first mode the natural frequency is $\omega_1 = 13.2$ rad/s, for the second mode the natural frequency is $\omega_2 = 18.1$ rad/s, for the third mode the natural frequency is $\omega_3 = 18.8$ rad/s, and for the fourth mode the natural frequency is $\omega_4 = 24.3$ rad/s.

Example 2.5. The Matlab code for this example is in Appendix B. For the simple system from Fig. 1.1 determine the natural frequencies and modes, the system transfer function, and transfer functions of each mode. Also determine the system impulse response and the impulse responses of each mode. Assume the system masses $m_1 = m_2 = m_3 = 1$, stiffnesses $k_1 = k_2 = k_3 = 3$, $k_4 = 0$, and the damping matrix proportional to the stiffness matrix, $D = 0.01K$ or $d_i = 0.01k_i$, i = 1, 2, 3, 4. There is a single input force at mass 3 and a single output: velocity of mass 1.

We determine the transfer function from (2.27), using data from Example 2.2. The magnitude and phase of the transfer function are plotted in Fig. 2.5. The magnitude plot shows resonance peaks at natural frequencies $\omega_1 = 0.7708$ rad/s, $\omega_2 = 2.1598$ rad/s, and $\omega_3 = 3.1210$ rad/s. The phase plot shows a 180-degree phase change at each resonant frequency.

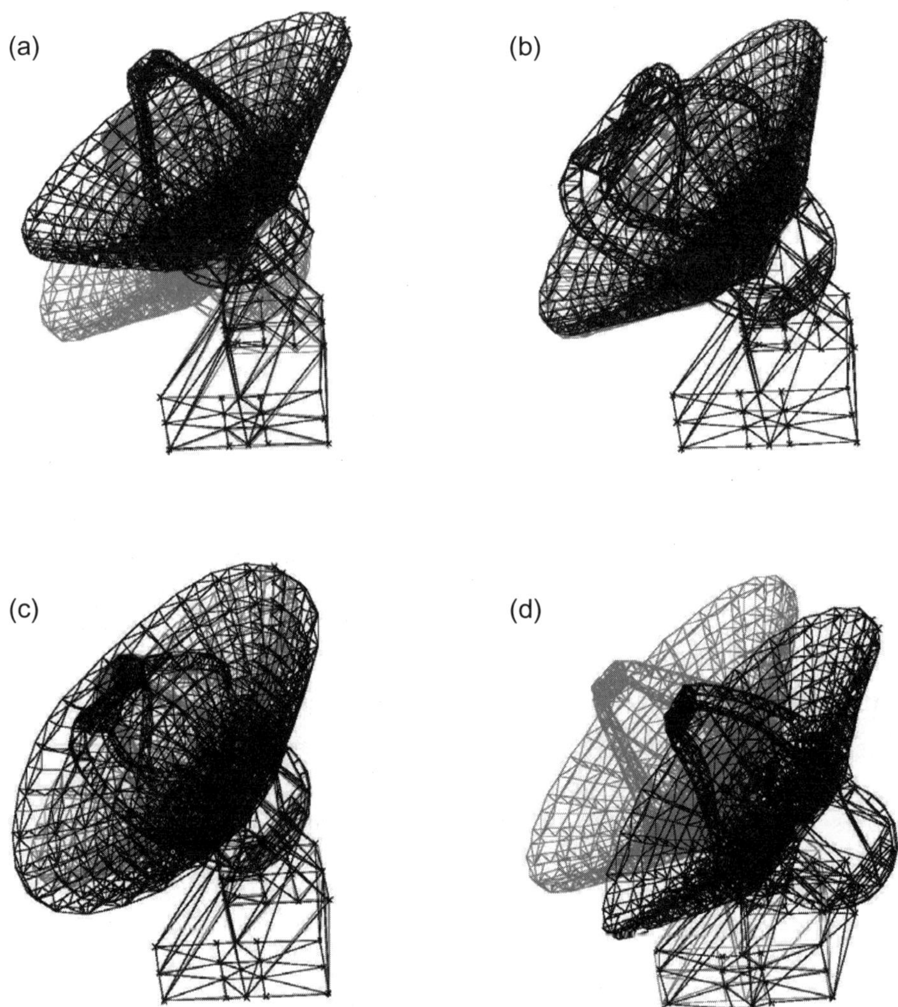

Figure 2.4. Antenna modes: (a) First mode (of natural frequency 2.10 Hz); (b) second mode (of natural frequency 2.87 Hz); (c) third mode (of natural frequency 2.99 Hz); and (d) fourth mode (of natural frequency 3.87 Hz). For each mode the nodal displacements are sinusoidal, have the same frequency, and the displacements are shown at their extreme values. Gray color denotes undeformed state.

We determine the transfer functions of modes 1, 2, and 3 from (2.28), and their magnitudes and phases are shown in Fig. 2.6. According to Property 2.1, the transfer function of the entire structure is a sum of the modal transfer functions, and this is shown in Fig. 2.6, where the transfer function of the structure was constructed as a sum of transfer functions of individual modes.

The impulse response of the structure is shown in Fig. 2.7; it was obtained from (2.19). It consists of three harmonics (or responses of three modes) of natural frequencies $\omega_1 = 0.7708$ rad/s, $\omega_2 = 2.1598$ rad/s, and $\omega_3 = 3.1210$ rad/s. The harmonics are shown on the impulse response plot, but are more explicit at the impulse response spectrum plot, Fig. 2.7, as the spectrum peaks at these frequencies.

Impulse response is the time-domain associate of the transfer function (through the Parseval theorem); therefore, Property 2.1 can be written in time domain as

$$h(t) = \sum_{i=1}^{n} h_i(t)$$

where $h(t)$ is the impulse response of a structure and $h_i(t)$ is the impulse response of the ith mode. Thus, the structural impulse response is a sum of modal responses. This is illustrated in Fig. 2.8, where impulse responses of modes 1, 2, and 3 are plotted. Clearly the total response as in Fig. 2.7 is a sum of the individual responses. Note that each response is a sinusoid of frequency equal to the natural frequency, and of exponentially decayed amplitude, proportional to the modal damping ζ_i. Note also that the higher-frequency responses decay faster.

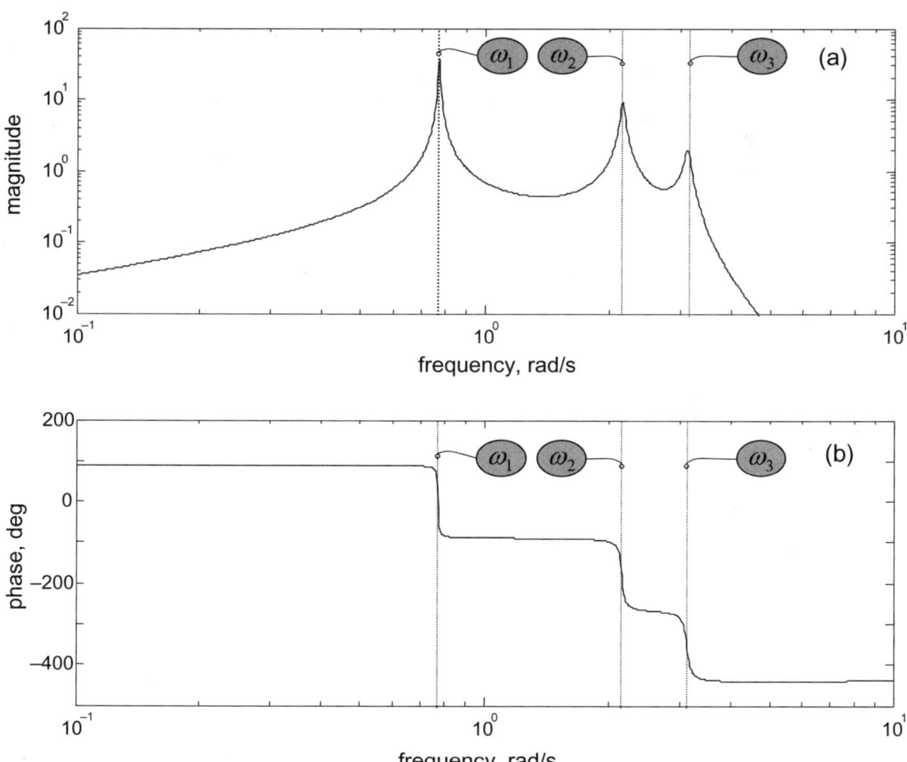

Figure 2.5. Transfer function of a simple system: (a) Magnitude shows three resonance peaks; and (b) phase shows three shifts of 180 degrees; ω_1, ω_2, and ω_3 denote the natural frequencies.

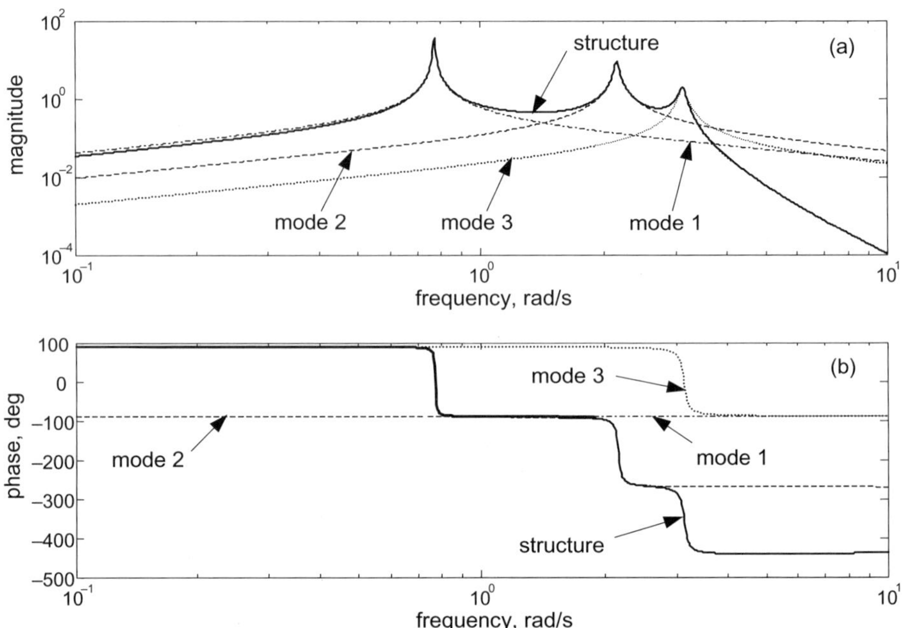

Figure 2.6. The transfer functions of single modes and of the structure: (a) Magnitudes; and (b) phases. The plots illustrate that the structure transfer function is a sum of modal transfer functions.

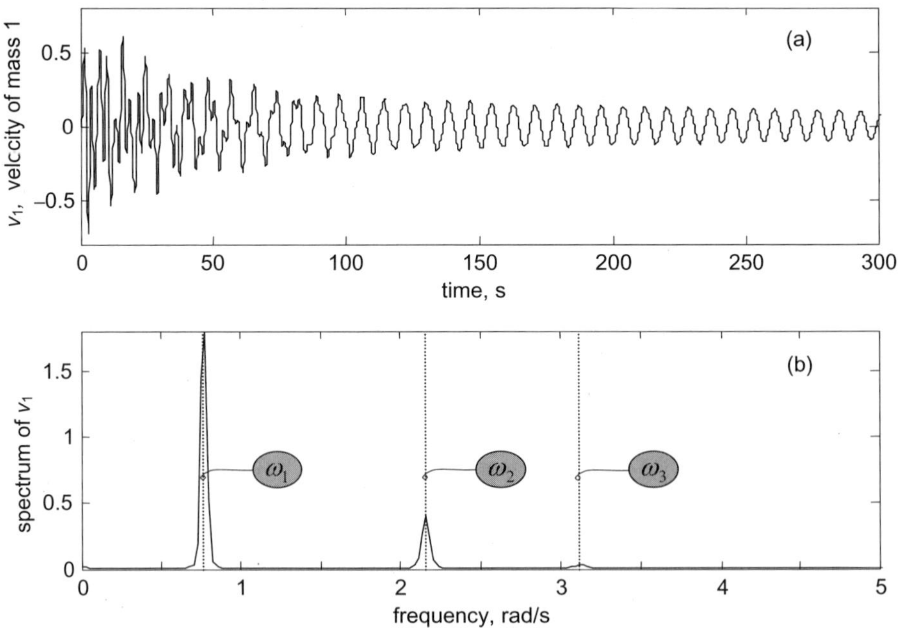

Figure 2.7. Impulse response of the simple system: (a) Time history; and (b) its spectrum. Both show that the response is composed of three harmonics.

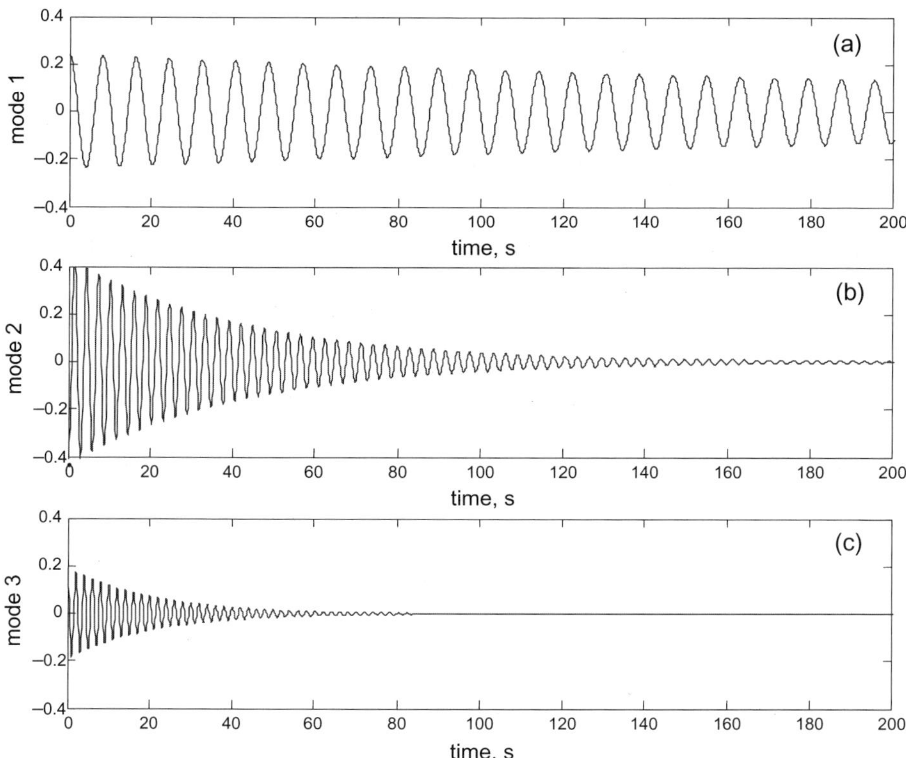

Figure 2.8. Impulse responses of (a) first mode; (b) second mode; and (c) third mode. All show single frequency time histories.

2.3 State-Space Structural Models

For the purposes of structural dynamic simulations, control system analysis, and design, it is convenient to represent the flexible structure equations in a state-space form, as in (2.1). Recall that a set of the three state-space parameters (A, B, C) is called the state-space representation, and x is the state vector, u is the input, and y is the output. Again, the representation depends on the choice of the state vector, while the input and output remain invariant. It makes a difference what state representation is chosen for system analysis or for controller design. It is shown in this book that modal representation is specifically useful for the purpose of dynamics and control of flexible structures.

2.3.1 Nodal Models

In order to obtain a state representation from the nodal model as in (2.7) we rewrite the latter equation as follows (assuming that the mass matrix is nonsingular):

$$\ddot{q} + M^{-1}D\dot{q} + M^{-1}Kq = M^{-1}B_o u,$$
$$y = C_q q + C_v \dot{q}. \tag{2.33}$$

A state is a vector that contains the minimal number of physical variables that enable us to calculate uniquely the output using the applied input. For a structure, nodal displacements and velocities allow for such determination of the outputs. Therefore, we define the state vector x as a combination of the structural displacements, q, and velocities, $\dot{q}$, i.e.,

$$x = \begin{Bmatrix} x_1 \\ x_2 \end{Bmatrix} = \begin{Bmatrix} q \\ \dot{q} \end{Bmatrix}. \tag{2.34}$$

In this case, (2.33) can be rewritten as follows:

$$\dot{x}_1 = x_2,$$
$$\dot{x}_2 = -M^{-1}Kx_1 - M^{-1}Dx_2 + M^{-1}B_o u,$$
$$y = C_{oq}x_1 + C_{ov}x_2.$$

Combining the above equations into one, we obtain the state equations as in (2.1), with the following state-space representation:

$$A = \begin{bmatrix} 0 & I \\ -M^{-1}K & -M^{-1}D \end{bmatrix}, \qquad B = \begin{bmatrix} 0 \\ M^{-1}B_o \end{bmatrix}, \qquad C = \begin{bmatrix} C_{oq} & C_{ov} \end{bmatrix}, \tag{2.35}$$

where A is $N \times N$, B is $N \times s$, and C is $r \times N$. The dimension of the state model N is twice the number of degrees of freedom of the system n_d, i.e., $N = 2n_d$.

Example 2.6. Determine the nodal state-space model for the simple structure from Example 2.1.

From (2.35) we obtain

$$A = \left[\begin{array}{ccc:ccc} 0 & 0 & 0 & 1 & 0 & 0 \\ 0 & 0 & 0 & 0 & 1 & 0 \\ 0 & 0 & 0 & 0 & 0 & 1 \\ \hdashline -6 & 3 & 0 & -0.06 & 0.03 & 0 \\ 3 & -6 & 3 & 0.03 & -0.06 & 0.03 \\ 0 & 3 & -3 & 0 & 0.03 & -0.03 \end{array}\right], \qquad B = \left[\begin{array}{c} 0 \\ 0 \\ 0 \\ \hdashline 0 \\ 0 \\ 1 \end{array}\right],$$

$$C = \left[\begin{array}{ccc|ccc} 1 & 0 & 0 & 0 & 0 & 0 \\ 0 & 0 & 0 & 1 & 0 & 0 \\ 0 & 0 & 0 & 0 & 0 & 1 \end{array}\right].$$

2.3.2 Models in Modal Coordinates

Frequently the order of the nodal representation is unacceptably high. For example, it is not uncommon that the number of degrees of freedom of the finite-element model exceeds 1000. Therefore, the nodal state representation is rarely used in structural dynamics. An alternative approach is to obtain the state-space representation using the modal coordinates and the second-order modal form (2.19), where the number of equations is significantly lower, while the accuracy of the analysis has not suffered. In this subsection we introduce three different state-space models using modal coordinates q_m.

We obtain the first model by defining the following state variables as $x_1 = \Omega q_m$ and $x_2 = \dot{q}_m$, that is,

$$x = \begin{Bmatrix} x_1 \\ x_2 \end{Bmatrix} = \begin{Bmatrix} \Omega q_m \\ \dot{q}_m \end{Bmatrix}. \tag{2.36}$$

In this case (2.19) is presented as a set of the following first-order equations:

$$\begin{aligned} \dot{x}_1 &= \Omega x_2, \\ \dot{x}_2 &= -\Omega x_1 - 2Z\Omega x_2 + B_m u, \\ y &= C_{mq}\Omega^{-1} x_1 + C_{mv} x_2. \end{aligned} \tag{2.37}$$

These equations are presented in the state-space form as in (2.1), with the state triple as follows:

$$A = \begin{bmatrix} 0 & \Omega \\ -\Omega & -2Z\Omega \end{bmatrix}, \qquad B = \begin{bmatrix} 0 \\ B_m \end{bmatrix}, \qquad C = \begin{bmatrix} C_{mq}\Omega^{-1} & C_{mv} \end{bmatrix}. \tag{2.38}$$

The second state-space model in modal coordinates we obtain by transforming the state-space representation (2.38) using (2.4) and the following transformation:

$$R = \begin{bmatrix} I & 0 \\ -Z & I \end{bmatrix}. \tag{2.39}$$

Applying the above transformation to the state vector (2.36) we obtain a new state variable

$$x = \begin{Bmatrix} \Omega q_m \\ Z\Omega q_m + \dot{q}_m \end{Bmatrix}. \tag{2.40}$$

The corresponding state representation is

$$A = \begin{bmatrix} -Z\Omega & \Omega \\ -\Omega - Z^2\Omega & -Z\Omega \end{bmatrix}, \qquad B = \begin{bmatrix} 0 \\ B_m \end{bmatrix}, \qquad C = \begin{bmatrix} C_{mq}\Omega^{-1} - C_{mv}Z & C_{mv} \end{bmatrix}. \tag{2.41}$$

For small Z (i.e., such that $Z^2 \cong 0$) it simplifies to

$$A = \begin{bmatrix} -Z\Omega & \Omega \\ -\Omega & -Z\Omega \end{bmatrix}, \qquad B = \begin{bmatrix} 0 \\ B_m \end{bmatrix}, \qquad C = \begin{bmatrix} C_{mq}\Omega^{-1} - C_{mv}Z & C_{mv} \end{bmatrix}. \tag{2.42}$$

Comparing (2.38) and (2.42) we see that, although the state matrices A and the output matrices C are different, they actually are very close to each other. Indeed, the transformation matrix from (2.36) to (2.40) is

$$R = \begin{bmatrix} I & 0 \\ Z & I \end{bmatrix}, \tag{2.43}$$

and it differs from the identity matrix by a small off-diagonal element Z.

In the third model the state vector consists of modal displacements and velocities, $x_1 = q_m$, and $x_2 = \dot{q}_m$. This is the most straightforward approach and it has direct physical interpretation, therefore it is the most popular model; however, its properties are not so useful as the first and second models, as will be seen later.

The state vector of the third model is as follows:

$$x = \begin{Bmatrix} x_1 \\ x_2 \end{Bmatrix} = \begin{Bmatrix} q_m \\ \dot{q}_m \end{Bmatrix}; \tag{2.44}$$

therefore, (2.19) is presented now as a set of first-order equations

$$\begin{aligned} \dot{x}_1 &= x_2, \\ \dot{x}_2 &= -\Omega^2 x_1 - 2Z\Omega x_2 + B_m u, \\ y &= C_{mq} x_1 + C_{mv} x_2, \end{aligned}$$

which is equivalent to the state-space form (2.1), with the state triple as follows:

$$A = \begin{bmatrix} 0 & I \\ -\Omega^2 & -2Z\Omega \end{bmatrix}, \qquad B = \begin{bmatrix} 0 \\ B_m \end{bmatrix}, \qquad C = \begin{bmatrix} C_{mq} & C_{mv} \end{bmatrix}. \tag{2.45}$$

The transformation from the third to the first model is given by

$$R = \begin{bmatrix} \Omega & 0 \\ 0 & I \end{bmatrix}. \tag{2.46}$$

The dimension of the modal models is the most obvious advantage over the nodal state-space models. The dimension of the modal state-space representation is $2n$, while the nodal state-space representation, as in (2.35), is $2n_d$, and typically we have $n \ll n_d$, i.e., the order of the model in modal coordinates is much lower than the model in nodal coordinates.

Another advantage of the models in modal coordinates is their definition of damping properties. While the mass and stiffness matrices are, as a rule, derived in the nodal coordinates (e.g., from a finite-element model), the damping matrix is commonly not known, but is conveniently evaluated in the modal coordinates. Usually, the damping estimation is more accurate in modal coordinates.

In Appendix A the Matlab functions *modal1m.m* and *modal1n.m* determine the model 1 in modal coordinates using the modal data or nodal data; and functions *modal2m.m* and *modal2n.m* determine the model 2 in modal coordinates using the modal data or nodal data.

Example 2.7. Obtain the third state-space model in modal coordinates for the simple structure from Example 2.2.

From (2.45) we have

$$A = \left[\begin{array}{ccc|ccc} 0 & 0 & 0 & 1 & 0 & 0 \\ 0 & 0 & 0 & 0 & 1 & 0 \\ 0 & 0 & 0 & 0 & 0 & 1 \\ \hline -9.7409 & 0 & 0 & -0.0974 & 0 & 0 \\ 0 & -4.6649 & 0 & 0 & -0.0466 & 0 \\ 0 & 0 & -0.5942 & 0 & 0 & -0.0059 \end{array}\right],$$

$$B = \begin{bmatrix} 0 \\ 0 \\ 0 \\ \hline 0.328 \\ -0.591 \\ 0.737 \end{bmatrix},$$

$$C = \left[\begin{array}{ccc|ccc} 0.591 & 0.737 & 0.328 & 0 & 0 & 0 \\ 0 & 0 & 0 & 0.591 & 0.737 & 0.328 \\ 0 & 0 & 0 & 0.328 & -0.591 & 0.737 \end{array}\right].$$

Example 2.8. For a simple system from Example 2.7 find the second state-space model in modal coordinates, as in (2.42).

Applying the transformation (2.39) to the representation from Example 2.7 we obtain

$$A = \left[\begin{array}{ccc|ccc} -0.0487 & 0 & 0 & 3.1210 & 0 & 0 \\ 0 & -0.0233 & 0 & 0 & 2.1598 & 0 \\ 0 & 0 & -0.0030 & 0 & 0 & 0.7708 \\ \hline -3.1203 & 0 & 0 & -0.0487 & 0 & 0 \\ 0 & -2.1596 & 0 & 0 & -0.0233 & 0 \\ 0 & 0 & -0.7708 & 0 & 0 & -0.0030 \end{array}\right],$$

$$B = \begin{bmatrix} 0 \\ 0 \\ 0 \\ \hline 0.3280 \\ -0.5910 \\ 0.7370 \end{bmatrix},$$

and

$$C = \left[\begin{array}{ccc|ccc} 0.1894 & 0.3412 & 0.4255 & 0 & 0 & 0 \\ -0.0092 & -0.0080 & -0.0013 & 0.5910 & 0.7370 & 0.3280 \\ -0.0051 & 0.0064 & -0.0028 & 0.3280 & -0.5910 & 0.7370 \end{array}\right].$$

2.3.3 Modal Models

Although the above representations were derived using modal displacements, q_m, they are not considered modal state representations. The modal state-space representation is a triple (A_m, B_m, C_m) characterized by the block-diagonal state matrix, A_m,

$$A_m = \text{diag}(A_{mi}) = \left[\begin{array}{cc|cc|cc|cc} \times & \times & 0 & 0 & \cdots & \cdots & 0 & 0 \\ \times & \times & 0 & 0 & \cdots & \cdots & 0 & 0 \\ \hline 0 & 0 & \times & \times & \cdots & \cdots & 0 & 0 \\ 0 & 0 & \times & \times & \cdots & \cdots & 0 & 0 \\ \hline \cdots & \cdots & \cdots & \cdots & \cdots & \cdots & \cdots & \cdots \\ \cdots & \cdots & \cdots & \cdots & \cdots & \cdots & \cdots & \cdots \\ \hline 0 & 0 & 0 & 0 & \cdots & \cdots & \times & \times \\ 0 & 0 & 0 & 0 & \cdots & \cdots & \times & \times \end{array}\right], \quad i = 1, 2, \ldots, n, \tag{2.47}$$

where A_{mi} are 2×2 blocks (their nonzero elements are marked with $\times$), and the modal input and output matrices are divided, correspondingly,

$$B_m = \begin{bmatrix} B_{m1} \\ B_{m2} \\ \vdots \\ B_{mn} \end{bmatrix}, \qquad C_m = \begin{bmatrix} C_{m1} & C_{m2} & \cdots & C_{mn} \end{bmatrix}, \tag{2.48}$$

where B_{mi} and C_{mi} are $2 \times s$ and $r \times 2$ blocks, respectively.

The state x of the modal representation consists of n independent components, x_i, that represent a state of each mode

$$x = \begin{Bmatrix} x_1 \\ x_2 \\ \vdots \\ x_n \end{Bmatrix}, \tag{2.49}$$

and each component consists of two states

$$x_i = \begin{Bmatrix} x_{i1} \\ x_{i2} \end{Bmatrix}. \tag{2.50}$$

The ith component, or mode, has the state-space representation (A_{mi}, B_{mi}, C_{mi}) independently obtained from the state equations

$$\dot{x}_i = A_{mi} x_i + B_{mi} u,$$
$$y_i = C_{mi} x_i, \tag{2.51}$$
$$y = \sum_{i=1}^{n} y_i.$$

This decomposition is justified by the block-diagonal form of the matrix A_m, and is illustrated in Fig. 2.9 for $n = 2$. In generic coordinates each state depends on itself (through the gain A_{mi} shown in Fig. 2.9 with a solid line) and on other states (through the gains A_{mij} shown in Fig. 2.9 with a dashed line). In modal coordinates the cross-coupling gains A_{mij} are zero, thus each state is independent and depends only on itself.

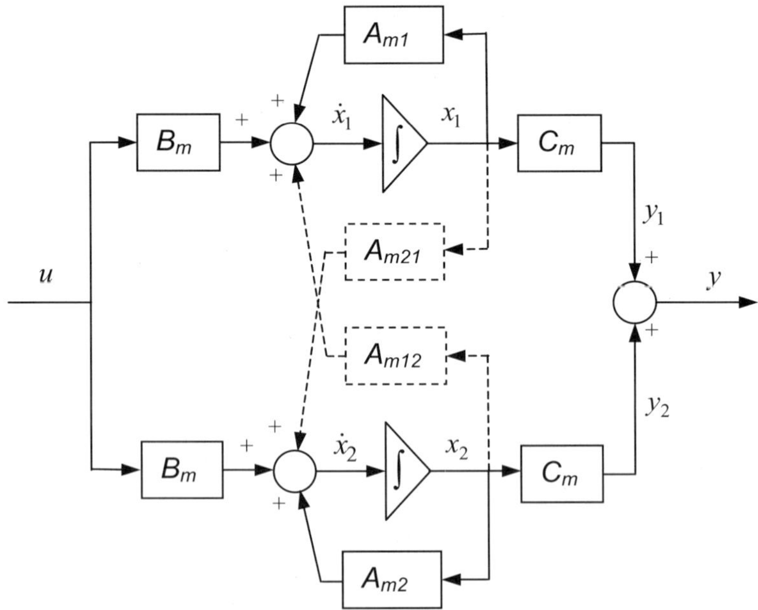

Figure 2.9. Block diagram of the state-space representation of a structure with two modes: The modal cross coupling (marked with a dashed line) is nonexistent.

We consider three modal representations in this book. The blocks A_{mi}, B_{mi}, and C_{mi} of these models are as follows:

- Modal model 1:

$$A_{mi} = \begin{bmatrix} 0 & \omega_i \\ -\omega_i & -2\zeta_i\omega_i \end{bmatrix}, \qquad B_{mi} = \begin{bmatrix} 0 \\ b_{mi} \end{bmatrix}, \qquad C_{mi} = \begin{bmatrix} \dfrac{c_{mqi}}{\omega_i} & c_{mvi} \end{bmatrix}; \tag{2.52}$$

- Modal model 2:

$$A_{mi} = \begin{bmatrix} -\zeta_i\omega_i & \omega_i \\ -\omega_i & -\zeta_i\omega_i \end{bmatrix}, \qquad B_{mi} = \begin{bmatrix} 0 \\ b_{mi} \end{bmatrix}, \qquad C_{mi} = \begin{bmatrix} \dfrac{c_{mqi}}{\omega_i} - c_{mvi}\zeta_i & c_{mvi} \end{bmatrix}; \tag{2.53}$$

- Modal model 3:

$$A_{mi} = \begin{bmatrix} 0 & 1 \\ -\omega_i^2 & -2\zeta_i\omega_i \end{bmatrix}, \qquad B_{mi} = \begin{bmatrix} 0 \\ b_{mi} \end{bmatrix}, \qquad C_{mi} = \begin{bmatrix} c_{mqi} & c_{mvi} \end{bmatrix}. \tag{2.54}$$

The ith state component for the first modal model is as follows:

$$x_i = \begin{Bmatrix} \omega_i q_{mi} \\ \dot{q}_{mi} \end{Bmatrix}, \tag{2.55}$$

for the second modal model it is

$$x_i = \begin{Bmatrix} \omega_i q_{mi} \\ \zeta_i\omega_i q_{mi} + \dot{q}_{mi} \end{Bmatrix}, \tag{2.56}$$

and for the third modal model it is

$$x_i = \begin{Bmatrix} q_{mi} \\ \dot{q}_{mi} \end{Bmatrix}, \tag{2.57}$$

where q_{mi} and $\dot{q}_{mi}$ are the ith modal displacement and velocity, as defined in (2.18). Note that each component consists of modal displacement and velocity which, by (2.18), gives the original (nodal) displacement q and velocity $\dot{q}$. Note also that eigenvalues of A_{mi} are the complex conjugate poles given by (2.32).

We obtain the modal models 1, 2, and 3 from the corresponding state-space representations in modal coordinates as in (2.38), (2.42), and (2.45), by simply rearranging the columns of A and C and the rows of A and B. Consider, for example, the third representation, with the state vector $x^T = \begin{bmatrix} q_m^T & \dot{q}_m^T \end{bmatrix}$, consisting of modal displacements followed by modal rates. We transform it to the new state defined as follows:

$$x = \begin{Bmatrix} q_{m1} \\ \dot{q}_{m1} \\ \hdashline q_{m2} \\ \dot{q}_{m2} \\ \hdashline \vdots \\ \hdashline q_{mn} \\ \dot{q}_{mn} \end{Bmatrix} = \begin{Bmatrix} x_1 \\ x_2 \\ \vdots \\ x_n \end{Bmatrix}, \tag{2.58}$$

where the modal displacement for each mode stays next to its rate. The variable x_i in the above equation is defined by (2.57). Formally, this state-space representation is derived using the transformation matrix R in the form

$$R = \begin{bmatrix} e_1 & 0 \\ 0 & e_1 \\ \hdashline e_2 & 0 \\ 0 & e_2 \\ \hdashline \vdots & \vdots \\ \hdashline e_n & 0 \\ 0 & e_n \end{bmatrix} = \begin{bmatrix} E_1 \\ E_2 \\ \vdots \\ E_n \end{bmatrix}, \qquad \text{where} \qquad E_i = \begin{bmatrix} e_i & 0 \\ 0 & e_i \end{bmatrix}, \tag{2.59}$$

while e_i is an n row vector with all elements equal to zero except the ith which is equal to one, and 0 denotes an n row vector of zeros (actually, we simply rearrange the coordinates).

We obtain the modal models 1 and 2 in a similar manner by rearranging the states in the state vectors (2.36) and (2.40). The new state vectors for these representations are as follows:

$$x = \begin{Bmatrix} \omega_1 q_{m1} \\ \dot{q}_{m1} \\ \hdashline \omega_2 q_{m2} \\ \dot{q}_{m2} \\ \hdashline \vdots \\ \hdashline \omega_n q_{mn} \\ \dot{q}_{mn} \end{Bmatrix} = \begin{Bmatrix} x_1 \\ x_2 \\ \vdots \\ x_n \end{Bmatrix}, \qquad x = \begin{Bmatrix} \omega_1 q_{m1} \\ \zeta_1 \omega_1 q_{m1} + \dot{q}_{m1} \\ \hdashline \omega_2 q_{m2} \\ \zeta_2 \omega_2 q_{m2} + \dot{q}_{m2} \\ \hdashline \vdots \\ \hdashline \omega_n q_{mn} \\ \zeta_n \omega_n q_{mn} + \dot{q}_{mn} \end{Bmatrix} = \begin{Bmatrix} x_1 \\ x_2 \\ \vdots \\ x_n \end{Bmatrix}. \tag{2.60}$$

In the above equation x_i for modal models 1 and 2 are defined by (2.55).

We obtain the state representations from the representations as in (2.38) and (2.42), respectively, by rearranging the columns of A and C and the rows of A and B.

Note that the modal models (2.52), (2.53), and (2.54) are not unique in the sense that for the same matrix A_i, one can obtain different matrices B_i and C_i, as explained in Appendix A.1. In particular, the first component of the input matrix B_i might not necessarily be zero, unless one finds a unique transformation that preserves the zero entry. For details see Appendix A.1.

In Appendix A the Matlab functions *modal1m.m* and *modal1n.m* determine the modal model 1 using the modal or nodal data; and functions *modal2m.m* and *modal2n.m* determine the modal model 2 using the modal or nodal data. Also, the functions *modal1.m* and *modal2.m* determine modal model 1 or 2 using an arbitrary state-space representation.

Example 2.9. Obtain modal model 2 from the model in Example 2.8.

Using transformation (2.59) we find

$$A_m = \left[\begin{array}{cc|cc|cc} -0.0487 & 3.1210 & 0 & 0 & 0 & 0 \\ -3.1203 & -0.0487 & 0 & 0 & 0 & 0 \\ \hline 0 & 0 & -0.0233 & 2.1598 & 0 & 0 \\ 0 & 0 & -2.1596 & -0.0233 & 0 & 0 \\ \hline 0 & 0 & 0 & 0 & -0.0030 & 0.7708 \\ 0 & 0 & 0 & 0 & -0.7708 & -0.0030 \end{array}\right],$$

$$B_m = \left[\begin{array}{c} 0 \\ 0.3280 \\ \hline 0 \\ -0.5910 \\ \hline 0 \\ 0.7370 \end{array}\right] = \begin{bmatrix} b_{m1} \\ b_{m2} \\ b_{m3} \end{bmatrix},$$

and

$$C_m = \left[\begin{array}{cc|cc|cc} 0.1894 & 0 & 0.3412 & 0 & 0.4255 & 0 \\ -0.0092 & 0.5910 & -0.0080 & 0.7370 & -0.0013 & 0.3280 \\ -0.0051 & 0.3280 & 0.0064 & -0.5910 & -0.0028 & 0.7370 \end{array}\right] = \begin{bmatrix} c_{m1} & c_{m2} & c_{m3} \end{bmatrix}.$$

The transfer function of a structure is defined in (2.5). In modal coordinates it is, of course,

$$G(s) = C_m (sI - A_m)^{-1} B_m. \tag{2.61}$$

As was said before, the transfer function is invariant under the coordinate transformation; however, its internal structure is different from the generic transfer function (2.6). In modal coordinates the matrix $sI - A_m$ is block-diagonal, and it can be decomposed into a sum of transfer functions for each mode, therefore,

Property 2.2. *Transfer Function in Modal Coordinates.* *The structural transfer function is a composition of modal transfer functions:*

$$G(\omega) = \sum_{i=1}^{n} G_{mi}(\omega) = \sum_{i=1}^{n} \frac{(c_{mqi} + j\omega c_{mvi}) b_{mi}}{\omega_i^2 - \omega^2 + 2j\zeta_i \omega_i \omega}, \tag{2.62}$$

where

$$G_{mi}(\omega) = C_{mi}(j\omega I - A_{mi})^{-1} B_{mi} = \frac{(c_{mqi} + j\omega c_{mvi}) b_{mi}}{\omega_i^2 - \omega^2 + 2j\zeta_i \omega_i \omega}, \quad i = 1, \ldots, n, \tag{2.63}$$

is the transfer function of the ith mode. The value of the transfer function at the ith resonant frequency is approximately equal to the value of the ith mode transfer function at this frequency:

$$G(\omega_i) \cong G_{mi}(\omega_i) = \frac{(-jc_{mqi} + \omega_i c_{mvi}) b_{mi}}{2\zeta_i \omega_i^2}. \tag{2.64}$$

Proof. Introducing A, B, and C as in (2.45) to the definition of the transfer function we obtain

$$G(\omega) = C_m (j\omega I - A_m)^{-1} B_m = \sum_{i=1}^{n} C_{mi}(j\omega I - A_{mi})^{-1} B_{mi} = \sum_{i=1}^{n} G_{mi}(\omega),$$

which proves the first part. The second part follows from the first part by noting that, for flexible structures with distinct natural frequencies and low damping, $\|G_{mj}(\omega_i)\|_2 \ll \|G_{mi}(\omega_i)\|_2$ for $i \neq j$. ▣

3 Special Models

↳ *how to describe less-common structures*

> Do not quench your inspiration and your imagination;
> do not become the slave of your model.
> —*Vincent van Gogh*

Models described in the previous chapter include typical structural models, which are continuous-time, stable, and with proportional damping. In this chapter we consider models that are not typical in the above sense but, nevertheless, often used in engineering practice. Thus, we will consider models with rigid-body modes (which are unstable), models with nonproportional damping, discrete-time structural models, models with acceleration measurements, and generalized structural models. The latter include two kinds of inputs: controlled (or test) inputs and disturbance inputs, and also two kinds of outputs: measured outputs and outputs where the system performance is evaluated.

3.1 Models with Rigid-Body Modes

Many structures are "free" or unrestrained—they are not attached to a base. An example is the Deep Space Network antenna structure shown in Fig. 1.5: if uncontrolled, it can rotate freely with respect to the azimuth (vertical) axis and its dish can freely rotate with respect to the elevation (horizontal) axis. Modal analysis for such structures shows that they have zero natural frequency, and that the corresponding natural mode shows structural displacements without flexible deformations. A mode without flexible deformations is called a rigid-body mode. Corresponding zero frequency implies that the zero frequency harmonic excitation (which is a constant force or torque) causes rigid-body movement of the structure. Structural analysts sometimes ignore this mode, as there is no deformation involved. However, it is of crucial importance for a control engineer, since this mode is the one that allows a controller to move a structure and track a command.

We obtain the rigid-body modes by solving the same eigenvalue problem as presented for the standard models. Since the natural frequency is zero, the modal equation (2.10) becomes

$$\det(K) = 0, \tag{3.1}$$

i.e., the stiffness matrix becomes singular. The corresponding rigid-body mode ϕ_{rb} is the one that satisfies the equation

$$K\phi_{rb} = 0. \tag{3.2}$$

We obtain the modal equations for the rigid-body modes from (2.26) by assuming $\omega_i = 0$, i.e.,

$$\begin{aligned} \ddot{q}_{mi} &= b_{mi}u, \\ y_i &= c_{mqi}q_{mi} + c_{mvi}\dot{q}_{mi}, \\ y &= \sum_{i=1}^{n} y_i. \end{aligned} \tag{3.3}$$

The state-space modal model for a rigid-body mode exists only in form 3, namely,

$$A_{mi} = \begin{bmatrix} 0 & 1 \\ 0 & 0 \end{bmatrix}, \qquad B_{mi} = \begin{bmatrix} 0 \\ b_{mi} \end{bmatrix}, \qquad C_{mi} = \begin{bmatrix} c_{mqi} & c_{mvi} \end{bmatrix}, \tag{3.4}$$

and we obtain it from (2.54) by assuming $\omega_i = 0$. Finally, we obtain the transfer function of a rigid-body mode from (2.62) for $\omega_i = 0$,

$$G_{mi}(\omega) = \frac{-(c_{mqi} + j\omega c_{mvi})b_{mi}}{\omega^2}. \tag{3.5}$$

For an experienced engineer rigid-body frequency and mode are not difficult to determine: rigid-body frequency is known in advance, since it is always zero, and rigid-body mode can be predicted as a structural movement without deformation. The importance of distinguishing it from "regular" modes is the fact that they make a system unstable, thus a system that requires special attention.

Example 3.1. Find natural frequencies and modes of a simple system from Example 2.1 assuming that $k_1 = 0$. The latter assumption causes the structure to float (there are no springs k_1 and k_4 that attach the structure to the base).

For this system, the nodal model consists of the mass matrix M, $M = \text{diag}(m_1,\ m_2,\ m_3) = I_3$, and the stiffness matrix:

$$K = \begin{bmatrix} k_2 & -k_2 & 0 \\ -k_2 & k_2 + k_3 & -k_3 \\ 0 & -k_3 & k_3 \end{bmatrix} = \begin{bmatrix} 3 & -3 & 0 \\ -3 & 6 & -3 \\ 0 & -3 & 3 \end{bmatrix}.$$

By solving the eigenvalue problem we find that the natural frequency matrix is

$$\Omega = \begin{bmatrix} 0.0000 & 0 & 0 \\ 0 & 1.7321 & 0 \\ 0 & 0 & 3.0000 \end{bmatrix},$$

and the modal matrix is

$$\Phi = \begin{bmatrix} 0.5774 & 0.7071 & 0.4083 \\ 0.5774 & 0.0000 & -0.8165 \\ 0.5774 & -0.7071 & 0.4083 \end{bmatrix}. \quad \text{(a)}$$

The modes are shown in Fig. 3.1. Note that the first mode does not have flexible deformations (springs are neither expanded nor compressed). This is the rigid-body mode. Note also that the corresponding natural frequency is zero.

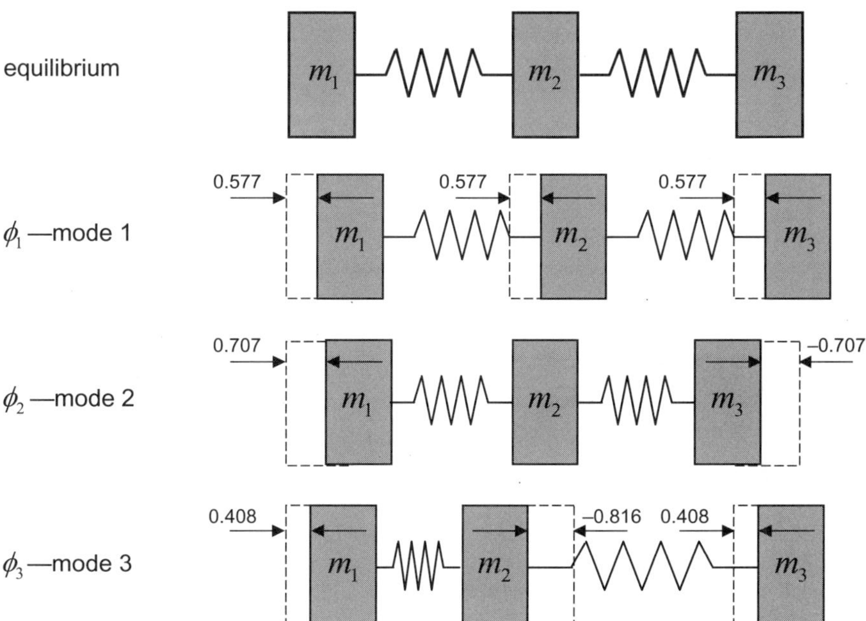

Figure 3.1. Modes of a simple system (see (a)): The first mode is the rigid-body mode (without spring deformation).

Example 3.2. Determine the rigid-body modes of the Deep Space Network antenna.

The antenna has actually two rigid-body modes: rigid-body rotation with respect to the azimuth (vertical) axis, and rigid-body rotation with respect to the elevation (horizontal) axis. Figure 3.2 shows the azimuth rigid-body mode. Figure 3.2(a) presents the initial position from the side view, Fig. 3.2(b) presents the modal displacement (rigid-body rotation with respect to the azimuth axis) from the side view, Fig. 3.2(c) presents the initial position from the top view, and Fig. 3.2(d) presents the modal displacement from the top view. Note that there are no structural deformations, only displacements.

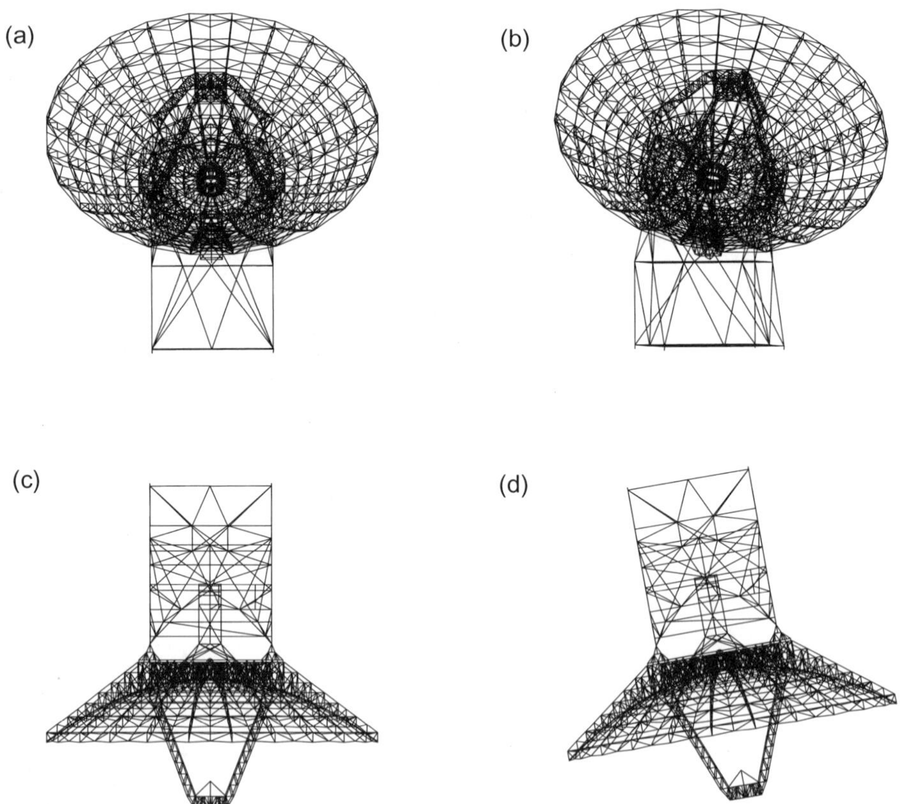

Figure 3.2. Antenna in neutral position (a) and (c); and rigid-body mode of the antenna (b) and (d); where no flexible deformations are observed.

3.2 Models with Accelerometers

Accelerometers are frequently used as structural sensors due to their simplicity, and because they do not require a reference frame. However, they amplify high-frequency parasitic noise. In this section we will discuss the modeling of structures instrumented with accelerometers. The acceleration output was not an option in the standard structural model, in the second order, or in the state-space model, cf. (2.7) and (2.35). In both models the output is composed of structural displacements and/or velocities.

3.2.1 State-Space Representation

Since accelerations are not state variables of the system, they cannot be expressed solely as a linear combination of the states, that is, the output cannot just be $y = Cx$, as in (2.1).

Consider the state-space model with the output being a linear combination of nodal accelerations, i.e.,

$$y = C_a \ddot{q}. \tag{3.6}$$

This output is not a part of the state vector x. Note, however, that it is a part of the state derivative, since

$$\dot{x} = \begin{Bmatrix} \dot{q} \\ \ddot{q} \end{Bmatrix}. \tag{3.7}$$

Combining (3.6) and (3.7) we find that the acceleration output is expressed through the state derivative

$$y = \begin{bmatrix} 0 & C_a \end{bmatrix} \dot{x}. \tag{3.8}$$

Introducing $\dot{x}$ from (2.1) we obtain

$$y = \begin{bmatrix} 0 & C_a \end{bmatrix} Ax + \begin{bmatrix} 0 & C_a \end{bmatrix} Bu$$

or, using A from (2.35), we arrive at

$$y = \begin{bmatrix} -C_a M^{-1} K & -C_a M^{-1} D \end{bmatrix} x + C_a M^{-1} B_o u,$$

which is, in short notation,

$$y = Cx + Du, \tag{3.9}$$

where

$$C=\left[-C_aM^{-1}K \quad -C_aM^{-1}D\right]=-C_a\left[M^{-1}K \quad M^{-1}D\right] \tag{3.10}$$

and

$$D=C_aM^{-1}B_o. \tag{3.11}$$

Equations (3.10), (3.11) are the output equations for the state-space model of a structure with acceleration output.

In this way, we obtained an extended state-space model with four (*A*,*B*,*C*,*D*) rather than three (*A*,*B*,*C*) parameters. Its corresponding state equations are as follows:

$$\begin{aligned}\dot{x}&=Ax+Bu,\\ y&=Cx+Du.\end{aligned} \tag{3.12}$$

The matrix *D* is called a feed-through matrix, since it represents this part of the output that is proportional to the input. Note that *D* denotes the feed-through matrix and the damping matrix. They should not be confused.

Note from (3.11) that matrix *D* in the acceleration measurements is zero if the accelerometer location matrix C_a and the excitation force matrix B_o are orthogonal with respect to matrix M^{-1}. In particular, for a diagonal matrix *M*, the matrix *D* is zero if accelerometers and excitation forces are not collocated.

We can obtain a similar model in modal coordinates, where the acceleration output is a linear combination of modal accelerations, namely,

$$y=C_{ma}\ddot{q}_m, \tag{3.13}$$

where $C_{ma}=C_a\Phi$. Introducing the state derivative from (2.1) we obtain

$$y=\left[0 \quad C_{ma}\right]\dot{x}=\left[0 \quad C_{ma}\right]Ax+\left[0 \quad C_{ma}\right]Bu.$$

Next, using the state-space representation from (2.45) we rewrite the above equation as follows:

$$y=\left[-C_{ma}\Omega^2 \quad -2C_{ma}Z\Omega\right]x+C_{ma}B_mu.$$

Thus, in short, we again obtain $y=Cx+Du$, where

$$\begin{aligned}C&=\left[-C_{ma}\Omega^2 \quad -2C_{ma}Z\Omega\right]=-C_{ma}\left[\Omega^2 \quad 2Z\Omega\right],\\ D&=C_{ma}B_m.\end{aligned} \tag{3.14}$$

Example 3.3. The Matlab code for this example is in Appendix B. Determine the transfer function and the impulse response of the same system as in Example 2.5. The output is an acceleration of mass 1 rather than the velocity of the same mass.

The state-space representation of the structure has the same A and B matrices as the system in Example 2.5. We determine the C and D matrices from (3.10) and (3.11). For this case $C_a = \begin{bmatrix} 1 & 0 & 0 \end{bmatrix}$ and $B_o^T = \begin{bmatrix} 0 & 0 & 1 \end{bmatrix}$. The mass matrix is an unit matrix, $M = I_3$, and the stiffness matrix is

$$K = \begin{bmatrix} 6 & -3 & 0 \\ -3 & 6 & -3 \\ 0 & -3 & 3 \end{bmatrix}.$$

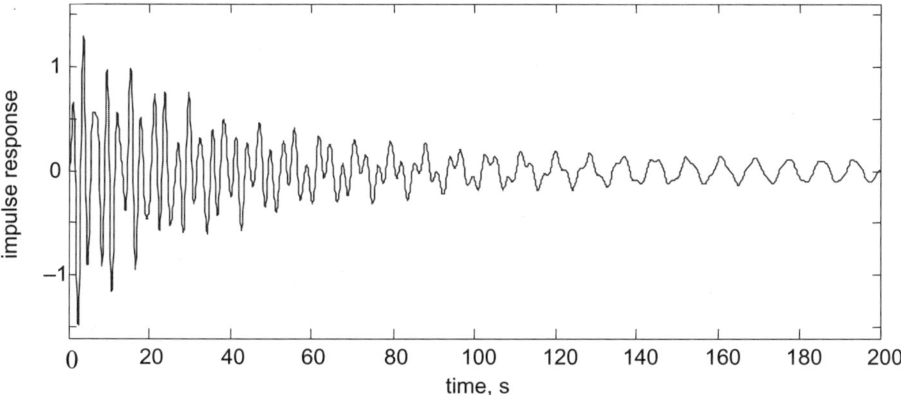

Figure 3.3. Impulse response of a simple structure.

For these data we obtained $D = C_a M^{-1} B_o = 0$, since C_a and B_o are orthogonal, and $C = [-C_a M^{-1} K \quad -C_a M^{-1} D] = [-6 \ 3 \ 0 \ 0 \ 0 \ 0]$. It is interesting to note from the obtained output matrix C that the acceleration of the first mass ($\ddot{q}_1$) is a linear combination of displacements of the first mass (q_1) and the second mass (q_2), that is, $\ddot{q}_1 = -6q_1 + 3q_2$. The impulse response for the structure is shown in Fig. 3.3, and the magnitude and phase of the transfer function in Fig. 3.4. Comparing the velocity response from Example 2.5 (Figs. 2.5 and 2.7) and the acceleration response (Figs. 3.3 and 3.4), note that the higher modes are more visible in the acceleration impulse response, even for $t > 100$ s, and that the resonance peaks in the acceleration transfer function are higher for higher modes; the phase of the transfer function for the acceleration sensor is 90 degrees higher than for the velocity sensor.

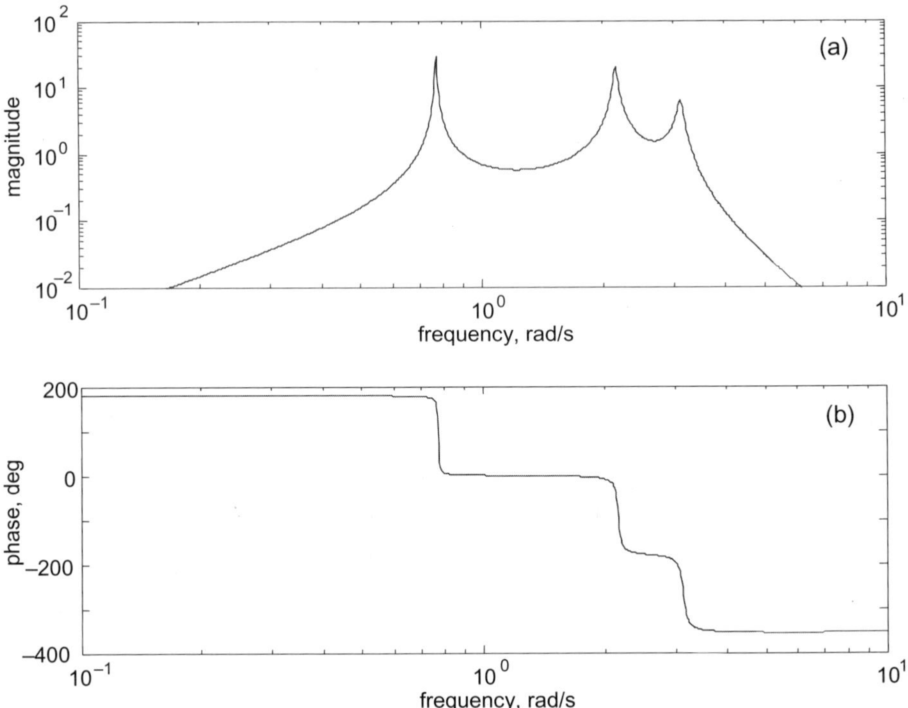

Figure 3.4. Transfer function of a simple structure: (a) Amplitude; and (b) phase.

3.2.2 Second-Order Representation

For a structure with accelerometers we obtain the second-order equations similar to (2.7). Note that from this equation we obtain the acceleration as

$$\ddot{q} = -M^{-1}D\dot{q} - M^{-1}Kq + M^{-1}B_o u. \tag{3.15}$$

Let the accelerometer locations be defined by the output matrix C_a, that is,

$$y = C_a \ddot{q}; \tag{3.16}$$

thus, introducing (3.15) to the above equation yields

$$y = -C_a M^{-1}D\dot{q} - C_a M^{-1}Kq + C_a M^{-1}B_o u$$

or

$$y = C_v \dot{q} + C_q q + D_a u, \tag{3.17}$$

where

$$C_v = -C_a M^{-1} D,$$
$$C_q = -C_a M^{-1} K, \qquad (3.18)$$
$$D_a = C_a M^{-1} B_o.$$

Similar equations can be obtained in modal coordinates. Namely, using (2.19), we arrive at the following acceleration output equation:

$$y = C_m \ddot{q}_m = C_{mv} \dot{q}_m + C_{mq} q_m + D_{ma} u, \qquad (3.19)$$

where

$$C_{mq} = -C_{ma} \Omega^2, \qquad C_{mv} = -2C_{ma} Z\Omega, \qquad (3.20)$$

and the feed-through term does not depend on the coordinate system

$$D_{ma} = C_{ma} B_m = C_a M^{-1} B_o = D_a, \qquad (3.21)$$

while

$$C_{ma} = C_a \Phi. \qquad (3.22)$$

For a single mode we use (2.26), and in this case the acceleration output is as follows:

$$y_i = c_{mvi} \dot{q}_{mi} + c_{mqi} q_{mi},$$
$$y = \sum_{i=1}^{n} y_i + D_a u, \qquad (3.23)$$

and c_{mqi}, c_{mvi} are the ith column of C_{mq}, C_{mv}, respectively.

3.2.3 Transfer Function

A structure with the accelerometers can be considered as a structure with rate sensors cascaded with differentiating devices (the derivative of a rate gives acceleration). For simplicity of notation we consider a structure with a single accelerometer. Denote (A_r, B_r, C_r) and $G_r = C_r(sI - A_r)^{-1} B_r$ as the state-space triple and as the transfer function, respectively, of the structure with a rate sensor. The transfer function G_a of the structure with an accelerometer is therefore

$$G_a = j\omega G_r. \tag{3.24}$$

3.3 Models with Actuators

A flexible structure in testing, or in a closed-loop configuration, is equipped with actuators. Does their presence impact the dynamics of a flexible structure? We answer this question for proof-mass actuators, and inertial actuators attached to a structure.

3.3.1 Model with Proof-Mass Actuators

Proof-mass actuators are widely used in structural dynamics testing. In many cases, however, the actuator dynamics are not included in the model. The proof-mass actuator consists of mass m and a spring with stiffness k, and they are attached to a structure at node n_a. This is a reaction-type force actuator, see [144], [57]. It generates a force by reacting against the mass m, thus force f_o acts on the structure, and $-f$ acts on the mass m (Fig. 3.5 at position n_a). Typically, the stiffness of the proof-mass actuator is much smaller than the dynamic stiffness of the structure (often it is zero).

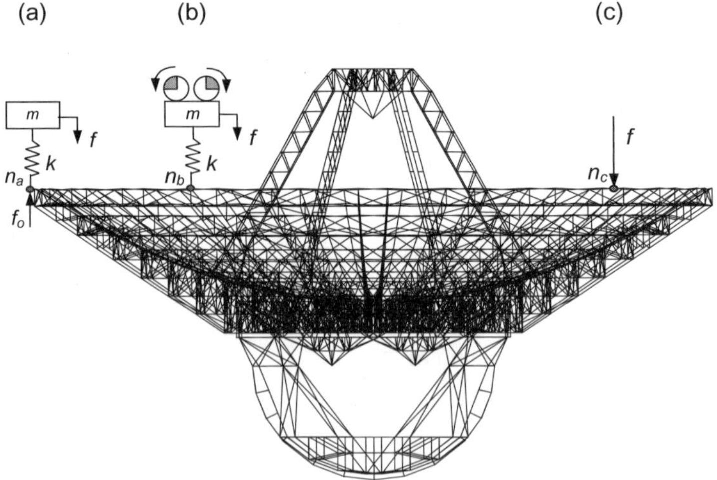

Figure 3.5. A structure with different actuators: (a) With proof-mass actuator; (b) with inertial actuator; and (c) with ideal actuator.

Let M_s, D_s, and K_s be the mass, damping, and stiffness matrices of the structure, respectively, and let B_s be the matrix of the actuator location,

$$B_s = \begin{bmatrix} 0 & 0 & \dots & 0 & 1 & 0 & \dots & 0 \end{bmatrix}^T,$$

with a nonzero term at the actuator location n_a. Denote $G_s(\omega) = -\omega^2 M_s + j\omega D_s + K_s$ and $j = \sqrt{-1}$, then the dynamic stiffness of a structure at the actuator location is defined as

$$k_s = \frac{1}{B_s^T G_s^{-1} B_s}. \tag{3.25}$$

The dynamic stiffness is the inverse of the frequency response function at the actuator location. At zero frequency it is reduced to the stiffness constant at the actuator location.

The structural dynamics of an "ideal" actuator are excited by the force f, as in Fig. 3.5, at node n_c. In contrast, the force generated by the proof-mass actuator consists of an additional force f_a, which is a reaction force from the actuator mount, see Fig. 3.6. Thus, the total force f_o acting on the structure is

$$f_o = f + f_a. \tag{3.26}$$

From Fig. 3.6 we find

$$m\ddot{q} + k(q - q_s) = -f,$$
$$f_a = k(q - q_s),$$

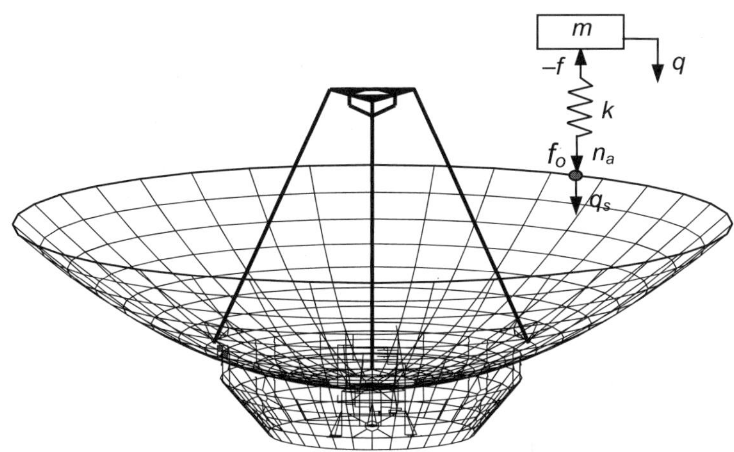

Figure 3.6. Displacements and forces at the proof-mass actuator mounting.

and the structural displacement q_s at node n_a is obtained from the dynamic stiffness of the structure

$$k_s q_s = f + f_a.$$

Combining the last three equations together, after a little algebra, we arrive at the following relationship:

$$f_a = \frac{\rho^2 - \beta}{1 + \beta - \rho^2} f, \tag{3.27}$$

where

$$\rho = \frac{\omega_o}{\omega}, \qquad \omega_o = \sqrt{\frac{k}{m}}, \qquad \beta = \frac{k}{k_s}. \tag{3.28}$$

Introducing (3.27) and (3.28) into (3.26) we obtain the following relationship between the proof-mass actuator force (f_o) and the ideal actuator force (f),

$$f_o = \alpha_c f, \qquad \alpha_c = \frac{1}{1 + \beta - \rho^2}. \tag{3.29}$$

It follows from the above equations that the actuator force, f_o, approximately reproduces the ideal force f if $\alpha_c \cong 1$. This is obtained if

$$\beta \ll 1 \qquad \text{and} \qquad \rho \ll 1 \tag{3.30}$$

The above conditions are satisfied when the actuator stiffness is small (compared with the structural stiffness), and the actuator mass is large enough, such that the actuator natural frequency is smaller than the structural principal frequency. Hence, we replace the above conditions with the following ones:

$$k \ll k_s \qquad \text{and} \qquad \omega \gg \omega_o. \tag{3.31}$$

If these conditions are satisfied, we obtain $f \cong f_o$ and, consequently the transfer function of the system with the proof-mass actuator is approximately equal to the transfer function of the system without the proof-mass actuator.

In addition to conditions (3.31), consider the following ones:

$$\omega_o \ll \omega_1 \qquad \text{and} \qquad k \ll \min_i k_{si}, \tag{3.32}$$

where ω_1 is the fundamental (lowest) frequency of the structure. These conditions say that the actuator natural frequency should be significantly lower than the fundamental frequency of the structure, and that the actuator stiffness should be much smaller than the dynamic stiffness of the structure at any frequency of interest. If the aforementioned conditions are satisfied, we obtain $\alpha_{ci} \cong 1$ for $i = 1,\ldots,n$; thus, the forces of the structure with the proof-mass actuator are equal to the forces of the structure without the proof-mass actuator. Note also that for many cases, whenever the first condition of (3.31) is satisfied, the second condition (3.32) is satisfied too.

Example 3.4. Compare the transfer functions of the 3D truss, Fig. 1.3, with and without proof-mass actuator. The force input is at node 21, acting in the *y*-direction, and the rate output is measured at node 14 in the *y*-direction.

The magnitude of its transfer function for this force is shown in Fig. 3.7 as a solid line. The proof-mass actuator was attached to node 21 to generate the input force. The mass of the proof-mass actuator is $m = 0.1$ Ns^2/cm, and its stiffness is $k = 1$ N/cm. Its natural frequency is $\omega_o = 3.1623\,\text{rad/s}$, much lower than the truss fundamental frequency, $\omega_1 = 32.8213\,\text{rad/s}$. The plot of the magnitude of the transfer function for the truss with the proof-mass actuator is shown in Fig. 3.7 as a dashed line. The figure shows perfect overlapping of the transfer functions for $f \gg f_o$, where $f_o = \omega_o / 2\pi = 0.5033\,\text{Hz}$.

3.3.2 Model with Inertial Actuators

In the inertial actuator, force is proportional to the square of the excitation frequency. It consists of mass *m* and a spring with stiffness *k*, and they are attached to a structure at node, say, *na*. The force acts on mass *m* exclusively (Fig. 3.5 at position n_b). It is assumed that the stiffness of the actuator is much smaller than the dynamic stiffness of the structure.

This configuration is shown in Fig. 3.5, position n_b. The force acting on mass *m* is proportional to the squared frequency

$$f = \kappa \omega^2, \tag{3.33}$$

where κ is a constant. The relationship between transfer functions of a structure, without (G_s) and with (G_c) an inertial actuator, is as follows:

$$G_c = \alpha_c G_s, \qquad \alpha_c = \frac{\kappa \omega_o}{1 + \beta - \rho^2}, \tag{3.34}$$

which are derived from the actuator equations

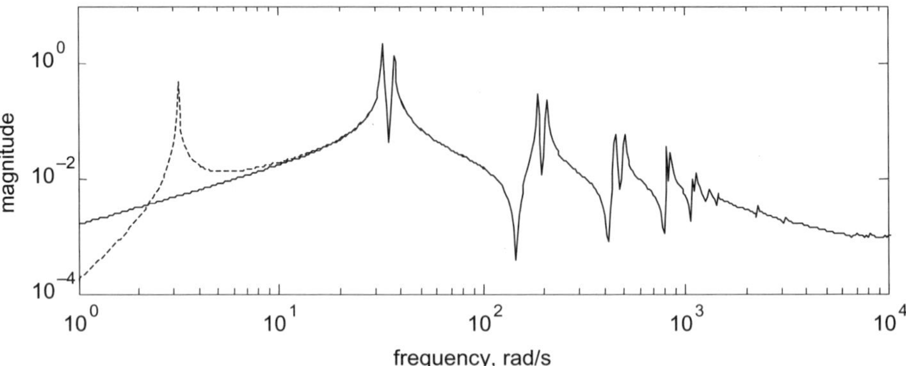

Figure 3.7. 3D truss without and with the proof-mass actuator: Magnitudes of the transfer function without (solid line) and with (dashed line) the proof-mass actuator; they are identical except for the resonant peak of the proof-mass actuator itself.

$$m\ddot{q} + k(q - q_s) = \kappa\omega^2,$$
$$f_o = f_a = k(q - q_s).$$

For these equations we obtain

$$f_o = \alpha_c f, \qquad \alpha_c = \frac{\kappa\omega^2}{1 + \beta - \rho^2}. \tag{3.35}$$

The above result shows that the structural transfer function with the inertial actuator is proportional to the structural transfer function without the actuator, and the proportionality coefficient depends on frequency.

In applications, measurement noise is unavoidable, thus care should be taken in the selection of the scaling factor. For instance, if the scaling factor is too small, modes with small norms cannot be detected, and the reduction procedure could be biased.

3.4 Models with Small Nonproportional Damping

The damping properties of structures are often assumed in the modal form, i.e., they are introduced as damping coefficients ζ_i in the modal equations (2.19) or (2.26). This is done not only for the sake of analytical simplicity, but also because it is the most convenient way to measure or estimate it. This is the way, for example, to estimate the material damping in the finite-element analysis of large flexible structures, where the modal analysis is executed, the low-frequency modes retained, and modal damping for these modes assumed. The resulting damping is a proportional one. In another approach, a damping matrix proportional either to the mass or to the stiffness matrix, or to both, is introduced. This technique produces proportional damping as well.

However, in many practical problems structural damping is not proportional. In this book we analyze structural dynamics and design controllers for the proportional damping only. The question arises: Are the analysis and the design procedures valid for the case of nonproportional damping? In this section we show that small nonproportional damping can be replaced with proportional damping without causing a significant error.

Several authors analyzed the replacement of nonproportional damping with proportional damping; including Cronin [23], Chung and Lee [16], Bellos and Inman [8], Yae and Inman [141], Nicholson [112], and Felszeghy [35]. The simplest and most common approach to the problem is to replace the full modal damping matrix with a diagonal one by neglecting the off-diagonal terms of the nonproportional damping matrix. Several researchers studied the error bounds generated by this simplified approach; see, for example, Shahruz and Ma [123], Uwadia and Esfandiari [130], Hwang and Ma [76], Bhaskar [11], and Gawronski and Sawicki [63].

In order to analyze the impact of nonproportional damping on system dynamics we consider (2.19)—the second-order modal equation of a structure. We replace the proportional damping matrix $2Z\Omega$ with the full matrix D. This matrix can be decomposed into the diagonal $(2Z\Omega)$ and off-diagonal (D_o) components $D = 2Z\Omega + D_o$, so that the equation of motion is as follows:

$$\ddot{q} + 2Z\Omega\dot{q} + D_o\dot{q} + \Omega^2 q = Bu. \tag{3.36}$$

Now, q is the displacement of the nonproportionally damped structure, and q_m is the displacement of the proportionally damped structure—a solution of (2.19).

Denote by e_i the ith modal error between nonproportionally and proportionally damped structures, i.e., $e_i = q_i - q_{mi}$. Subtracting (2.19) from (3.36) we obtain, for the ith mode,

$$\ddot{e}_i + 2\zeta_i\omega_i\dot{e}_i + \omega_i^2 e_i = -d_{oi}\dot{q}, \tag{3.37}$$

where d_{oi} is the ith row of D_o. A question arises as to when the error is small (compared to the system displacement q) so that the nonproportional part, D_o, can be ignored.

We show the following property:

Property 3.1. ***Error of a Mode with Nonproportional Damping.*** *For nonclustered natural frequencies the error e_i of the ith mode is limited as follows:*

$$\|e_i\|_2 \ll \|q\|_2, \qquad \text{for } i = 1,\ldots,n. \tag{3.38}$$

Proof. We rewrite (3.37) written in the frequency domain as

$$e_i(\omega) = g_i(\omega) d_{oi} q_o(\omega), \tag{3.39}$$

where

$$g_i(\omega) = \frac{-j\omega}{\omega_i^2 - \omega^2 + 2j\zeta_i\omega_i\omega},$$

and q_o is equal to a displacement vector q except the ith component, which is equal to zero. From (3.39) it follows that

$$\begin{aligned}
\|e_i\|_2^2 &= \frac{1}{4\pi^2}\int_{-\infty}^{\infty} |e_i|^2\, d\omega = \frac{1}{4\pi^2}\int_0^{\infty} |g_i(\omega)|^2 \operatorname{tr}(d_{oi} q_o(\omega) q_o^*(\omega) d_{oi}^T)\, d\omega \\
&= \|d_{oi}\|_2^2 \frac{1}{4\pi^2}\int_0^{\infty} |g_i(\omega)|^2 \operatorname{tr}(q_o(\omega) q_o^*(\omega))\, d\omega \\
&\le \|d_{oi}\|_2^2 |g_{i\max}|^2 \frac{1}{4\pi^2}\int_0^{\infty} \operatorname{tr}(q_o(\omega) q_o^*(\omega))\, d\omega \\
&= \|d_{oi}\|_2^2 |g_i(\omega_i)|^2 \|q_o\|_2^2 .
\end{aligned}$$

However,

$$|g_i(\omega_i)| = \frac{1}{2\zeta_i\omega_i} \qquad \text{and} \qquad \|q_o\|_2 \ll \|q\|_2;$$

therefore,

$$\|e_i\|_2^2 \ll \frac{\|d_{oi}\|_2^2}{4\zeta_i^2\omega_i^2}\|q\|_2^2 \le \|q\|_2^2$$

since $d_{oik} \le 2\zeta_i\omega_i$. ▣

The above property implies that for separate natural frequencies the off-diagonal elements of the damping matrix can be neglected regardless of their values.

The following example illustrates the insignificance of the nonproportional damping terms:

Example 3.5. Examine the impact of nonproportional damping on the dynamics of a flexible truss as in Fig. 1.2. The truss damping matrix is proportional to the stiffness matrix, however we added damping at node 5, see Fig. 3.8.

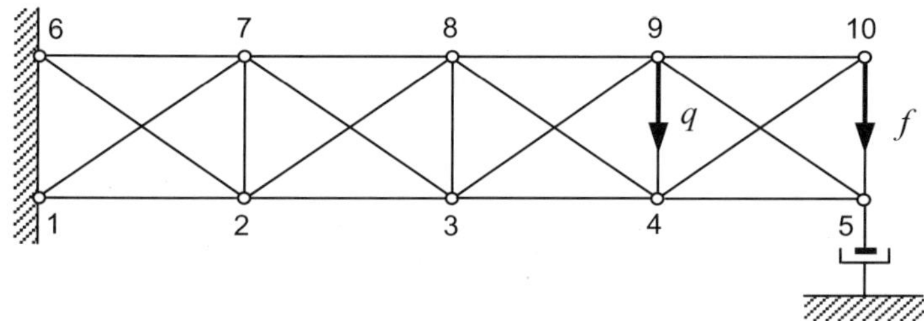

Figure 3.8. The 2D truss with added damper to create nonproportional damping.

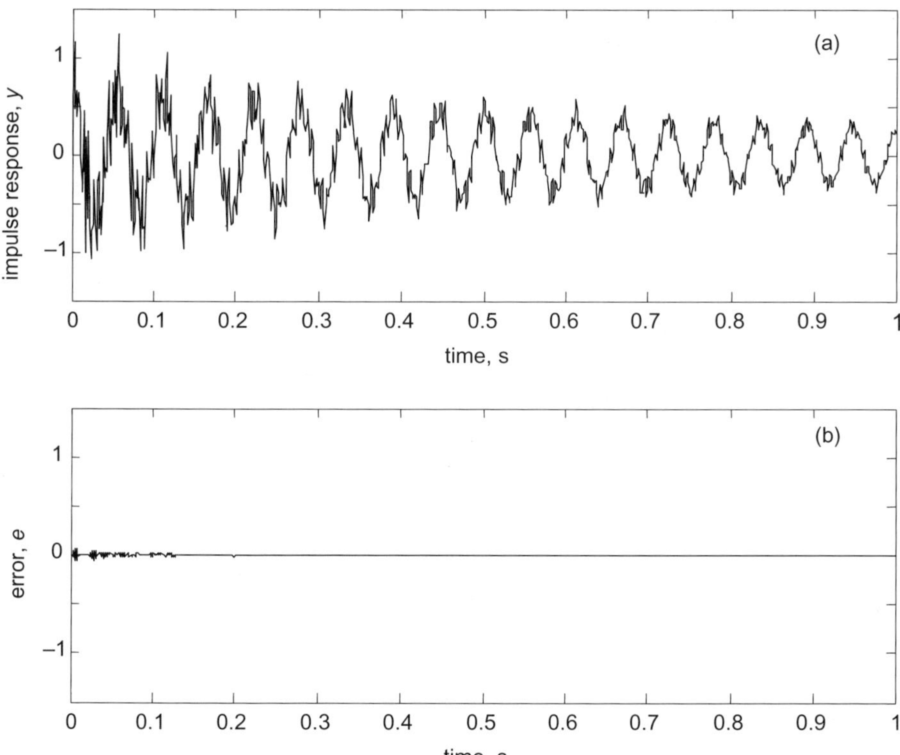

Figure 3.9. The impulse responses: (a) Of the nonproportionally damped system; and (b) of the response error of the system with the equivalent proportional damping. The error is small when compared with the response itself.

This addition makes the damping nonproportional. The structural damping is about 1.5% of the critical damping, and the concentrated viscous damping in a vertical direction at node 5 is ten times larger than the structural damping at this location.

We apply the impulse force at node 10, measure the response at node 9 in a vertical direction, and compare the exact solution (i.e., with full damping matrix) and the approximate solution obtained by neglecting off-diagonal terms in the damping matrix. The exact response (y) is shown in Fig. 3.9(a), while the difference, $e = y - y_d$, between the exact response and the response of the system with the proportional damping (y_d) is shown in Fig. 3.9(b). This difference is small, since the approximate solution has the error $\|y - y_d\|_2 / \|y\|_2 = 0.019$. However, the difference between the full and diagonal matrices is not small. Indeed, let D and D_d stand for the full damping matrix and the diagonal part of the same matrix, then the damping matrix diagonality index, defined as $\|D - D_d\|_2 / \|D\|_2$, is not a small number in this case, it is equal to 0.760.

3.5 Generalized Model

The system models, so far considered, have multiple inputs and outputs. Although multiple, these inputs and outputs could be grouped into a single input vector and a single output vector. For example, a system to be controlled is equipped with sensors and actuators. The independent variables produced by actuators are system inputs, while variables registered by sensors are system outputs. In reality, the situation is more complex. When a system is under testing or control, we can distinguish two kinds of inputs and two kinds of outputs that cannot (or it is not advisable to) be grouped into a single input or output vector. In a controlled system, for example, there is one kind of input: those accessed by a controller and without controller access, such as external disturbances and commands. The same system can have two kinds of outputs: those measured by the controller and those that characterize system performance and often cannot be sensed by the controller. Similarly, in a system under dynamic test, the first kind of inputs are actuator signals, and the second kind are disturbances. The first kind of outputs are sensor signals and the second kind are sensor and environmental noises, or system performance outputs that cannot be accessed by available sensors. This distinction is a basis for the definition of a generalized model.

The generalized model is a two-port system, which consists of two *kinds* of inputs, denoted u and w, and two *kinds* of outputs, denoted y and z, see Fig. 3.10.

The inputs to the generalized model consist of two vector signals:

- the actuator vector, denoted u, which consists of all inputs handled by the controller, or applied as test inputs; and
- the disturbance vector, w, noises and disturbances, which are not manipulated by the controller, or are not a part of the test input.

The outputs of the generalized model consist of two vector signals:

- the sensor vector, y, used for the controller for feedback purposes, or the measured test signals; and
- the performance vector, z, the outputs to be controlled, or to evaluate test performance.

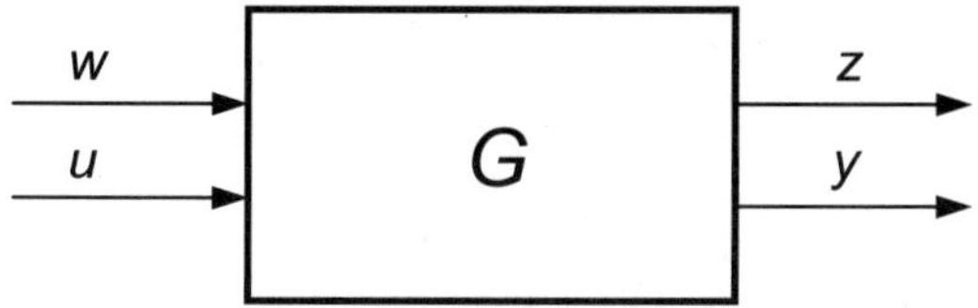

Figure 3.10. The generalized model consists of two inputs (u—actuator and w—disturbance) and two outputs (y—sensor and z—performance).

3.5.1 State-Space Representation

For a generalized structure let A be the state matrix of the system, B_u, B_w represent the input matrices of u and w, respectively, and C_y, C_z represent the output matrices of y and z, respectively. For simplicity, we assume the feed-through terms D_{uy}, D_{wz}, etc., to be equal to zero. Then the state-space representation of the generalized model is as follows:

$$\begin{aligned} \dot{x} &= Ax + B_u u + B_w w, \\ y &= C_y x, \\ z &= C_z z. \end{aligned} \tag{3.40}$$

3.5.2 Transfer Function

We can write the transfer function of the generalized structure as follows

$$\begin{Bmatrix} y \\ z \end{Bmatrix} = \begin{bmatrix} G_{uy} & G_{wy} \\ G_{uz} & G_{wz} \end{bmatrix} \begin{Bmatrix} u \\ w \end{Bmatrix} \tag{3.41}$$

or, equivalently, as

$$
\begin{aligned}
y &= G_{uy}u + G_{wy}w, \\
z &= G_{uz}u + G_{wz}w.
\end{aligned}
\tag{3.42}
$$

The transfer functions are related to the state-space representation as follows:

$$
\begin{aligned}
G_{uy} &= C_y(sI - A)^{-1}B_u, \\
G_{wy} &= C_y(sI - A)^{-1}B_w, \\
G_{uz} &= C_z(sI - A)^{-1}B_u, \\
G_{wz} &= C_z(sI - A)^{-1}B_w.
\end{aligned}
\tag{3.43}
$$

The block diagram of the decomposed standard system that corresponds to (3.42) is shown in Fig. 3.11.

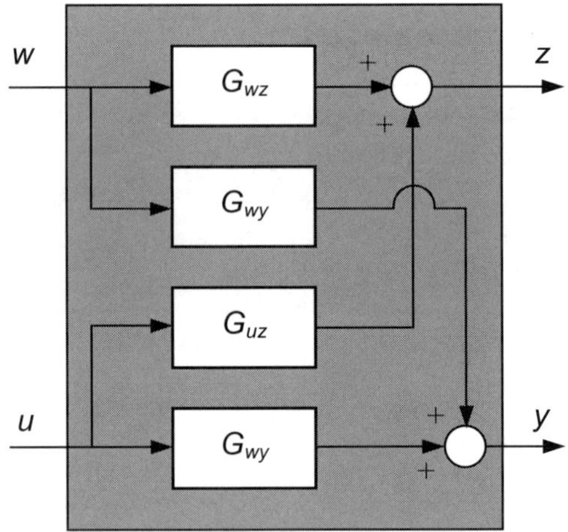

Figure 3.11. Decomposed generalized structure.

3.6 Discrete-Time Models

Continuous-time models are predominantly utilized in the analysis of structural dynamics. However, when implementation is considered, such as structural testing or control, discrete-time models are primarily required. The reason for using the digitalized models in structural testing is that the test data are collected in a digital form (sampled data), and in structural control most controllers are implemented digitally. Therefore, we analyze the discrete-time models along with the continuous-time models.

3.6.1 State-Space Representation

We assume a continuous-time model in the form of the state-space representation (A,B,C,D). The discrete-time sequences of this model are sampled continuous-time signals, i.e.,

$$x_k = x(k\Delta t), \qquad u_k = u(k\Delta t), \qquad \text{and} \qquad y_k = y(k\Delta t), \tag{3.44}$$

for $k = 1, 2, 3 \ldots$. The corresponding discrete-time representation for the sampling time Δt is (A_d, B_d, C_d, D_d), where

$$A_d = e^{A\Delta t}, \qquad B_d = \int_0^{\Delta t} e^{A\tau} B \; d\tau, \qquad C_d = C, \qquad \text{and} \qquad D_d = D, \tag{3.45}$$

and the corresponding state-space equations are

$$\begin{aligned} x_{k+1} &= A_d x_k + B_d u_k, \\ y_k &= C x_k + D u_k. \end{aligned} \tag{3.46}$$

The discretization can be carried out numerically using the *c2d* command of Matlab.

Similarly to the continuous-time models the discrete-time models can also be presented in modal coordinates. Assume small damping and that the sampling rate is sufficiently fast, such that the Nyquist sampling theorem is satisfied (i.e., $\omega_i \Delta t \le \pi$ for all i), see, for example, [37, p. 111], then the state matrix in modal coordinates A_{dm} is block-diagonal,

$$A_{dm} = \operatorname{diag}(A_{dmi}), \qquad i = 1, \ldots, n. \tag{3.47}$$

The 2×2 blocks A_{dmi} are in the form, see [98],

$$A_{dmi} = e^{-\zeta_i \omega_i \Delta t} \begin{bmatrix} \cos(\omega_i \Delta t) & -\sin(\omega_i \Delta t) \\ \sin(\omega_i \Delta t) & \cos(\omega_i \Delta t) \end{bmatrix}, \tag{3.48}$$

where ω_i and ζ_i are the ith natural frequency and the ith modal damping, respectively. The modal input matrix B_{dm} consists of $2 \times s$ blocks B_{dmi},

$$B_{dm} = \begin{bmatrix} B_{dm1} \\ B_{dm2} \\ \vdots \\ B_{dmn} \end{bmatrix}, \tag{3.49}$$

where

$$B_{dmi} = S_i B_{mi},$$
$$S_i = \frac{1}{\omega_i}\begin{bmatrix} \sin(\omega_i \Delta t) & -1+\cos(\omega_i \Delta t) \\ 1-\cos(\omega_i \Delta t) & \sin(\omega_i \Delta t) \end{bmatrix}, \tag{3.50}$$

and B_{mi} is part of the continuous-time modal representation, see (2.42) and (2.51). The discrete-time modal matrix C_{dm} is the same as the continuous-time modal matrix C_m.

The poles of the matrix A_{dm} are composed of the poles of matrices A_{dmi}, $i = 1,\ldots, n$. For the ith mode the poles of A_{dmi} are

$$s_{1,2} = e^{-\zeta_i \omega_i \Delta t}(\cos(\omega_i \Delta t) \pm \sin(\omega_i \Delta t)). \tag{3.51}$$

The location of the poles is shown in Fig. 3.12, which is quite different from the continuous-time system, cf., Fig. 2.1. For a stable system they should be inside the unit circle, which is the case of small damping ζ_i.

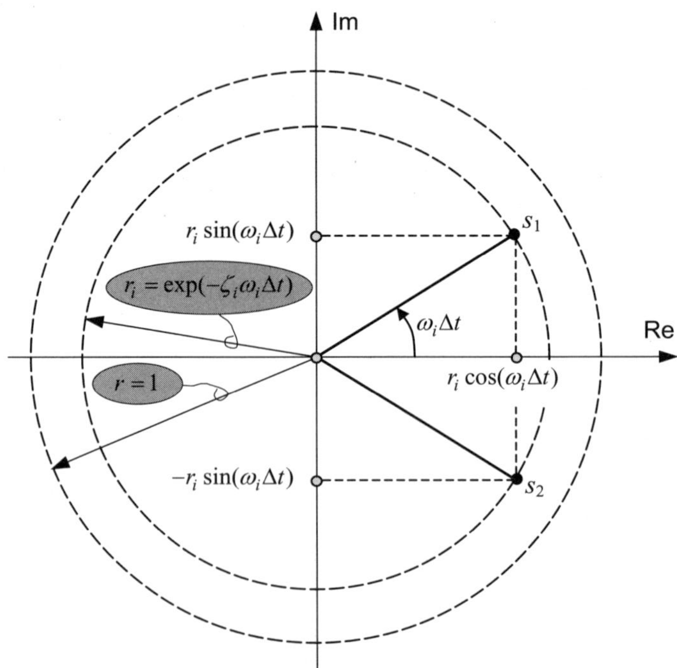

Figure 3.12. Pole location of the *i*th mode of a lightly damped structure in discrete time: It is a complex pair with angle proportional to natural frequency and magnitude close to 1.

The question arises how to choose the sampling time Δt. Note that from the Nyquist criterion the ith natural frequency is recovered if the sampling rate is at least twice the natural frequency in Hz ($f_i = \omega_i / 2\pi$), i.e., if

$$\frac{1}{\Delta t} \geq 2 f_i$$

or, if

$$\omega_i \Delta t \leq \pi \qquad \text{or} \qquad \Delta t \leq \frac{\pi}{\omega_i}. \tag{3.52}$$

Considering all modes, the sampling time will be smaller than the smallest π / ω_i,

$$\Delta t \leq \frac{\pi}{\max\limits_i(\omega_i)}. \tag{3.53}$$

Example 3.6. Obtain a discrete-time model for a simple system from Example 2.9, modal form.

Using the Matlab *c2d* command we obtain

$$A_d = \left[\begin{array}{cc:cc:cc} 0.9970 & 0.0770 & 0 & 0 & 0 & 0 \\ -0.0770 & 0.9964 & 0 & 0 & 0 & 0 \\ \hdashline 0 & 0 & 0.9768 & 0.2138 & 0 & 0 \\ 0 & 0 & -0.2138 & 0.9722 & 0 & 0 \\ \hdashline 0 & 0 & 0 & 0 & 0.9519 & 0.3055 \\ 0 & 0 & 0 & 0 & 0.3056 & 0.9423 \end{array}\right], \qquad B_d = \left[\begin{array}{c} 0.0046 \\ 0.1206 \\ \hdashline 0.0644 \\ -0.0084 \\ \hdashline -0.0337 \\ 0.0064 \end{array}\right],$$

and the discrete-time output matrix is equal to the continuous-time output matrix, $C_d = C$.

3.6.2 Transfer Function

We obtain the transfer function from the state-space representation (3.46) by introducing the shift operator z such that $x_{i+1} = z x_i$, and we find that

$$G_d(z) = C_d (zI - A_d)^{-1} B_d + D_d. \tag{3.54}$$

We use this equation to obtain the transfer function for the ith mode. Using A_d as in (3.48), we determine

$$(zI - A_{dmi})^{-1} = \frac{1}{d}\begin{bmatrix} z - e^{-\zeta\omega_i\Delta t}\cos(\omega_i\Delta t) & e^{-\zeta\omega_i\Delta t}\sin(\omega_i\Delta t) \\ -e^{-\zeta\omega_i\Delta t}\sin(\omega_i\Delta t) & z - e^{-\zeta\omega_i\Delta t}\cos(\omega_i\Delta t) \end{bmatrix}, \tag{3.55}$$

where d is the characteristic polynomial $d = z^2 - 2ze^{-\zeta\omega_i\Delta t}\cos(\omega_i\Delta t) + e^{-2\zeta\omega_i\Delta t}$.

Next, using B_d as in (3.50) and noting that $B_{mi} = \begin{bmatrix} 0 \\ b_{oi} \end{bmatrix}$, we arrive at

$$G_{di}(z) = \frac{C_{mi}}{\omega_i d}\begin{bmatrix} (1 - \cos(\omega_i\Delta t))(z + e^{-\zeta\omega_i\Delta t}) \\ \sin(\omega_i\Delta t)(z - e^{-\zeta\omega_i\Delta t}) \end{bmatrix} b_{oi}, \tag{3.56}$$

which is the transfer function of the ith mode. Note that C_{mi} and b_{oi} in the above equation are the output and input matrices of the continuous-time model.

4
Controllability and Observability

how to excite and monitor a structure

It is the theory which decides
what we can observe
—*Albert Einstein*

Controllability and observability are structural properties that carry information useful for structural testing and control, yet they are not fully utilized by structural engineers. The usefulness can be found by reviewing the definitions of the controllability and observability of a structure. A structure is controllable if the installed actuators excite all its structural modes. It is observable if the installed sensors detect the motions of all the modes. This information, although essential in many applications (e.g., in the placement of sensors and actuators), is too limited. It answers the question of mode excitation or detection in terms of yes or no. The more quantitative answer is supplied by the controllability and observability grammians, which represent a degree of controllability and observability of each mode.

In this chapter we discuss the controllability and observability properties of flexible structures. The fundamental property of a flexible structure in modal coordinates consists of a set of uncoupled modes, as shown in Property 2.1. It allows us to treat the properties of each individual mode separately and to combine them into a property of the entire structure. This also refers to the controllability and observability properties of the whole system, which are combined out of the properties of individual modes. These controllability and observability properties are used later in this book in the evaluation of structural testing and in control analysis and design.

4.1 Definition and Properties

The controllability and observability properties of a linear time-invariant system can be heuristically described as follows. The system dynamics described by the state variable (x) is excited by the input (u) and measured by the output (y). However, the input may not be able to excite all states (or, equivalently, to move them in an arbitrary direction). In this case we cannot fully control the system. Also, not all states may be represented at the output (or, equivalently, the system states cannot be recovered from a record of the output measurements). In this case we cannot fully observe the system. However, if the input excites all states, the system is controllable, and if all the states are represented in the output, the system is observable. More precise definitions follow.

4.1.1 Continuous-Time Systems

Controllability, as a measure of interaction between the input and the states, involves the system matrix A and the input matrix B. A linear system, or the pair (A, B), is controllable at t_o if it is possible to find a piecewise continuous input $u(t)$, $t \in [t_o, t_1]$, that will transfer the system from the initial state, $x(t_o)$, to the origin $x(t_1) = 0$, at finite time $t_1 > t_o$. If this is true for all initial moments t_o and all initial states $x(t_o)$ the system is completely controllable. Otherwise, the system, or the pair (A, B), is uncontrollable.

Observability, as a measure of interaction between the states and the output, involves the system matrix A and the output matrix C. A linear system, or the pair (A, C), is observable at t_o if the state $x(t_o)$ can be determined from the output $y(t)$, $t \in [t_o, t_1]$, where $t_1 > t_o$ is some finite time. If this is true for all initial moments t_o and all initial states $x(t_o)$ the system is completely observable. Otherwise, the system, or the pair (A, C), is unobservable. Note that neither the controllability nor observability definition involves the feed-through term D.

There are many criteria to determine system controllability and observability; see [88], [143]. We consider two of them. First, a linear time-invariant system (A, B, C), with s inputs is completely controllable if and only if the $N \times sN$ matrix

$$\mathcal{C} = \begin{bmatrix} B & AB & A^2B & \ldots & A^{N-1}B \end{bmatrix} \tag{4.1}$$

has rank N. A linear time-invariant system (A, B, C) with r outputs is completely observable if and only if the $rN \times N$ matrix of

$$\mathcal{O} = \begin{bmatrix} C \\ CA \\ CA^2 \\ \vdots \\ CA^{N-1} \end{bmatrix} \tag{4.2}$$

has rank N.

The above criteria, although simple, have two serious drawbacks. First, they answer the controllability and observability question in yes and no terms. Second, they are useful only for a system of small dimensions. The latter can be visible if we assume, for example, that the system is of dimension $N = 100$. In order to answer the controllability and observability question we have to find powers of A up to 99. Finding A^{99} for a 100×100 matrix is a numerical task that easily results in numerical overflow.

The alternative approach uses grammians to determine the system properties. Grammians are nonnegative matrices that express the controllability and observability properties qualitatively, and are free of the numerical difficulties mentioned above. The controllability and observability grammians are defined as follows, see, for example, [88]:

$$\begin{aligned} W_c(t) &= \int_0^t \exp(A\tau)BB^T \exp(A^T\tau)\ d\tau, \\ W_o(t) &= \int_0^t \exp(A^T\tau)C^T C \exp(A\tau)\ d\tau. \end{aligned} \tag{4.3}$$

We can determine them alternatively and more conveniently from the following differential equations:

$$\begin{aligned} \dot{W}_c &= AW_c + W_c A^T + BB^T, \\ \dot{W}_o &= A^T W_o + W_o A + C^T C. \end{aligned} \tag{4.4}$$

The solutions $W_c(t)$ and $W_o(t)$ are time-varying matrices. At the moment we are interested in the stationary, or time-invariant, solutions (the time-varying case is discussed later). For a stable system, we obtain the stationary solutions of the above equations by assuming $\dot{W}_c = \dot{W}_o = 0$. In this case, the differential equations (4.4) are replaced with the algebraic equations, called Lyapunov equations,

$$\begin{aligned} AW_c + W_c A^T + BB^T &= 0, \\ A^T W_o + W_o A + C^T C &= 0. \end{aligned} \tag{4.5}$$

For stable A, the obtained grammians W_c and W_o are positive definite.

The grammians depend on the system coordinates, and for a linear transformation of a state x into a new state x_n, such that $x_n = Rx$, they are transformed to new grammians W_{cn} and W_{on} as follows:

$$W_{cn} = R^{-1} W_c R^{-T},$$
$$W_{on} = R^T W_o R. \tag{4.6}$$

The eigenvalues of the grammians change during the coordinate transformation. However, the eigenvalues of the grammian product are invariant. It can be shown as follows:

$$\lambda_i(W_{cn} W_{on}) = \lambda_i(R^{-1} W_c R^{-T} R^T W_o R) = \lambda_i(R^{-1} W_c W_o R) = \lambda_i(W_c W_o). \tag{4.7}$$

These invariants are denoted γ_i,

$$\gamma_i = \sqrt{\lambda_i(W_c W_o)}, \qquad i = 1, \ldots, N, \tag{4.8}$$

and are called the Hankel singular values of the system.

4.1.2 Discrete-Time Systems

Consider now a discrete-time system as given by (3.46). For the sampling time Δt the controllability matrix $\mathcal{C}_k$ is defined similarly to the continuous-time systems, as follows:

$$\mathcal{C}_k = \begin{bmatrix} B & AB & \cdots & A^{k-1}B \end{bmatrix}. \tag{4.9}$$

The controllability grammian $W_c(k)$ over the time interval $[0, k\Delta t]$ is defined as

$$W_c(k) = \sum_{i=0}^{k} A^i B B^T (A^i)^T. \tag{4.10}$$

Unlike the continuous-time systems we can use the controllability matrix of the discrete-time system to obtain the discrete-time controllability grammian $W_c(k)$. Namely,

$$W_c(k) = \mathcal{C}_k \mathcal{C}_k^T. \tag{4.11}$$

The stationary grammian (for $k \to \infty$) satisfies the discrete-time Lyapunov equation

$$W_c - A W_c A^T = B B^T, \tag{4.12}$$

but can still be obtained from (4.11) using large enough *k*, since $A^k \to 0$ for $k \to \infty$. Similarly the observability matrix $\mathcal{O}_k$ is defined as follows:

$$\mathcal{O}_k = \begin{bmatrix} C \\ CA \\ \vdots \\ CA^{k-1} \end{bmatrix}, \tag{4.13}$$

and the discrete-time observability grammian $W_o(k)$ for the time interval $[0, k\Delta t]$ is defined as

$$W_o(k) = \sum_{i=0}^{k} (A^i)^T C^T C A^i, \tag{4.14}$$

which is obtained from the observability matrix

$$W_o(k) = \mathcal{O}_k^T \mathcal{O}_k. \tag{4.15}$$

For $k \to \infty$ (stationary solution) the observability grammian satisfies the following Lyapunov equation:

$$W_o - A^T W_o A = C^T C. \tag{4.16}$$

Similarly to the continuous-time grammians, the eigenvalues of the discrete-time grammian product are invariant under linear transformation. These invariants are denoted γ_i,

$$\gamma_i = \sqrt{\lambda_i(W_c W_o)}, \qquad i = 1, ..., N, \tag{4.17}$$

and are called the Hankel singular values of the discrete-time system.

4.1.3 Relationship Between Continuous- and Discrete-Time Grammians

Let (A, B, C) be the state-space representation of a discrete-time system. From the definitions (4.10) and (4.14) of the discrete-time controllability and observability grammians we obtain

$$\begin{aligned} W_c &= BB^T + ABB^T A^T + A^2 BB^T (A^2)^T + \cdots, \\ W_o &= C^T C + A^T C^T CA + (A^2)^T C^T CA + \cdots. \end{aligned} \tag{4.18}$$

We show that the discrete-time controllability and observability grammians do not converge to the continuous-time grammians when the sampling time approaches zero, see [109]. Indeed, consider the continuous-time observability grammian

$$W_{ocont} = \int_0^\infty e^{At} BB^T e^{A^T t}\, dt.$$

This can be approximated in discrete time, at time moments $t = 0, \Delta t, 2\Delta t, ...,$ as

$$W_{ocont} = \sum_{i=0}^{\infty} e^{iA^T \Delta t} C^T C e^{iA\Delta t} \Delta t = \sum_{i=0}^{\infty} (A_d^i)^T C^T C A_d^i \Delta t.$$

Introducing the second equation of (4.18) one obtains

$$W_{ocont} = \lim_{\Delta t \to 0} \Delta t\, W_{odiscr}. \tag{4.19}$$

Obtaining the controllability grammians is similar. First note that for a small sampling time one has

$$B_d \cong \Delta t B_c. \tag{4.20}$$

Indeed, from the definition of B_d, one obtains

$$B_d = \int_0^{\Delta t} e^{A_c \tau} B_c\, d\tau = \int_0^{\Delta t} (I + A_c \tau + \tfrac{1}{2} A_c^2 \tau^2 + \cdots) B_c\, d\tau$$

$$= B_c \Delta t + \tfrac{1}{2} A_c B_c \Delta t^2 + \tfrac{1}{2} A_c^2 B_c \Delta t^3 + \cdots \cong \Delta t B_c.$$

Now, from the definition of the continuous-time controllability grammian, the following holds:

$$W_{ccont} = \int_0^\infty e^{A_c \tau} B_c B_c^T e^{A_c^T \tau}\, d\tau = \lim_{\Delta t \to 0} \sum_{i=0}^{\infty} e^{iA_c \Delta t} B_c B_c^T e^{iA_c^T \Delta t} \Delta t.$$

Using (4.20) and $A_d = e^{A_c \Delta t}$ we obtain

$$W_{ccont} = \lim_{\Delta t \to 0} \frac{1}{\Delta t} \sum_{i=0}^{\infty} A_d^i B_d B_d^T (A_d^i)^T = \lim_{\Delta t \to 0} \frac{1}{\Delta t} W_{cdiscr};$$

hence,

$$W_{ccont} = \lim_{\Delta t \to 0} \frac{1}{\Delta t} W_{cdiscr}. \tag{4.21}$$

Note, however, from (4.19) and (4.21), that the product of the discrete-time controllability and observability grammians converge to the continuous-time grammians,

$$W_{ccont}W_{ocont} = \lim_{\Delta t \to 0}(W_{cdiscr}W_{odiscr}); \tag{4.22}$$

therefore, the discrete-time Hankel singular values converge to the continuous-time values, as the sampling time approaches zero:

$$\gamma_{icont} = \lim_{\Delta t \to 0}\gamma_{idiscr}. \tag{4.23}$$

4.2 Balanced Representation

Consider a case when controllability and observability grammians are equal and diagonal. The diagonality means that each state has its own and independent measure of controllability and observability (which is the diagonal value of the grammians). The equality of grammians means that each state is equally controllable and observable or, in terms of structures, each mode is equally controllable and observable (excited to the same degree as it is sensed). The equality and diagonality of grammians is a feature of special usefulness—this allows us to evaluate each state (or mode) separately, and to determine their values for testing and for control purposes. Indeed, if a state is weakly controllable and, at the same time, weakly observable, it can be neglected without impacting the accuracy of analysis, dynamic testing, or control design procedures. On the other hand, if a state is strongly controllable and strongly observable, it must be retained in the system model in order to preserve accuracy of analysis, test, or control system design. Knowing the importance of the diagonal and equal grammians, we proceed to their definition and determination.

The system triple (A,B,C) is open-loop balanced, if its controllability and observability grammians are equal and diagonal, as defined by Moore in [109],

$$\begin{aligned} &W_c = W_o = \Gamma, \\ &\Gamma = \operatorname{diag}(\gamma_1,...,\gamma_N), \\ &\gamma_i \geq 0, \qquad i = 1,...,N. \end{aligned} \tag{4.24}$$

The matrix Γ is diagonal, and its diagonal entries γ_i are called Hankel singular values of the system (which were earlier introduced as eigenvalues of the product of the controllability and observability grammians).

A generic representation (A,B,C) can be transformed into the balanced representation (A_b, B_b, C_b), using the transformation matrix R, such that

$$\begin{aligned} A_b &= R^{-1} A R, \\ B_b &= R^{-1} B, \\ C_b &= C R. \end{aligned} \tag{4.25}$$

The matrix R is determined as follows, see [53],

$$R = PU\Gamma^{-1/2}. \tag{4.26}$$

Its inverse is conveniently determined as

$$R^{-1} = \Gamma^{-1/2} V^T Q. \tag{4.27}$$

The matrices Γ, V, and U are obtained from the singular value decomposition of the matrix H,

$$H = V\Gamma U^T \qquad \text{and} \qquad V^T V = I, \qquad U^T U = I, \tag{4.28}$$

where H is obtained as a product of the matrices P and Q,

$$H = QP \tag{4.29}$$

and P, Q, in turn, are obtained from the decomposition of the controllability and observability grammians, respectively,

$$\begin{aligned} W_c &= PP^T, \\ W_o &= Q^T Q. \end{aligned} \tag{4.30}$$

The singular value decomposition can be used to decompose W_c and W_o.

The algorithm is proved as follows. Using (4.6), (4.27), and (4.30) we obtain

$$W_{cb} = R^{-1} W_c R^{-T} = \Gamma^{-1/2} V^T Q P P^T Q^T V \Gamma^{-1/2}.$$

Now, introducing (4.29) to the above equation, we obtain

$$W_{cb} = \Gamma^{-1/2} V^T H H^T V \Gamma^{-1/2},$$

and introducing (4.28) to the above equation we have

$$W_{cb} = \Gamma^{-1/2} V^T V \Gamma U^T U \Gamma V^T V \Gamma^{-1/2}.$$

Taking into account that $V^TV = I$ and $U^TU = I$ we determine that

$$W_{cb} = \Gamma^{-1/2}\Gamma^2\,\Gamma^{-1/2} = \Gamma.$$

In a similar way we find that $W_{ob} = \Gamma$, thus the system is balanced.

The Matlab function *balan2.m*, which transforms a representation (*A*,*B*,*C*) to the open-loop balanced representation, is given in Appendix A.8. Also, the Matlab function *balreal* of the Control System Toolbox performs the continuous-time system balancing. For the discrete-time systems the transformation to the balanced representation is obtained similarly to the continuous-time case. The Matlab function *balreal* performs discrete-time system balancing as well.

4.3 Balanced Structures with Rigid-Body Modes

Some plants (e.g., tracking systems) include rigid-body modes that allow structures to move freely, as they are not attached to a base. In analytical terms a structure with rigid-body modes has *m* poles at zero ($m \le 6$), and it is observable and controllable (see [29]). It is also assumed that a system matrix *A* is nondefective (cf. [73]), i.e., that the geometric multiplicity of poles at zero is *m*. Grammians for structures with rigid-body modes do not exist (since they reach the infinity value), although the structures with rigid-body modes are controllable and observable. Here we show how to represent the grammians so that the infinite values of some of their components do not prevent us performing the analysis of the remaining measurable part.

A system with rigid-body modes (A, B, C) can be represented by the following state-space representation:

$$A = \begin{bmatrix} 0_m & 0_{m\times(N-m)} \\ 0_{(N-m)\times m} & A_o \end{bmatrix}, \qquad B = \begin{bmatrix} B_r \\ B_o \end{bmatrix}, \qquad C = \begin{bmatrix} C_r & C_o \end{bmatrix}, \qquad (4.31)$$

where $0_{m\times N}$ is an $m \times N$ zero matrix. Matrix *A* in the form as above always exists since it has *m* poles at zero, and matrices *B*, *C* exist too, due to the nondefectiveness of A_o.

Similar to *A* in (4.31), the system matrix of the balanced structures with rigid-body modes is also block-diagonal, $A = \text{diag}(0_m, A_{bo})$, and the balanced representation (A_b, B_b, C_b) is obtained by the transformation *R*, using (4.25). However, the transformation *R* is in the form $R = \text{diag}(I_m, R_o)$ where I_m is an identity matrix of order *m* (that remains the rigid-body part unchanged), and R_o balances the inner matrix A_o, i.e.,

$$A_{bo} = R_o^{-1} A_o R_o. \tag{4.32}$$

In summary, to balance structures with rigid-body modes and to obtain the Hankel singular values we need to represent (A, B, C) in the form as in (4.31), and then to balance the triple (A_o, B_o, C_o) that has no poles at zero. The obtained Hankel singular values add to the infinity Hankel singular values that correspond to the poles at zero. Let γ_o represent the vector of Hankel singular values of (A_o, B_o, C_o), then

$$\gamma = \{\inf, \gamma_o\} \tag{4.33}$$

represents the vector of Hankel singular values of (A,B,C), where $\inf = \{\infty, \infty, \ldots, \infty\}$ contains m values at infinity.

4.4 Input and Output Gains

In this section we will consider the controllability and observability properties that are specific to structures. We have found already that the input and output matrices of a system contain the information on the system controllability and observability. Norms of these matrices could serve as crude information on the controllability and observability. We will present these matrices in modal coordinates, where they possess special properties. Namely, the 2-norms of the input and output matrices in modal coordinates contain information on the structural controllability and observability, and are called input and output gains, respectively.

In the following the approximate relationships are used and denoted by the equality sign "$\cong$". They are applied in the following sense: Two variables, x and y, are approximately equal ($x \cong y$) if $x = y + \varepsilon$ and $\|e\| \ll \|y\|$. For example, if D is a diagonal matrix, then S is diagonally dominant, $S \cong D$ if the terms s_{ij} satisfy the condition $s_{ij} + \varepsilon = d_{ij}$, $i, j = 1,..., n,$ and ε is small when compared to d_{ii}.

Input and Output Gains of a Structure. The 2-norms $\|B_m\|_2$ and $\|C_m\|_2$ of the input and output matrices in modal coordinates, as in (2.38), are called the input and output gains of the system

$$\begin{aligned} \|B_m\|_2 &= \left(\operatorname{tr}(B_m B_m^T)\right)^{1/2}, \\ \|C_m\|_2 &= \left(\operatorname{tr}(C_{mq}\Omega^{-2}C_{mq}^T + C_{mv}C_{mv}^T)\right)^{1/2}. \end{aligned} \tag{4.34}$$

Each mode in the model representation has its own gains.

Input and Output Gains of a Mode. The 2-norms $\|b_{mi}\|_2$ and $\|c_{mi}\|_2$ of the b_{mi} and c_{mi} matrices in modal coordinates, as in (2.52), are called the input and output gains of the ith mode

$$\begin{aligned} \|B_{mi}\|_2 &= \left(\mathrm{tr}(B_{mi}B_{mi}^T)\right)^{1/2} = \left(\mathrm{tr}(b_{mi}b_{mi}^T)\right)^{1/2}, \\ \|C_{mi}\|_2 &= \left(\mathrm{tr}(C_{mi}C_{mi}^T)\right)^{1/2} = \left(tr(\frac{c_{mqi}c_{mqi}^T}{\omega_i^2} + c_{mvi}c_{mvi}^T) \right)^{1/2}. \end{aligned} \tag{4.35}$$

In the second modal representation the small term in the output gain, which is proportional to the damping ratio, was neglected.

It is important to notice that the gains of a structure are the root-mean-square (rms) sum of the modal gains

$$\|B_m\|_2 = \sqrt{\sum_{i=1}^{n} \|B_{mi}\|_2^2} \quad \text{and} \quad \|C_m\|_2 = \sqrt{\sum_{i=1}^{n} \|C_{mi}\|_2^2}. \tag{4.36}$$

The modal output matrix C_m is defined as a combination of the displacement and rate matrices

$$C_m = \left[C_{mq}\Omega^{-1} \quad C_{mv} \right]. \tag{4.37}$$

Consider the common case when the displacement and rate sensors are not collocated. In this case the output gain has the following property:

$$\|C_m\|_2^2 = \left\|C_{mq}\Omega^{-1}\right\|_2^2 + \|C_{mv}\|_2^2. \tag{4.38}$$

For a single mode the output matrix c_{mi} is given as

$$C_{mi} = \begin{bmatrix} \dfrac{c_{mqi}}{\omega_i} & c_{mvi} \end{bmatrix}, \tag{4.39}$$

and in the case when the displacement and rate sensors are not collocated the output gain of the ith mode is

$$\|C_{mi}\|_2^2 = \left\|\frac{c_{mqi}}{\omega_i}\right\|_2^2 + \|c_{mvi}\|_2^2. \tag{4.40}$$

For a structure with an accelerometer the output gain is

$$\|C_m\|_2 \cong \|C_{ma}\Omega\|_2 . \tag{4.41}$$

It is derived from the gain definition, (4.35) and (3.20),

$$C_m = \begin{bmatrix} C_{mq}\Omega^{-1} & C_{mv} \end{bmatrix} = \begin{bmatrix} -C_{ma}\Omega & -2C_{ma}Z\Omega \end{bmatrix} = -C_{ma}\begin{bmatrix} I & 2Z \end{bmatrix}\Omega.$$

However, $\zeta_i \ll 1$, $i = 1,\ldots n$; thus, $Z \ll I$; hence,

$$C_m \cong -C_{ma}\begin{bmatrix} I & 0 \end{bmatrix}\Omega,$$

which gives $\|C_m\|_2 \cong \|C_{ma}\Omega\|_2$.

Gains for the individual modes are obtained similarly

$$\|c_{mi}\|_2 \cong \|c_{mai}\omega_i\|_2 . \tag{4.42}$$

Example 4.1. Obtain its input and output gains for the modal model of form 2 given in Example 2.9.

The input and output gains are: For the first mode $\|B_{m1}\|_2 = 0.3280$, $\|C_{m1}\|_2 = 0.6760$; for the second mode $\|B_{m2}\|_2 = 0.5910$, $\|C_{m2}\|_2 = 0.9448$; and for the third mode $\|B_{m3}\|_2 = 0.7370$, $\|C_{m3}\|_2 = 0.8067$. The structural gains are $\|B_m\|_2 = 1.0000$ and $\|C_m\|_2 = 1.4143$.

4.5 Controllability and Observability of a Structural Modal Model

In the following we consider the controllability and observability properties that apply to structures. The modal state-space representation of flexible structures has specific controllability and observability properties, and its grammians are of specific form.

4.5.1 Diagonally Dominant Grammians

Assuming small damping, such that $\zeta \ll 1$, where $\zeta = \max(\zeta_i)$, i = 1, ..., n, the balanced and modal representations of flexible structures are closely related. One indication of this relationship is expressed in the grammian form. The balanced grammians are equal and diagonal; similarly, the grammians in modal coordinates are diagonally dominant, and by using appropriate scaling, they are approximately equal. This is expressed in the following property:

Property 4.1. ***Diagonally Dominant Grammians in Modal Coordinates.*** *In modal coordinates controllability and observability grammians are diagonally dominant, i.e.,*

$$\begin{aligned} W_c &\cong \operatorname{diag}(w_{ci} I_2), \\ W_o &\cong \operatorname{diag}(w_{oi} I_2), \qquad i = 1, \ldots, n, \end{aligned} \tag{4.43}$$

where $w_{ci} > 0$ and $w_{oi} > 0$ are the modal controllability and observability factors. The approximate Hankel singular values are obtained as a geometric mean of the modal controllability and observability factors,

$$\gamma_i \cong \sqrt{w_{cii} w_{oii}}\,. \tag{4.44}$$

Proof. Consider a flexible structure in the modal representation 1, as in (2.47), (2.48), or (2.52). We prove the diagonal dominance of the controllability and observability grammians by the introduction of this modal representation to the Lyapunov equations (4.5). By inspection, for $\zeta \to 0$ we obtain finite values of the off-diagonal terms w_{cij} and w_{oij} for $i \neq j$, i.e., $\lim_{\zeta \to 0} w_{cij} < \infty$ and $\lim_{\zeta \to 0} w_{oij} < \infty$, while the diagonal terms tend to infinity, $\lim_{\zeta \to 0} w_{cii} = \infty$ and $\lim_{\zeta \to 0} w_{oii} = \infty$. Moreover, the difference between the diagonal terms in each block is finite, $\lim_{\zeta \to 0} | w_{cii} - w_{ci+1,i+1} | < \infty$; thus, for small ζ the grammians in modal coordinates are diagonally dominant, having 2×2 blocks on the diagonal with almost identical diagonal entries of each block. Equation (4.44) is a direct consequence of the diagonal dominance of W_c and W_o, and the fact that the eigenvalues of the grammian product are invariant.

The second part, (4.44), follows from the invariance of the eigenvalues of the grammian product. ▣

The profiles of grammians and system matrix A in modal coordinates are drawn in Fig. 4.1. The controllability grammian is, at the same time, a covariance matrix of the states x excited by the white noise input, i.e., $W_c = E(xx^T)$. Thus, the diagonal dominance results in the following conclusion that, under white noise or impulse excitation, modes are almost independent or almost orthogonal.

Example 4.2. Determine the grammians and Hankel singular values (exact and approximate) for the modal model from Example 2.9.

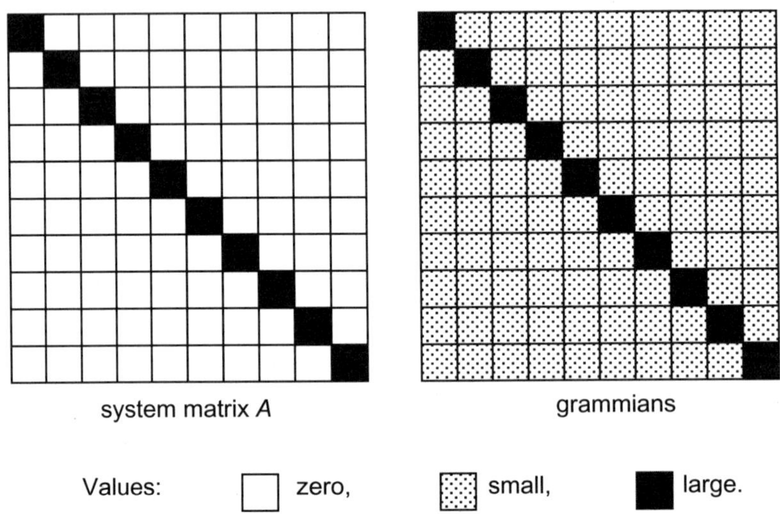

Figure 4.1. Profiles of the system matrix A (diagonal) and the grammians (diagonally dominant) in the modal coordinates.

Using the transformation R as in (4.26) or Matlab's *balreal.m* function we obtain

$$W_{mc} = \left[\begin{array}{cc|cc|cc} 0.6095 & 0.0095 & 0.0122 & -0.0946 & 0.0350 & 0.0048 \\ 0.0095 & 0.6086 & 0.1370 & 0.0114 & -0.0014 & 0.1418 \\ \hline 0.0122 & 0.1370 & 4.5464 & 0.0491 & 0.1489 & 0.0016 \\ -0.0946 & 0.0114 & 0.0491 & 4.5475 & -0.0013 & 0.4171 \\ \hline 0.0350 & -0.0014 & 0.1489 & -0.0013 & 122.6197 & -0.4726 \\ 0.0048 & 0.1418 & 0.0016 & 0.4171 & -0.4726 & 122.0863 \end{array}\right],$$

$$W_{mo} = \left[\begin{array}{cc|cc|cc} 2.2922 & -0.0305 & -0.0057 & -0.1227 & -0.0373 & -0.0008 \\ 0.0305 & 2.2952 & 0.1517 & -0.0065 & -0.0007 & -0.0903 \\ \hline -0.0057 & 0.1517 & 8.9050 & -0.0740 & 0.0223 & 0.0003 \\ -0.1227 & -0.0065 & -0.0740 & 8.9039 & 0.0002 & -0.0417 \\ \hline -0.0373 & -0.0007 & 0.0223 & 0.0002 & 26.0878 & 0.0568 \\ -0.0008 & -0.0903 & 0.0003 & -0.0417 & 0.0568 & 26.0863 \end{array}\right].$$

Indeed, the grammians are diagonally dominant. The approximate grammians were obtained from (4.43),

$$w_{c1} = 0.6095, \qquad w_{c2} = 4.5464, \qquad w_{c3} = 122.6197,$$
$$w_{o1} = 2.2922, \qquad w_{o2} = 8.9050, \qquad w_{o3} = 26.0878,$$

and are close to the actual ones,

$$w_{c1} = 0.6090, \qquad w_{c2} = 4.5469, \qquad w_{c3} = 122.6243,$$
$$w_{o1} = 2.2937, \qquad w_{o2} = 8.9045, \qquad w_{o3} = 26.0871.$$

The approximate Hankel singular values for each mode are obtained from (4.44),

$$\Gamma_{ap} = \text{diag}(1.1818,\ 1.1818,\ 6.3625,\ 6.3625,\ 56.5579,\ 56.5579)$$

and are close to the actual Hankel singular values

$$\Gamma = \text{diag}(1.1756,\ 1.1794,\ 6.3575,\ 6.3735,\ 56.5110,\ 56.5579).$$

4.5.2 Closed-Form Grammians

Next, we show that for flexible structures the grammians of each mode can be expressed in a closed form. This allows for their speedy determination for structures with a large number of modes, and allows for the insight into the grammian physical interpretation. Let B_{mi} and C_{mi} be the $2 \times s$ and $r \times 2$ blocks of modal B_m and C_m, respectively, and the latter matrices are part of the modal representation, (2.38), (2.42), and (2.45). In this case the following property is valid:

Property 4.2. *Closed-Form Controllability and Observability Grammians.* *In modal coordinates the diagonal entries of the controllability and observability grammians, as in* (4.43), *are as follows:*

$$w_{ci} = \frac{\|B_{mi}\|_2^2}{4\zeta_i\omega_i}, \qquad w_{oi} = \frac{\|C_{mi}\|_2^2}{4\zeta_i\omega_i}, \tag{4.45}$$

and the approximate Hankel singular values are obtained from

$$\gamma_i \cong \frac{\|B_{mi}\|_2 \|C_{mi}\|_2}{4\zeta_i\omega_i}. \tag{4.46}$$

Proof. By inspection, introducing the modal representation (2.38), (2.42), and (2.45) to the Lyapunov equations (4.5). ▣

The above equations show that the controllability grammian of the ith mode is proportional to the square of the ith modal input gain, and inversely proportional to the ith modal damping and modal frequency. Similarly, the observability grammian is proportional to the square of the ith modal output gain, and inversely proportional to the ith modal damping and modal frequency. Finally, the Hankel singular value is the geometric mean of the previous two, and is proportional to the ith input and output gains and is inversely proportional to the ith modal damping and modal frequency.

Example 4.3. Determine the closed-form Hankel singular values for the modal model from Example 2.9.

From (4.46) we obtain $\gamma_1 = 1.1817$, $\gamma_2 = 6.3627$, and $\gamma_3 = 56.5585$, which are close to the actual values obtained in Example 4.2.

Example 4.4. Compare the exact and approximate Hankel singular values of the International Space Station structure. The input is a force at node 8583 (marked by the white circle at the top of Fig. 1.9), and the output is a rate at this node.

The results in Fig. 4.2 show good coincidence between the exact and approximate Hankel singular values.

4.5.3 Approximately Balanced Structure in Modal Coordinates

By comparing the grammians in the balanced and modal representations we noticed that the balanced and modal representations are close to each other. The closeness of the balanced and modal representations can also be observed in the closeness of the system matrix A in both representations. It was shown in Chapter 2 that the matrix A in modal coordinates is diagonal (with a 2×2 block on the diagonal). We show that the system matrix A in the balanced representation is diagonally dominant.

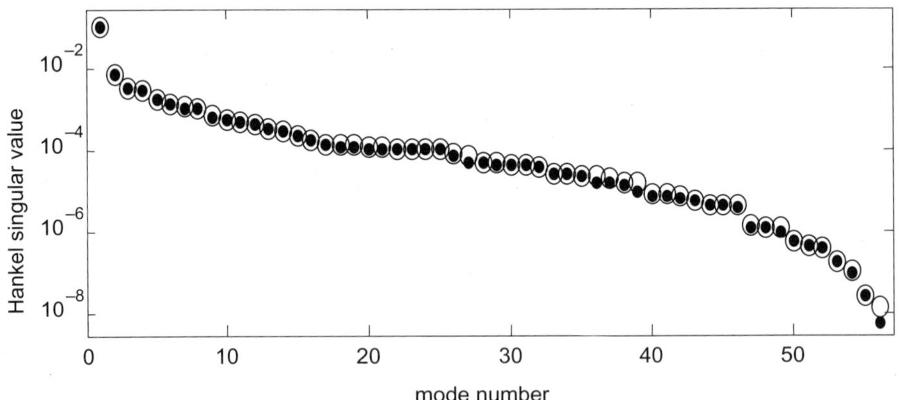

Figure 4.2. The exact (○) and approximate (•) Hankel singular values for the International Space Station structure are almost identical.

Property 4.3. ***Diagonal Dominance of the System Matrix in the Balanced Coordinates.*** *In the balanced representations the system matrix A is block diagonally dominant with 2×2 blocks on the diagonal, and B, C are divided into $2 \times s$ and $r \times 2$ blocks*

$$A \cong \operatorname{diag}(A_i), \qquad B = \begin{bmatrix} B_1 \\ B_2 \\ \vdots \\ B_n \end{bmatrix}, \qquad C = \begin{bmatrix} C_1 & C_2 & \cdots & C_n \end{bmatrix}, \quad (4.47)$$

where $i = 1, \ldots, n$, *and*

$$A_i = \begin{bmatrix} -\zeta_i \omega_i & \omega_i \\ -\omega_i & -\zeta_i \omega_i \end{bmatrix}. \quad (4.48)$$

Proof. Since the grammians in modal coordinates are diagonally dominant, the transformation matrix R from the modal to the balanced coordinates is diagonally dominant itself. The system matrix in the balanced coordinates is $A = R^{-1} A_m R$; therefore, it is diagonally dominant. ▣

The profiles of the grammians and a system matrix A are drawn in Fig. 4.3.

Example 4.5. Find the balanced representation of a simple system from Example 2.5, and check if the matrix A is diagonally dominant.

We use transformation R from (4.26) to obtain the following balanced representation:

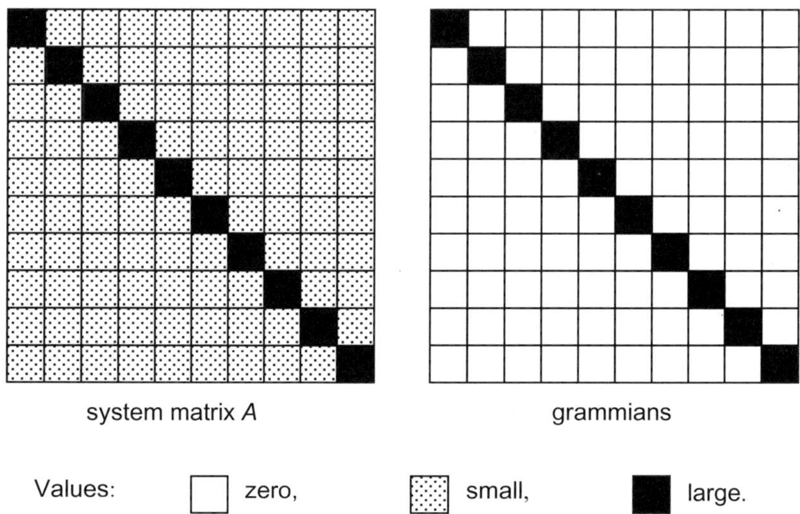

Figure 4.3. Profiles of the system matrix A (diagonally dominant) and the grammians (diagonal) in the balanced coordinates.

$$A_b = \left[\begin{array}{cc:cc:cc} -0.0029 & 0.7685 & 0.0003 & 0.0015 & -0.0032 & 0.0006 \\ -0.7731 & -0.0030 & 0.0014 & 0.0004 & 0.0028 & -0.0018 \\ \hdashline -0.0059 & 0.0058 & -0.0263 & -2.1400 & 0.0243 & -0.0047 \\ 0.0050 & -0.0052 & 2.1807 & -0.0203 & -0.0174 & 0.0105 \\ \hdashline -0.0044 & 0.0044 & -0.0441 & 0.0380 & -0.0805 & -3.0883 \\ 0.0020 & -0.0020 & 0.0190 & -0.0179 & 3.1523 & -0.0170 \end{array}\right],$$

$$B_b = \left[\begin{array}{c} 0.5772 \\ -0.5822 \\ \hdashline 0.5790 \\ -0.5074 \\ \hdashline 0.4357 \\ -0.2001 \end{array}\right],$$

$$C_b = \left[\begin{array}{cc:cc:cc} 0.2707 & 0.2703 & -0.1649 & -0.2029 & 0.0414 & 0.1222 \\ 0.2082 & -0.2088 & -0.4373 & 0.3568 & 0.3814 & -0.1318 \\ 0.4654 & -0.4715 & 0.3418 & -0.2983 & 0.2064 & -0.0881 \end{array}\right],$$

showing the diagonally dominant matrix A_b.

Define a state-space representation as almost balanced if its grammians are almost equal and diagonally dominant, $\Gamma \cong W_c \cong W_o$. We saw that the modal grammians are diagonally dominant. The question arises: Can grammians in modal coordinates be almost equal? That is, can the modal representation be almost-balanced? The following property answers the questions:

Property 4.4. ***Approximate Balancing by Scaling the Modal Coordinates.*** *By scaling the modal representation (A_m, B_m, C_m) one obtains an almost-balanced representation (A_{ab}, B_{ab}, C_{ab}), such that its grammians are almost equal and diagonally dominant*

$$A_{ab} = A_m, \qquad B_{ab} = R_{ab}^{-1} B_m, \qquad C_{ab} = C_m R_{ab}, \tag{4.49}$$

and

$$x_m = R x_{ab}. \tag{4.50}$$

Above, R_{ab} is a diagonal matrix

$$R_{ab} = \operatorname{diag}(r_i I_2), \qquad i = 1, \ldots, n \tag{4.51}$$

and the ith entry is given by the ratio of the input and output gains

$$r_i = \left(\frac{w_{ci}}{w_{oi}} \right)^{1/4} = \sqrt{\frac{\|B_{mi}\|_2}{\|C_{mi}\|_2}}. \tag{4.52}$$

Note that the transformation matrix R_{ab} *leaves matrix* A_m *unchanged, and scales matrices* B_m *and* C_m.

Proof. We prove this property by inspection. Introducing (4.51) and (4.52) to (4.6) we obtain

$$\begin{aligned} \overline{W}_c &\cong \overline{W}_o \cong \Gamma = \operatorname{diag}(\gamma_i I_2), \\ \gamma_i &= \sqrt{w_{ci} w_{oi}}; \end{aligned} \tag{4.53}$$

hence, by using transformation (4.49), we obtain an approximately balanced representation (A_{ab}, B_{ab}, C_{ab}) from the modal representation (A_m, B_m, C_m). ▣

A similar property can be derived for the modal representation 2. The closeness of these two modal representations follows from the fact that the transformation R_{12} from modal representation 1 to 2 is itself diagonally dominant, as in (2.43). The closeness of the modal, balanced and almost-balanced, coordinates is illustrated in Fig. 4.4.

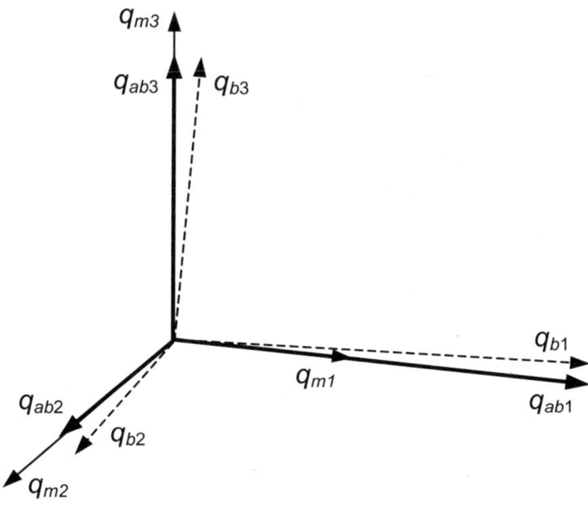

Figure 4.4. Modal, balanced, and almost-balanced coordinates: Almost-balanced coordinates are scaled modal coordinates to “fit” the balanced coordinates.

Similar to the modal representation, the almost-balanced state-vector x_{ab} is divided into n components,

$$x_{ab} = \begin{Bmatrix} x_{ab1} \\ x_{ab2} \\ \vdots \\ x_{abn} \end{Bmatrix}. \tag{4.54}$$

The components are independent, which is justified by the diagonal matrix A_{ab} of the almost-balanced representation. The state-space representation $(A_{abi}, B_{abi}, C_{abi})$ is associated with each component x_{abi}.

Consider the state-space representation $(A_{abi}, B_{abi}, C_{abi})$ of the ith-balanced mode, then $\|B_{abi}\|_2$ is its input gain and $\|C_{abi}\|_2$ is its output gain.

Property 4.5. *Input and Output Gains.* *In the almost-balanced representation the input and output gains are equal:*

$$\|B_{abi}\|_2 = \|C_{abi}\|_2. \tag{4.55}$$

Proof. This can be shown by introducing transformation R_{ab} as in (4.49)–(4.52), obtaining

$$\|B_{abi}\|_2 = \|C_{abi}\|_2 = \sqrt{\|B_{mi}\|_2 \|C_{mi}\|_2}.$$ ▣

In the almost-balanced representation the grammians are almost equal, $w_{ci} \cong w_{oi}$ or $\Gamma \cong W_c \cong W_o$, where the matrix of the Hankel singular values has the following form, $\Gamma \cong \operatorname{diag}(\gamma_1, \gamma_1, \gamma_2, \gamma_2, \ldots, \gamma_n, \gamma_n)$. The Hankel singular values are obtained as

$$\gamma_i \cong \frac{\|B_{abi}\|_2^2}{4\zeta_i \omega_i} = \frac{\|C_{abi}\|_2^2}{4\zeta_i \omega_i}. \tag{4.56}$$

Example 4.6. Obtain the almost-balanced state-space representation of the simple system in Example 2.9.

Starting from the state-space modal representation, as in Example 2.9, we find the transformation matrix R_{ab} as in (4.51) and (4.52), that is,

$$R_{ab} = \operatorname{diag}(0.7178,\ 0.7178,\ 0.8453,\ 0.8453,\ 1.4724,\ 1.4724).$$

The state matrix is equal to the modal matrix, $A_{ab} = A_m$, while the input and output matrices, B_{ab} and C_{ab}, are found from (4.49),

$$B_{ab} = \begin{bmatrix} -0.4798 \\ 0.0075 \\ \hdashline -0.7704 \\ -0.0083 \\ \hdashline -0.0025 \\ -0.8198 \end{bmatrix} = \begin{bmatrix} B_{ab1} \\ \hdashline B_{ab2} \\ \hdashline B_{ab3} \end{bmatrix},$$

$$C_{ab} = \left[\begin{array}{cc:cc:cc} -0.0020 & -0.1294 & -0.0028 & 0.2617 & 0.3825 & -0.0012 \\ -0.4038 & 0.0126 & 0.5653 & 0 & -0.0020 & -0.2948 \\ -0.2241 & 0.0070 & -0.4534 & 0 & -0.0045 & -0.6625 \end{array}\right] = \left[\begin{array}{c:c:c} C_{ab1} & C_{ab2} & C_{ab3} \end{array}\right].$$

In this representation (4.55) holds. Indeed,

$$\begin{aligned} \|B_{ab1}\|_2 &= \|C_{ab1}\|_2 = 0.4798, \\ \|B_{ab2}\|_2 &= \|C_{ab2}\|_2 = 0.7705, \\ \|B_{ab3}\|_2 &= \|C_{ab3}\|_2 = 0.8198. \end{aligned}$$

Finally, we found that the grammians obtained for this model are almost equal, i.e., $\Gamma \cong W_o \cong W_c = \operatorname{diag}(1.1817,\ 1.1817,\ 6.3627,\ 6.3627,\ 56.5585,\ 56.5585)$.

4.6 Controllability and Observability of a Second-Order Modal Model

In this section we present the controllability and observability properties of a structure given by the second-order model.

4.6.1 Grammians

The grammians and the balanced models are defined exclusively in the state-space representation, and they do not exist in the second-order form. This is a certain disadvantage since the second-order structural equations are popular forms of structural modeling. We will show, however, that for flexible structures one can find a second-order model which is almost balanced, and for which Hankel singular values can be approximately determined without using a state-space representation. First, we determine the grammians for the second-order modal model.

Property 4.6. ***Controllability and Observability Grammians of the Second-Order Modal Model.*** *The controllability* (w_c) *and observability* (w_o) *grammians of the second-order modal model are given as*

$$w_c = 0.25Z^{-1}\Omega^{-1}\text{diag}(B_m B_m^T),$$
$$w_o = 0.25Z^{-1}\Omega^{-1}\text{diag}(C_m^T C_m), \tag{4.57}$$

where $\text{diag}(B_m B_m^T)$ *and* $\text{diag}(C_m^T C_m)$ *denote the diagonal part of* $B_m B_m^T$ *and* $C_m^T C_m$, *respectively,* B_m *is given by* (2.23), $C_m = \begin{bmatrix} C_{mq}\Omega^{-1} & C_{mv} \end{bmatrix}$, *and* C_{mq}, C_{mv} *are defined as in* (2.24) *and* (2.25). *Therefore, the ith diagonal entries of* w_c *and* w_o *are*

$$w_{ci} = \frac{\|b_{mi}\|_2^2}{4\zeta_i\omega_i},$$
$$w_{oi} = \frac{\|c_{mi}\|_2^2}{4\zeta_i\omega_i}, \tag{4.58}$$

and b_{mi} *is the ith row of* B_m, *and* c_{mi} *is the ith column of* C_m.

Proof. In order to show this we introduce a state-space representation by defining the following state vector:

$$x_m = \begin{bmatrix} \Omega q_m \\ \dot{q}_m \end{bmatrix}.$$

The following state-space representation is associated with the above vector

$$A = \begin{bmatrix} 0 & \Omega \\ -\Omega & -2Z\Omega \end{bmatrix}, \qquad B = \begin{bmatrix} 0 \\ B_m \end{bmatrix}, \qquad C = \begin{bmatrix} C_{mq}\Omega^{-1} & C_{mv} \end{bmatrix}.$$

By inspection, for this representation, the grammians are diagonally dominant, in the form

$$W_c \cong \begin{bmatrix} w_c & 0 \\ 0 & w_c \end{bmatrix}, \qquad W_o \cong \begin{bmatrix} w_o & 0 \\ 0 & w_o \end{bmatrix},$$

where w_c and w_o are the diagonally dominant matrices, $w_c \cong \text{diag}(w_{ci})$ and $w_o \cong \text{diag}(w_{oi})$. Introducing the last two equations to the Lyapunov equations (4.5) we obtain (4.58). ▣

Having the grammians for the second-order models, the Hankel singular values are determined approximately from (4.58) as

$$\gamma_i \cong \sqrt{w_{ci} w_{oi}} = \frac{\|b_{mi}\|_2 \|c_{mi}\|_2}{4\zeta_i \omega_i}, \qquad i = 1,\ldots,n. \tag{4.59}$$

4.6.2 Approximately Balanced Structure in Modal Coordinates

Second-order modal models are not unique, since they are obtained using natural modes that are arbitrarily scaled. Hence we have a freedom to choose the scaling factor. By a proper choice of the scaling factors we introduce a model that is almost balanced, i.e., its controllability and observability grammians are approximately equal and diagonally dominant. The second-order almost-balanced model is obtained by scaling the modal displacement (q_m) as follows:

$$q_{ab} = R^{-1} q_m, \tag{4.60}$$

that is,

$$q_m = R q_{ab}, \tag{4.61}$$

and q_{ab} is the almost-balanced displacement.

The transformation R is obtained as follows. Denote $\|b_{mi}\|_2$ and $\|c_{mi}\|_2$ as the input and output gains, then

$$R = \operatorname{diag}(r_i), \qquad i = 1,\ldots,n, \tag{4.62}$$

and the ith entry r_i is defined as a square root of the gain ratio

$$r_i = \sqrt{\frac{\|b_{mi}\|_2}{\|c_{mi}\|_2}}. \tag{4.63}$$

Using (2.38) one obtains

$$\|c_{mi}\|_2^2 = \frac{\|c_{qi}\|_2^2}{\omega_i^2} + \|c_{vi}\|_2^2, \tag{4.64}$$

while

$$\|c_{qi}\|_2^2 = c_{qi}^T c_{qi} \qquad \text{and} \qquad \|c_{vi}\|_2^2 = c_{vi}^T c_{vi}. \tag{4.65}$$

In the above equations b_{mi} is the ith row of B_m, and c_{qi}, c_{vi}, c_{mi} are the ith columns of C_{mq}, C_{mv}, and C_m, respectively.

Introducing (4.60) and (4.62)–(4.65) to the modal equation (2.19) we obtain the almost-balanced second-order modal model

$$\ddot{q}_{ab} + 2Z\Omega\dot{q}_{ab} + \Omega^2 q_{ab} = B_{ab}u,$$
$$y = C_{abq}q_{ab} + C_{abv}\dot{q}_{ab}, \tag{4.66}$$

where

$$B_{ab} = R^{-1}B_m \tag{4.67}$$

and

$$C_{abq} = C_{mq}R, \qquad C_{abv} = C_{mv}R, \tag{4.68}$$

while the output matrix C_{ab} is defined as

$$C_{ab} = \begin{bmatrix} C_{abq}\Omega^{-1} & C_{abv} \end{bmatrix}. \tag{4.69}$$

This has the following property:

$$\|C_{ab}\|_2^2 = \left\|C_{abq}\Omega^{-1}\right\|_2^2 + \|C_{abv}\|_2^2. \tag{4.70}$$

A flexible structure in modal coordinates is described by its *natural modes*, ϕ_i, $i = 1,\ldots,n$. Similarly the almost-balanced modal representation is a modal representation with a unique scaling, and is described by the *almost-balanced modes*, ϕ_{abi}, $i = 1,\ldots,n$. The latter ones we obtain by rescaling the natural modes

$$\phi_{abi} = r_i\phi_i, \qquad i = 1,\ldots,n, \tag{4.71}$$

with the scaling factor r_i given by (4.63), and

$$\Phi_{ab} = \Phi R, \tag{4.72}$$

where $\Phi_{ab} = [\phi_{ab1}\ \phi_{ab2}\ \cdots\ \phi_{abn}]$, and Φ is a modal matrix, as in (2.12). In order to show this, note that from (2.18) one obtains $q = \Phi q_m$ or, equivalently

$$q = \sum_{i=1}^{n} \phi_i q_{mi}. \tag{4.73}$$

But, from (4.61) it follows that $q_{mi} = r_i q_{abi}$; thus, (4.73) is now

$$q = \sum_{i=1}^{n} r_i \phi_i q_{abi} = \sum_{i=1}^{n} \phi_{abi} q_{abi}, \tag{4.74}$$

where ϕ_{abi} is a balanced mode as in (4.71).

Property 4.7. *Grammians of the Almost-Balanced Model.* *In the almost-balanced model the controllability and observability grammians are approximately equal,*

$$w_{cabi} \cong w_{oabi} \cong \frac{\sqrt{\|b_{mi}\|_2 \|c_{mi}\|_2}}{4\zeta_i \omega_i} \cong \gamma_i. \tag{4.75}$$

Proof. From (4.58) we have

$$w_{cabi} = \frac{\|b_{abi}\|_2}{4\zeta_i \omega_i}, \qquad w_{oabi} = \frac{\|c_{abi}\|_2}{4\zeta_i \omega_i}. \tag{4.76}$$

However, from (4.49) and (4.52) it follows that

$$b_{abi} = b_{mi} \sqrt{\frac{\|c_{mi}\|_2}{\|b_{mi}\|_2}}, \qquad c_{abi} = c_{mi} \sqrt{\frac{\|b_{mi}\|_2}{\|c_{mi}\|_2}}. \tag{4.77}$$

Introducing the above equation to (4.76) we obtain approximately equal grammians as in (4.75). ▣

Define $\|b_{abi}\|_2$ and $\|c_{abi}\|_2$ as the input and output gains of the ith almost-balanced mode, respectively, and we find that these gains are equal.

Property 4.8. *Gains of the Almost-Balanced Model.* *In the second-order almost-balanced model, the input and output gains are equal,*

$$\|b_{abi}\|_2 = \|c_{abi}\|_2. \tag{4.78}$$

Proof. The transformation R as in (4.62) is introduced to (4.67) and (4.68) obtaining

$$\|b_{abi}\|_2 = \frac{1}{\|r_i\|} \|b_{mi}\|_2 = \sqrt{\frac{\|c_{mi}\|_2}{\|b_{mi}\|_2}} \|b_{mi}\|_2 = \sqrt{\|b_{mi}\|_2 \|c_{mi}\|_2}$$

and

$$\|c_{abi}\|_2 = \|c_{mi}\|_2\, r_i = \|c_{mi}\|_2 \sqrt{\frac{\|b_{mi}\|_2}{\|c_{mi}\|_2}} = \sqrt{\|b_{mi}\|_2 \|c_{mi}\|_2} = \|b_{abi}\|_2 \,. \qquad \blacksquare$$

Example 4.7. Determine the almost-balanced model of a simple structure from Example 2.2.

We obtain the transformation matrix R from (4.62) and (4.63) as R=diag(0.6836, 0.7671, 0.8989). Next, we find the almost-balanced input and output matrices from (4.67), (4.68), and (4.69), knowing from Example 2.2 that $\Omega = diag(3.1210, 2.1598, 0.7708)$; hence,

$$B_{ab} = \begin{bmatrix} 0.4798 \\ -0.7705 \\ 0.8198 \end{bmatrix} = \begin{bmatrix} b_{ab1} \\ b_{ab2} \\ b_{ab3} \end{bmatrix},$$

$$C_{abq}\Omega^{-1} = \begin{bmatrix} 0.1294 & 0.2617 & 0.3825 \\ 0 & 0 & 0 \\ 0 & 0 & 0 \end{bmatrix}, \qquad C_{abv} = \begin{bmatrix} 0 & 0 & 0 \\ 0.4040 & 0.5653 & 0.2948 \\ 0.2242 & -0.4534 & 0.6625 \end{bmatrix},$$

The output matrix C_{ab} is obtained by putting together $C_{abq}\Omega^{-1}$ and C_{abv}, such that the first column of $C_{abq}\Omega^{-1}$ is followed by the first column of C_{abv}, followed by the second column of $C_{abq}\Omega^{-1}$, followed by the second column of C_{abv}, etc., i.e.,

$$C_{ab} = \left[\begin{array}{cc|cc|cc} 0.1294 & 0 & 0.2617 & 0 & 0.3825 & 0 \\ 0 & 0.4040 & 0 & 0.5653 & 0 & 0.2948 \\ 0 & 0.2242 & 0 & -0.4534 & 0 & 0.6625 \end{array}\right] = \begin{bmatrix} c_{ab1} & c_{ab2} & c_{ab3} \end{bmatrix}.$$

The almost-balanced mode matrix is obtained from (4.72),

$$\Phi_{ab} = \begin{bmatrix} 0.4040 & 0.5653 & 0.2948 \\ -0.5038 & 0.2516 & 0.5313 \\ 0.2242 & -0.4534 & 0.6625 \end{bmatrix}.$$

Finally, it is easy to check that the input and output gains are equal,

$$\|b_{ab1}\|_2 = \|c_{ab1}\|_2 = 0.4798,$$
$$\|b_{ab2}\|_2 = \|c_{ab2}\|_2 = 0.7705,$$
$$\|b_{ab3}\|_2 = \|c_{ab3}\|_2 = 0.8198.$$

Also, from (4.76) we obtain $w_{c1} = w_{o1} = 1.1821$, $w_{c2} = w_{o2} = 6.3628$, and $w_{c3} = w_{o3} = 55.8920$, which shows that the model is almost balanced, since the exact Hankel singular values for this system are $\gamma_1 = 1.1794$, $\gamma_2 = 6.3736$, and $\gamma_3 = 56.4212$.

4.7 Three Ways to Compute Hankel Singular Values

Based on the above analysis one can see that there are three ways to obtain Hankel singular values for flexible structures in modal coordinates.

1. From the algorithm in Section 4.2. This algorithm gives the exact Hankel singular values. However, for large structures it could be time-consuming. Also, the relationship between the Hankel singular value and the natural mode it represents is not an obvious one: this requires one to examine the system matrix A in order to find the natural frequency related to the Hankel singular value in question.
2. From (4.43) and (4.44). This is an approximate value, and its determination can be time-consuming for large structures. However, there is a straightforward relationship between the Hankel singular values and natural frequencies (the Hankel singular value from (4.44) is found for the ith frequency).
3. From (4.46) or (4.59). This is an approximate value, but is determined fast, regardless of the size of the structure. Also, it is immediately known what mode it is associated with, and its closed-form allows for the parametric analysis and physical interpretation.

4.8 Controllability and Observability of the Discrete-Time Structural Model

Consider now a structure in modal coordinates. Similar to the continuous-time grammians the discrete-time grammians in modal coordinates are diagonally dominant,

$$
\begin{aligned}
W_c &\cong \operatorname{diag}(W_{c1}, W_{c2}, \ldots, W_{cn}), \\
W_o &\cong \operatorname{diag}(W_{o1}, W_{o2}, \ldots, W_{on}),
\end{aligned}
\tag{4.79}
$$

where W_{ci} and W_{oi} are 2×2 blocks, such that $W_{ci} = w_{ci} I_2$ and $W_{oi} = w_{oi} I_2$, see [98], where

$$w_{ci} = \frac{\|B_{mi}\|_2^2}{4\zeta_i\omega_i}\frac{2(1-\cos(\omega_i\Delta t)}{\omega_i^2\Delta t} = w_{ci\,cont}\frac{2(1-\cos(\omega_i\Delta t)}{\omega_i^2\Delta t} \tag{4.80}$$

and

$$w_{oi} = \frac{\|C_{mi}\|_2^2}{4\zeta_i\omega_i}\frac{1}{\Delta t} = w_{oi\,cont}\frac{1}{\Delta t} \tag{4.81}$$

In the above equations B_{mi} is the ith block of B_m in modal coordinates, and C_{mi} is the ith block of C_m in modal coordinates, where $C_m = [C_{mq}\Omega^{-1} \quad C_{mv}]$, see (2.42) for $Z \cong 0$. In the latter equation Ω is the diagonal matrix of natural frequencies, C_{mq} is the matrix of displacement measurements, and C_{mv} is the matrix of rate measurements. Also $w_{ci\,cont}$ and $w_{oi\,cont}$ denote the continuous-time controllability and observability grammians, respectively, cf. (4.45).

Note that the discrete-time controllability grammian deviates from the continuous-time controllability grammian by factor $\frac{2(1-\cos(\omega_i\Delta t)}{\omega_i^2\Delta t}$, while the discrete-time observability grammian deviates from the continuous-time observability grammian by factor $1/\Delta t$. Note also that the discrete-time grammians do not converge to the continuous-time grammians, but satisfy the following conditions:

$$\lim_{\Delta t\to 0}\frac{W_{ci}}{\Delta t} = W_{ci\,cont} \qquad \text{and} \qquad \lim_{\Delta t\to 0} W_{oi}\Delta t = W_{oi\,cont},$$

which is consistent with the Moore result; see [109] and Subsection 4.1.3 of this chapter.

The Hankel singular values are the square roots of the eigenvalues of the grammian products, $\Gamma = \lambda^{1/2}(W_c W_o)$. The approximate values of the Hankel singular values can be obtained from the approximate values of the grammians,

$$\gamma_i \cong \sqrt{w_{ci}w_{oi}} = \frac{\|B_{mi}\|_2\|C_{mi}\|_2}{4\zeta_i\omega_i}\frac{\sqrt{2(1-\cos(\omega_i\Delta t))}}{\omega_i\Delta t}. \tag{4.82}$$

Note that the discrete-time Hankel singular values differ from the continuous-time values by a coefficient k_i,

$$\gamma_i \cong k_i\gamma_{i\,cont}, \tag{4.83}$$

where

$$k_i = \frac{\sqrt{2(1-\cos\omega_i\Delta t)}}{\omega_i\Delta t}. \tag{4.84}$$

The plot of $k_i(\omega_i\Delta t)$ is shown in Fig. 4.5; this shows that for small sampling time, discrete- and continuous-time Hankel singular values are almost identical.

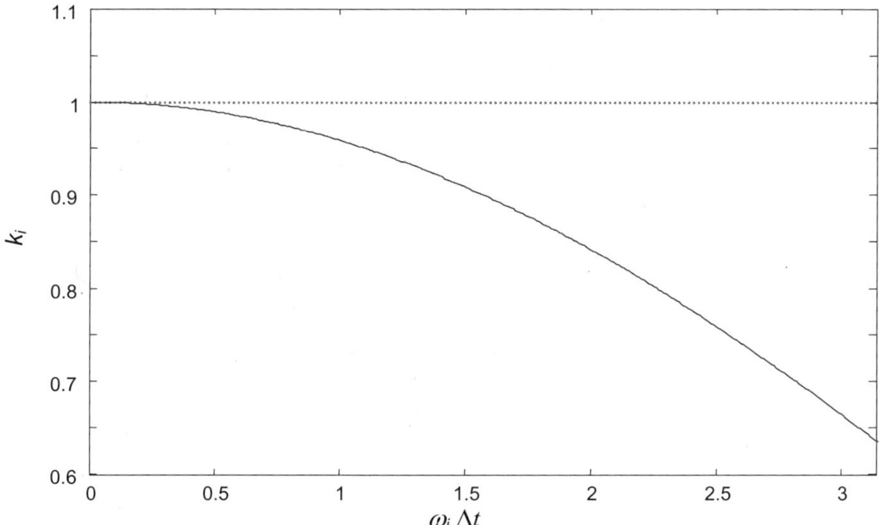

Figure 4.5. Plot of k_i versus $\omega_i\Delta t$: For fast sampling (i.e., small Δt) the k_i value is 1.

Note that if the sampling rate is high enough (or the sampling time small enough), the discrete-time Hankel singular values are very close to the continuous-time Hankel singular values. For example, if $\omega_i\Delta t \le 0.6$ the difference is less than 3%, if $\omega_i\Delta t \le 0.5$ the difference is less than 2%, and if $\omega_i\Delta t \le 0.35$ the difference is less than 1%. Note also that for a given sampling time the discrete-time Hankel singular values, corresponding to the lowest natural frequencies, are closer to the continuous-time Hankel singular values than the Hankel singular values corresponding to the higher natural frequencies.

Example 4.8. Consider the discrete-time simple system as in Example 2.9. For this system $k_1 = k_2 = k_3 = 3$, $k_4 = 0$, and $m_1 = m_2 = m_3 = 1$, while damping is proportional to the stiffness matrix, $D = 0.01K$. Determine its Hankel singular values for sampling time $\Delta t = 0.7$ s, and for $\Delta t = 0.02$ s, and compare with the continuous-time Hankel singular values.

The Hankel singular values for the continuous- and discrete-time structures with sampling times $\Delta t = 0.7$ s and $\Delta t = 0.02$ s are given in Table 4.1.

Table 4.1. Hankel singular values.

	Continuous time	Discrete time $\Delta t = 0.7$ s	Discrete time $\Delta t = 0.02$ s
Mode 1	20.342	20.138	20.342
	20.340	20.009	20.340
Mode 2	4.677	4.324	4.677
	4.671	4.225	4.670
Mode 3	0.991	0.848	0.991
	0.986	0.785	0.986

The table shows that for the sampling time $\Delta t = 0.7$ s the discrete-time Hankel singular values are smaller than the continuous-time values, especially for the third mode. In order to explain this, note that the natural frequencies are $\omega_1 = 0.771$ rad/s, $\omega_2 = 2.160$ rad/s, and $\omega_3 = 3.121$ rad/s. The sampling time must satisfy condition (3.52) for each mode. For the first mode $\pi/\omega_1 = 4.075$, for the second mode $\pi/\omega_2 = 1.454$, and for the third mode $\pi/\omega_3 = 1.007$. The sampling time satisfies the condition (3.52). However, for this sampling time, one obtains $\omega_1 \Delta t = 0.540$, $\omega_2 \Delta t = 1.512$, and $\omega_3 \Delta t = 2.185$. It is shown in Fig. 4.5 that the discrete-time reduction of the Hankel singular values with respect to continuous-time Hankel singular values is significant, especially for the third mode.

This is changed for the sampling time $\Delta t = 0.02$ s. In this case one obtains $\omega_1 \Delta t = 0.015$, $\omega_2 \Delta t = 0.043$, and $\omega_3 \Delta t = 0.062$. One can see from Fig. 4.5 that for these values of $\omega_i \Delta t$ the discrete-time Hankel singular values are almost equal to the continuous-time Hankel singular values.

Next, we verify the accuracy of the approximate relationship (4.83) between discrete- and continuous-time Hankel singular values. The accuracy is expressed with the coefficient k_i, (4.84). The Hankel singular values were computed for different sampling times, and compared with the continuous-time Hankel singular values. Their ratio determines the coefficient k_i. The plot of k_i obtained for all three modes and the plot of the approximate coefficient from (4.84) are shown in Fig. 4.6. The plot shows that the approximate curve and the actual curves are close, except for $\omega_i \Delta t$ very close to π.

4.9 Time-Limited Grammians

The steady-state grammians, defined over unlimited time integrals, are determined from the Lyapunov equations (4.5). The grammians over a finite-time interval $T = [t_1, t_2]$ (where $0 \le t_1 < t_2 < \infty$) are defined by (4.3), and can be obtained from the matrix differential equations (4.4). In many cases these equations cannot be conveniently solved, and the properties of their solutions are not readily visible. However, using the definitions from (4.3) we will derive the closed-form grammians over the finite interval T. Assume that a system is excited and its response measured within the time interval $T=[t_1, t_2]$. The grammians over this interval are defined as follows:

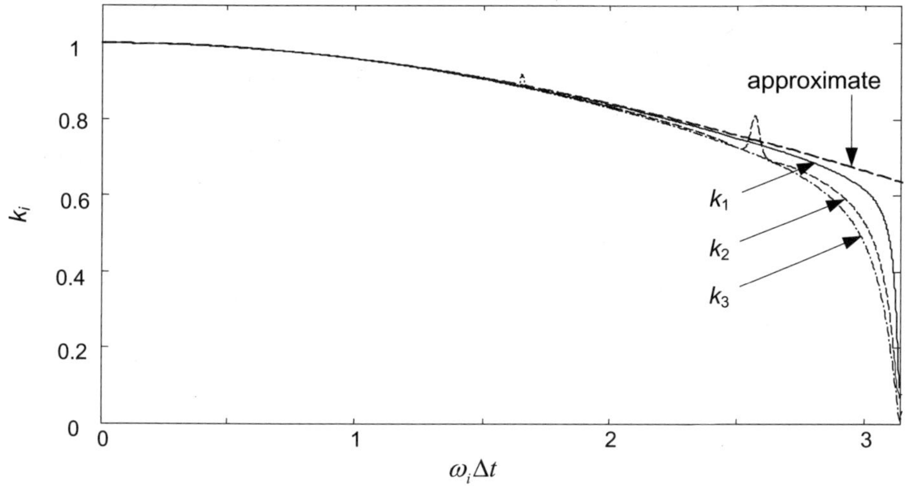

Figure 4.6. The exact and approximate coefficients (k) coincide except for $\omega_i\Delta t$ very close to π, which corresponds to the Nyquist frequency.

$$
\begin{aligned}
W_c(T) &= \int_{t_1}^{t_2} \exp(A\tau)BB^T \exp(A^T\tau)\,d\tau, \\
W_o(T) &= \int_{t_1}^{t_2} \exp(A^T\tau)C^T C\exp(A\tau)\,d\tau.
\end{aligned}
\tag{4.85}
$$

For a stable matrix A these grammians are positive-definite.

First, we express the controllability and observability grammians over the interval (0, t) through the infinite-time controllability grammian W_c.

Property 4.9(a). *Grammians in Time Interval (0, t).* *The controllability grammian $W_c(0,t)$ over the interval (0, t), and the observability grammian $W_o(0,t)$ over the interval (0, t), are obtained from the infinite-time controllability grammians W_c and W_o as follows:*

$$
\begin{aligned}
W_c(0,t) &= W_c - S(t)W_cS^T(t), \\
W_o(0,t) &= W_o - S^T(t)W_oS(t),
\end{aligned}
\tag{4.86}
$$

where

$$
S(t) = e^{At}. \tag{4.87}
$$

Proof. The controllability grammian $W_c = W_c(0,\infty)$,

$$W_c = \int_0^{\infty} e^{A\tau} BB^T e^{A^T\tau} \, d\tau$$

can be decomposed for $t < \infty$ as follows:

$$W_c = \int_0^{t} e^{A\tau} BB^T e^{A^T\tau} \, d\tau + \int_t^{\infty} e^{A\tau} BB^T e^{A^T\tau} \, d\tau.$$

In the second integral, the variable τ is replaced with $\theta = \tau - t$, yielding

$$\int_t^{\infty} e^{A\tau} BB^T e^{A^T\tau} \, d\tau = \int_0^{\infty} e^{A(\theta+t)} BB^T e^{A^T(\theta+t)} \, d\theta$$

$$= e^{At} \int_0^{\infty} e^{A\theta} BB^T e^{A^T\theta} \, d\theta \, e^{A^T t} = e^{At} W_c e^{A^T t};$$

therefore, combining the two latter equations one obtains

$$W_c = W_c(0,t) + e^{At} W_c e^{A^T t},$$

which, in turn, gives (4.86). The observability grammian is derived similarly. ▣

Note that equations (4.86) are the solutions of the Lyapunov differential equations (4.4). Indeed, from the first equation of (4.86) it follows that $\dot{W}_c(0,t) = -\dot{S} W_c S^T - S W_c \dot{S}^T$. Note also that $\dot{S} = \frac{d}{dt}(e^{At}) = Ae^{At} = AS$; thus, $\dot{W}_c(0,t) = -ASW_cS^T - SW_cS^T A^T$. Introducing the latter result and the first equation of (4.86) to (4.4) we obtain

$$-ASW_cS^T - SW_cS^T A^T = AW_c - ASW_cS^T + W_cA^T - SW_cS^T A^T + BB^T$$

or, after simplification,

$$0 = AW_c^T + W_cA^T + BB^T,$$

which is, of course, a steady-state Lyapunov equation, fulfilled for stable systems. Similarly, we can show that the observability grammian from (4.86) satisfies the second equation of (4.4).

Denote the time interval $T = [t_o, t_f]$ where $t_f > t_o$.

Property 4.9(b). ***Grammians in Time Interval (t_o, t_f).*** For $t_o > 0$ the following holds:

$$W_c(T) = S(t_o)W_cS^T(t_o) - S(t_f)W_cS^T(t_f) = W_c(t_o) - W_c(t_f),$$
$$W_o(T) = S^T(t_o)W_oS(t_o) - S^T(t_f)W_oS(t_f) = W_o(t_o) - W_o(t_f), \tag{4.88}$$

where

$$W_c(t) = S(t)W_cS^T(t),$$
$$W_o(t) = S^T(t)W_oS(t). \tag{4.89}$$

Proof. To prove the first part we begin with the definition

$$W_c(T) = \int_{t_o}^{t_f} e^{A\tau}BB^Te^{A^T\tau}\,d\tau,$$

and the introduction of the new variable $\theta = \tau - t_o$, obtaining

$$W_c(T) = \int_0^{t_f - t_o} e^{A(\theta + t_o)}BB^Te^{A^T(\theta + t_o)}d\theta$$
$$= e^{At_o}\int_0^{t_f - t_o} e^{A\theta}BB^Te^{A^T\theta}\,d\theta\, e^{A^Tt_o} = e^{At_o}\,W_c(t_f - t_o)e^{A^Tt_o},$$

which proves (4.88). The second part we prove by introducing (4.86) into (4.88),

$$W_c(T) = e^{At_o}(W_c - e^{A(t_f - t_o)}\,W_c\,e^{A^T(t_f - t_o)})e^{A^Tt_f} = e^{At_o}\,W_ce^{A^Tt_o} - e^{At_f}W_c\,e^{A^Tt_f}. \qquad \blacksquare$$

The time-limited grammians in modal coordinates have a simpler form, since the controllability grammian in modal coordinates is diagonally dominant and the matrix A is block-diagonal. The grammian block that corresponds to the ith mode has the form $w_{ci}(t_o, t_f)I_2$, and the matrix A block is

$$\begin{bmatrix} -\zeta_i\omega_i & \omega_i \\ -\omega_i & -\zeta_i\omega_i \end{bmatrix}.$$

Introducing it into (4.88) we obtain

$$w_{ci}(T) \cong w_{ci}e^{-2\zeta_i\omega_it_o}(1 - e^{-2\zeta_i\omega_i(t_f - t_o)}). \tag{4.90}$$

For the most practical case of $t_o = 0$ we find

$$w_{ci}(0,t) \cong w_{ci}(1-e^{-2\zeta_i\omega_i t}). \tag{4.91}$$

The latter equations show that for a stable system the modal grammians of limited time are positive definite, and that they grow exponentially, with the time constant $T_i = 1/2\zeta_i\omega_i$.

The observability grammian in modal coordinates for structures is diagonally dominant, and the matrix A is block-diagonal. The grammian block that corresponds to the ith mode has the form $w_{oi}(t_o,t_f)I_2$. Similar to the controllability grammian we get

$$w_{oi}(T) \cong w_{oi}e^{-2\zeta_i\omega_i t_o}(1-e^{-2\zeta_i\omega_i(t_f-t_o)}). \tag{4.92}$$

For the most practical case of $t_o = 0$ we obtain

$$w_{oi}(0,t) \cong w_{oi}(1-e^{-2\zeta_i\omega_i t}). \tag{4.93}$$

The latter equations show that for a stable system the modal grammians of limited time are positive definite, and that they grow exponentially, with the time constant $T_i = 1/2\zeta_i\omega_i$.

Define the Hankel singular values over the interval $T = (t_o, t_f)$ as follows:

$$\gamma_i(T) = \lambda_i^{1/2}\left(W_c(T)W_o(T)\right). \tag{4.94}$$

Introducing (4.90) and (4.92) to (4.94) we obtain

$$\gamma_i(T) \cong \gamma_i e^{-2\zeta_i\omega_i t_o}(1-e^{-2\zeta_i\omega_i(t_f-t_o)}) \tag{4.95}$$

or

$$\gamma_i(0,t) \cong \gamma_i(1-e^{-2\zeta_i\omega_i t}). \tag{4.96}$$

Example 4.9. Analyze a simple system with k_1=10, k_2=50, k_3=50, k_4=10, m_1=m_2=m_3=1, with proportional damping matrix, D=0.005K+0.1M. The input is applied to the third mass and the output is the velocity of this mass. Calculate the time limited Hankel singular values for $T = [0, t]$, and for t is varying from 0 to 25 s using the exact equations (4.88), (4.94), and the approximate equation (4.96).

The plots of the Hankel singular values for the system three modes are shown in Fig. 4.7. The plots show close approximation for the first two modes, and not-so-close for the third mode and for a short time span ($t < 1$ s).

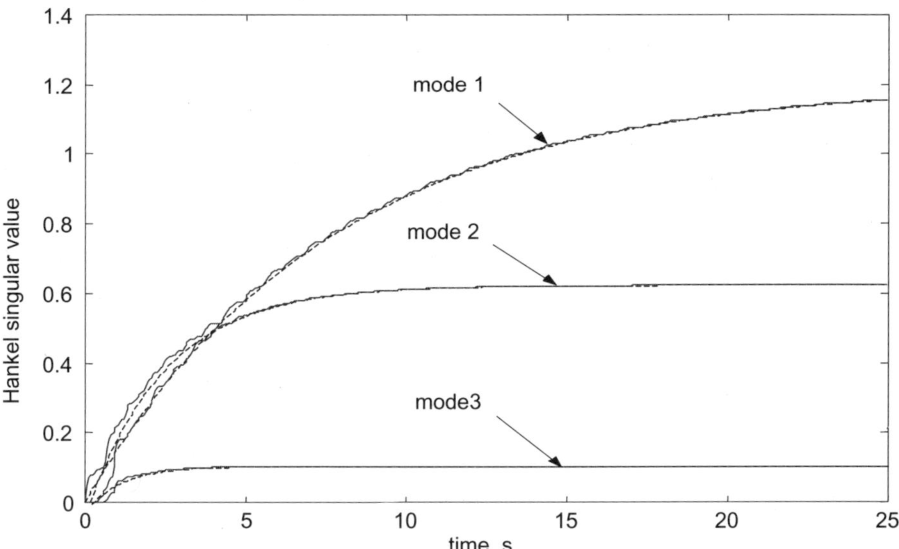

Figure 4.7. Hankel singular values versus time for the system modes: Exact (solid line) and approximate (dashed line).

4.10 Frequency-Limited Grammians

In this section we interpret the controllability and observability grammians in frequency domain. In order to do this, note that from the Parseval theorem the time integrals (4.3), for the time span $(0, \infty)$, can be substituted with the following integrals in the frequency domain:

$$\begin{aligned} W_c &= \frac{1}{2\pi}\int_{-\infty}^{\infty} H(\nu)BB^T H^*(\nu)d\nu, \\ W_o &= \frac{1}{2\pi}\int_{-\infty}^{\infty} H^*(\nu)C^T CH(\nu)d\nu, \end{aligned} \tag{4.97}$$

where

$$H(\nu) = (j\nu I - A)^{-1} \tag{4.98}$$

is the Fourier transform of e^{At}, and H^* is a complex-conjugate transpose of H.

The above grammians are defined over the entire frequency range. The frequency band $(-\infty, \infty)$ can be narrowed to $(-\omega, \omega)$ where $\omega < \infty$ by defining the grammians in the latter band as

$$W_c(\omega) = \frac{1}{2\pi} \int_{-\omega}^{\omega} H(\nu) B B^T H^*(\nu) d\nu,$$
$$W_o(\omega) = \frac{1}{2\pi} \int_{-\omega}^{\omega} H^*(\nu) C^T C H(\nu) d\nu. \tag{4.99}$$

We show the following property:

Property 4.10(a). ***Grammians in Frequency Interval (0, ω)*** are obtained from the following equations:

$$W_c(\omega) = W_c S^*(\omega) + S(\omega) W_c,$$
$$W_o(\omega) = S^*(\omega) W_o + W_o S(\omega), \tag{4.100}$$

where

$$S(\omega) = \frac{j}{2\pi} \ln\left((H^{-1})^* H\right) = \frac{j}{2\pi} \ln\left((A + j\omega I)(A - j\omega I)^{-1}\right) \tag{4.101}$$

and W_c is the controllability grammian obtained from the Lyapunov equation (4.5).

Proof. Note first that

$$A W_c + W_c A^T = -H^{-1} W_c - W_c (H^{-1})^*. \tag{4.102}$$

This can be shown by replacing A with H^{-1} as in (4.98). Next, introduce $BB^T = -AW_c - W_c A^T$ to (4.99), obtaining

$$W_c(\omega) = \frac{1}{2\pi} \int_{-\omega}^{\omega} H(\nu) B B^T H^*(\nu) d\nu = -\frac{1}{2\pi} \int_{-\omega}^{\omega} H(\nu)(A W_c + W_c A^T) H^*(\nu) d\nu.$$

Introducing (4.102) to the above equation we obtain

$$\begin{aligned} W_c(\omega) &= \frac{1}{2\pi} \int_{-\omega}^{\omega} H(\nu)\left(H^{-1}(\nu) W_c + W_c (H^{-1})^*(\nu)\right) H^*(\nu) d\nu \\ &= \frac{1}{2\pi} \int_{-\omega}^{\omega} \left(W_c H^*(\nu) + H(\nu) W_c\right) d\nu \\ &= W_c \frac{1}{2\pi} \int_{-\omega}^{\omega} H^*(\nu) d\nu + \frac{1}{2\pi} \int_{-\omega}^{\omega} H(\nu) d\nu \, W_c = W_c S^*(\omega) + S(\omega) W_c, \end{aligned}$$

since $S(\omega)$ in (4.101) is obtained as

$$S(\omega)=\frac{1}{2\pi}\int_{-\omega}^{\omega}H(\nu)d\nu. \tag{4.103}$$

The observability grammian is determined similarly. ▣

Define the frequency band $\Omega=[\omega_1,\ \omega_2]$, such that $\infty>\omega_2>\omega_1\geq 0$. It is easy to see that the grammians for the band Ω are obtained as

$$\begin{aligned}W_c(\Omega)&=W_c(\omega_2)-W_c(\omega_1),\\W_o(\Omega)&=W_o(\omega_2)-W_o(\omega_1).\end{aligned} \tag{4.104}$$

For this band the following property holds:

Property 4.11(a). ***Grammians in Frequency Interval*** **(ω_1, ω_2)** are obtained from the following equations:

$$\begin{aligned}W_c(\Omega)&=W_cS^*(\Omega)+S(\Omega)W_c,\\W_o(\Omega)&=S^*(\Omega)W_o+W_oS(\Omega),\end{aligned} \tag{4.105}$$

where

$$S(\Omega)=S(\omega_2)-S(\omega_1). \tag{4.106}$$

Proof. Introduce (4.100) to (4.104) to obtain (4.105). ▣

Next, we show the following property:

Property 4.12. Matrices A and $S(\omega)$ Commute.

$$AS(\omega)=S(\omega)A. \tag{4.107}$$

Proof. Note first that

$$AH(\omega)=H(\omega)A, \tag{4.108}$$

which we prove through the simple manipulations

$$\begin{aligned}AH(\omega)&=A(j\omega I-A)^{-1}=[(j\omega I-A)A^{-1}]^{-1}=[(j\omega A^{-1}-I)]^{-1}\\&=[A^{-1}(j\omega I-A)]^{-1}=(j\omega I-A)^{-1}A=H(\omega)A.\end{aligned}$$

Equation (4.107) follows directly from (4.108) and the definition (4.103) of $S(\omega)$. ▣

Using the above property we derive an alternative formulation for the frequency-limited grammians.

Property 4.10(b). ***Grammians in Frequency Interval (0, ω)*** are determined from the following equations:

$$\begin{aligned} &AW_c(\omega)+W_c(\omega)A^T+Q_c(\omega)=0, \\ &A^T W_o(\omega)+W_o(\omega)A+Q_o(\omega)=0, \end{aligned} \tag{4.109}$$

where

$$\begin{aligned} &Q_c(\omega)=S(\omega)BB^T+BB^T S^*(\omega), \\ &Q_o(\omega)=S^*(\omega)C^T C+C^T CS(\omega). \end{aligned} \tag{4.110}$$

Proof. Use (4.100) and apply the commutative Property 4.12 to obtain

$$\begin{aligned} AW_c(\omega)+W_c(\omega)A^T &= AW_cS^*+ASW_c+W_cA^TS^*+SW_cA^T, \\ S(W_cA^T+AW_c)+(W_cA^T+AW_c)S^* &= -SBB^T-BB^TS^*=-Q_c(\omega), \end{aligned}$$

i.e., the first of (4.109) is satisfied. The second (4.109) is proved similarly. ▣

Next we determine the grammians over the interval $\Omega=[\omega_1,\omega_2]$.

Property 4.11(b). ***Grammians in Frequency Interval Ω=(ω_1, ω_2)*** are obtained from the following equations:

$$\begin{aligned} &AW_c(\Omega)+W_c(\Omega)A^T+Q_c(\Omega)=0, \\ &A^T W_o(\Omega)+W_o(\Omega)A+Q_o(\Omega)=0, \end{aligned} \tag{4.111}$$

where

$$\begin{aligned} &Q_c(\Omega)=Q_c(\omega_2)-Q_c(\omega_1), \\ &Q_o(\Omega)=Q_o(\omega_2)-Q_o(\omega_1). \end{aligned} \tag{4.112}$$

Proof. Directly from (4.104) and (4.111). ▣

Example 4.10. Analyze a simple system as in Example 4.9, and obtain the frequency-limited grammians for $\Omega=[0,\omega]$, where ω is varying from 0 to 20 rad/s.

We obtain the grammians in modal coordinates from (4.111), and their plots for the three modes are shown in Fig. 4.8. The plots show that for $\omega > \omega_i$, i=1,2,3, (ω_i is the ith natural frequency, $\omega_1 = 2.55$ rad/s, $\omega_2 = 7.74$ rad/s, and $\omega_3 = 12.38$ rad/s), the grammians achieve constant value.

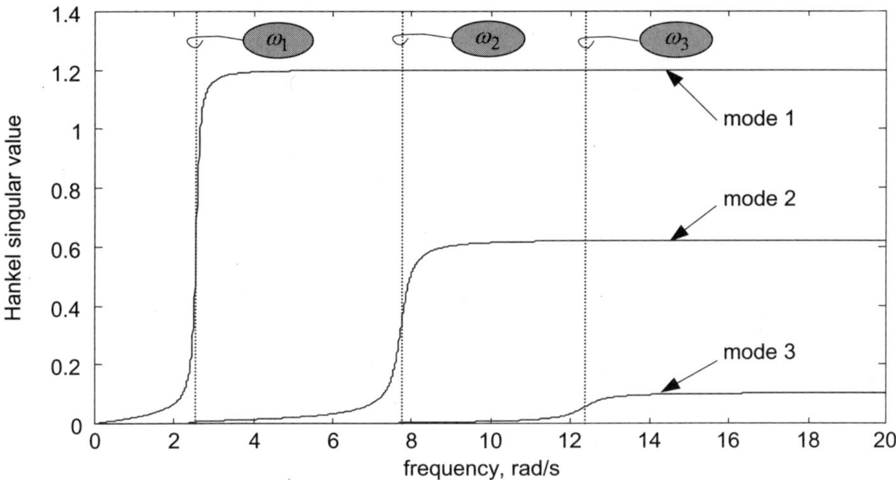

Figure 4.8. Hankel singular values versus frequency for the system modes.

4.11 Time- and Frequency-Limited Grammians

The time- and frequency-limited grammians characterize a system in a limited-time interval and in a limited-frequency window. They are obtained from the full time grammians using time and frequency transformation or vice versa. The results are identical in both cases, since the time and frequency transformations commute, as will be shown below.

Consider the controllability grammian in the finite-time interval, defined in (4.80). From the Parseval theorem, the grammian (4.97) in the infinite-time interval $T = [0, \infty)$ is equal to the grammian (4.97) in the infinite-frequency domain $\omega \in (-\infty, \infty)$. Assume now that $H(\nu)$ in (4.97) is measured within the finite-time interval and for $\nu \in [-\omega, \omega)$ so that W_c and W_o are determined from (4.100). Introducing (4.100) to (4.89) yields

$$\begin{aligned} W_c(t,\omega) &= S(t)W_c(\omega)S^T(t), \\ W_o(t,\omega) &= S^T(t)W_o(\omega)S(t), \end{aligned} \tag{4.113}$$

where $S(t)$ is given by (4.87) and $W_c(\omega)$, $W_o(\omega)$ by (4.100).

Consider now the grammians in frequency domain as in (4.100) and apply the Parseval theorem in time domain to obtain

$$\begin{aligned} W_c(t,\omega) &= W_c(t)S^*(\omega) + S(\omega)W_c(t), \\ W_o(t,\omega) &= W_o(t)S(\omega) + S^*(\omega)W_o(t), \end{aligned} \tag{4.114}$$

where $S(\omega)$ is given by (4.103) and S^* is the complex conjugate transposition of S.

We will show that grammians obtained from (4.113) and (4.114) are identical. Notice first from (4.87) and (4.103) that $S(t)$ and $S(\omega)$ commute, i.e., that

$$S(t)S(\omega) = S(\omega)S(t).$$

Next, using the above property, from (4.100) and (4.114) we obtain

$$\begin{aligned} W_c(t,\omega) &= S(t)W_cS^T(t)S^*(\omega) + S(\omega)S(t)W_cS^T(t), \\ W_o(t,\omega) &= S^T(t)W_oS(t)S(\omega) + S^*(\omega)S^T(t)W_oS(t). \end{aligned}$$

Introducing (4.89) to the above equations we prove the equality of (4.113) and (4.114).

As a consequence of the commuting property, the grammians over the finite-time interval T and finite-frequency interval Ω ($T = [t_1, t_2]$, $t_2 > t_1 \geq 0$, $\Omega = [\omega_1, \omega_2]$, and $\omega_2 > \omega_1 \geq 0$)) are determined from

$$\begin{aligned} W_c(T,\Omega) &= W_c(T,\omega_2) - W_c(T,\omega_1) = W_c(t_1,\Omega) - W_c(t_2,\Omega), \\ W_o(T,\Omega) &= W_o(T,\omega_2) - W_o(T,\omega_1) = W_o(t_1,\Omega) - W_o(t_2,\Omega), \end{aligned} \tag{4.115}$$

where

$$\begin{aligned} W_c(T,\omega) &= W_c(t_1,\omega) - W_c(t_2,\omega), \\ W_c(t,\Omega) &= W_c(t,\omega_2) - W_c(t,\omega_1), \end{aligned} \tag{4.116}$$

and

$$\begin{aligned} W_o(T,\omega) &= W_o(t_1,\omega) - W_o(t_2,\omega), \\ W_o(t,\Omega) &= W_o(t,\omega_2) - W_o(t,\omega_1), \end{aligned} \tag{4.117}$$

where

$$W_c(t,\omega) = S(t)W_c(\omega)S^T(t) = W_c(t)S^*(\omega) + S(\omega)W_c(t),$$
$$W_o(t,\omega) = S^T(t)W_o(\omega)S(t) = W_o(t)S(\omega) + S^*(\omega)W_o(t). \quad (4.118)$$

The Matlab program that computes the time-limited grammians, frequency-limited grammians, and time- and frequency-limited grammians is given in Appendix A.7.

Example 4.11. The Matlab code for this example is in Appendix B. Analyze a simple system as in Example 4.9, and obtain the time- and frequency-limited grammians for the time segment $T = [0, t]$, where t is varying from 0 to 25 s and for the band $\Omega = [0, \omega]$, where ω is varying from 0 to 20 rad/s.

The grammians in modal coordinates are obtained from (4.115), and their plots for the three modes are shown in Figs. 4.9(a),(b),(c). The plots show that grammians grow exponentially with time and that for $\omega > \omega_i$, $i = 1, 2, 3$ (ω_i is the ith natural frequency, $\omega_1 = 2.55$ rad/s, $\omega_2 = 7.74$ rad/s, and $\omega_3 = 12.38$ rad/s) the grammians achieve constant value.

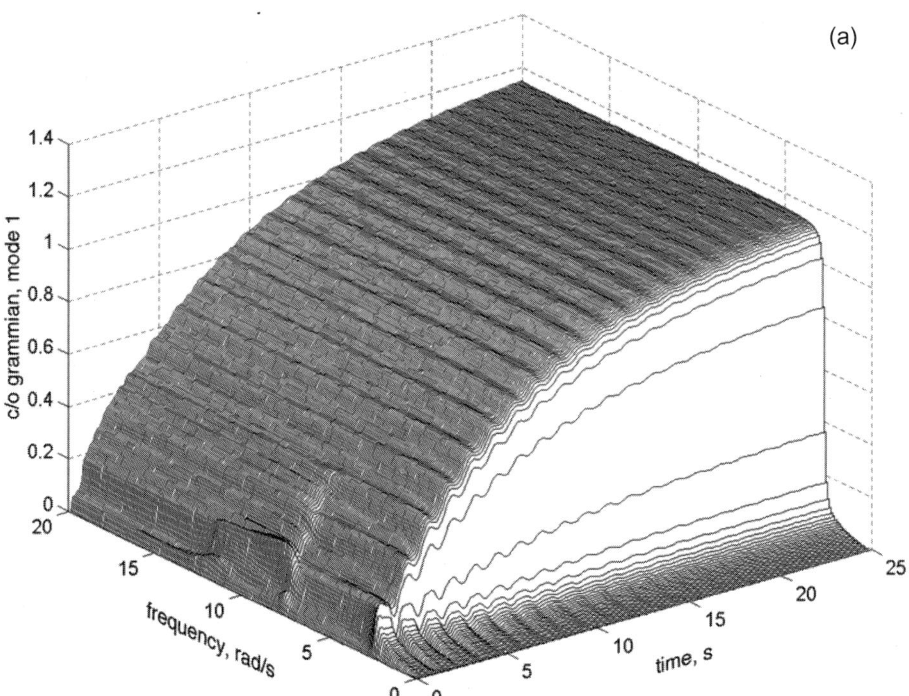

Figure 4.9. Hankel singular values versus time and frequency for (a) first mode.

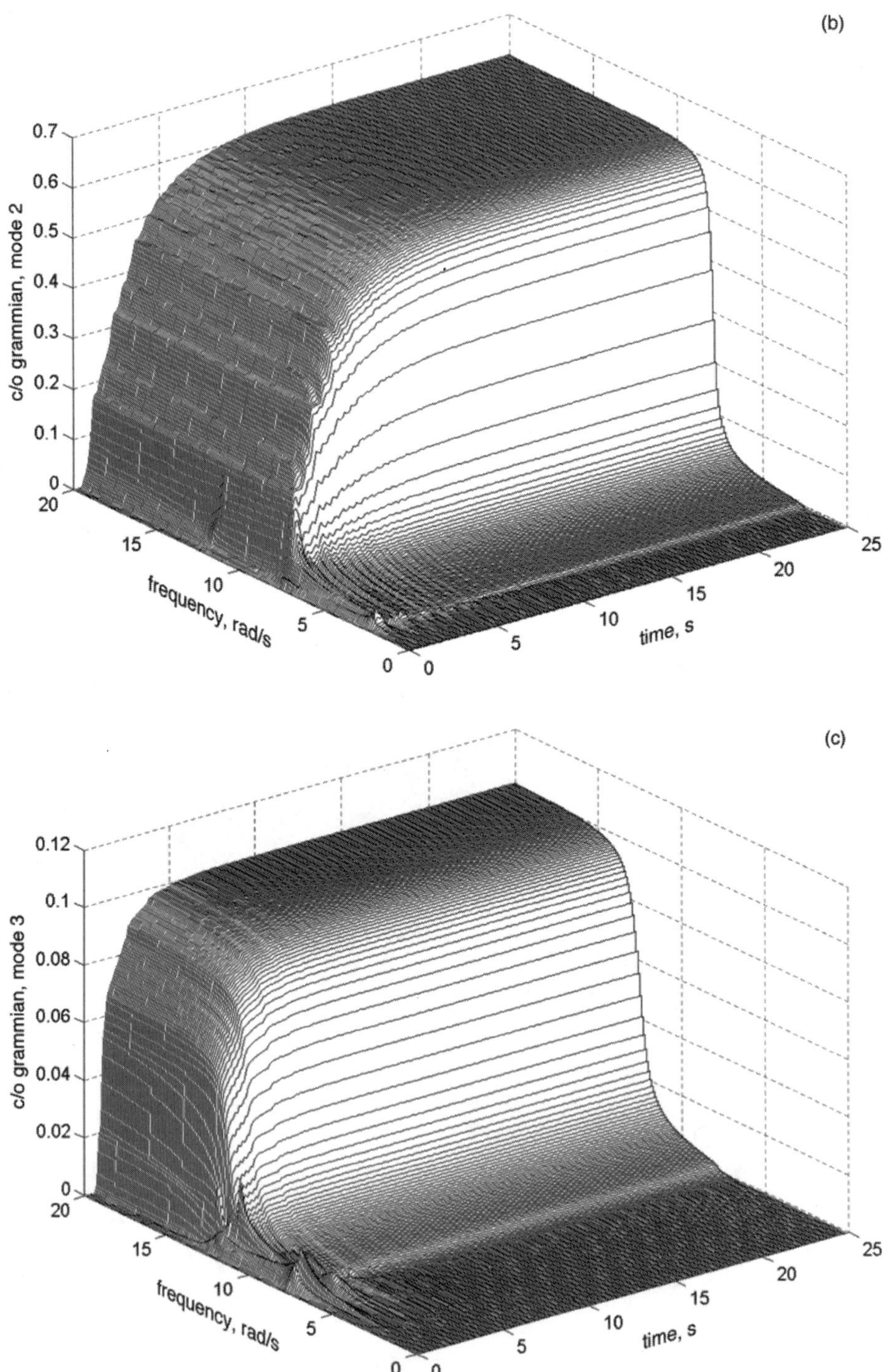

Figure 4.9. Hankel singular values versus time and frequency for (b) second mode; and (c) third mode.

4.12 Discrete-Time Grammians in Limited-Time and -Frequency Range

The above time- and frequency-limited grammians were determined for the continuous time and frequencies. If the time or frequency range is discrete, the grammians are determined differently.

Let the discrete-time state-space representation be (A,B,C), and let the sampling time be Δt. We obtain from (4.11) the discrete-time controllability grammian $W_c(k)$ over the time interval $[0, k\Delta t]$,

$$W_c(k) = \mathcal{C}_k \mathcal{C}_k^T \tag{4.119}$$

where $\mathcal{C}_k$ is the controllability matrix

$$\mathcal{C}_k = \begin{bmatrix} B & AB & \cdots & A^{k-1}B \end{bmatrix}. \tag{4.120}$$

Similarly we find the discrete-time observability grammian $W_o(k)$ for the time interval $[0, k\Delta t]$,

$$W_o(k) = \mathcal{O}_k^T \mathcal{O}_k, \tag{4.121}$$

where $\mathcal{O}_k$ is the observability matrix,

$$\mathcal{O}_k = \begin{bmatrix} C \\ CA \\ \vdots \\ CA^{k-1} \end{bmatrix}. \tag{4.122}$$

For the discrete-time system (A,B,C) of sampling frequency Δt the Nyquist frequency ω_n (in rad/s) is given as

$$\omega_n = \frac{\pi}{\Delta t}. \tag{4.123}$$

We determine the frequency-limited grammians, over the frequency interval $\Omega = [\omega_1, \omega_2]$ where $\omega_2 > \omega_1$, as follows; see [74]:

$$\begin{aligned} W_c(\Omega) &= W_c S^*(\Omega) + S(\Omega) W_c, \\ W_o(\Omega) &= W_o S(\Omega) + S^*(\Omega) W_o, \end{aligned} \tag{4.124}$$

where

$$S(\Omega) = S(\omega_2) - S(\omega_1) \tag{4.125}$$

and

$$S(\omega) = \frac{-1}{2\pi}\left(\frac{\pi\omega}{\omega_n} I + 2j\ln(e^{j\pi\omega/\omega_n} I - A)\right). \tag{4.126}$$

It is not difficult to check that for $\omega_1 = 0$ and $\omega_2 = \omega_n$ one obtains $S(\Omega) = S([0,\ \omega_n]) = \frac{1}{2}I$, which gives $W_c(\Omega) = W_c$ and $W_o(\Omega) = W_o$.

Example 4.12. Consider a discrete-time simple system from Fig. 1.1 with sampling time $\Delta t = 0.01$ s, with masses $m_1 = m_2 = m_3 = 1$, stiffnesses $k_1 = k_2 = k_3 = 3$, $k_4 = 0$, and a damping matrix proportional to the stiffness matrix, $D = 0.001K$ or $d_i = 0.01k_i$, i = 1, 2, 3, 4. There is a single input force at mass 3 and velocity of mass 1 is the output. Find the Hankel singular values for the frequency ranges $[0, \omega]$, where ω varies from 0 to 4 rad/s.

The system natural frequencies are $\omega_1 = 0.77$ rad/s, $\omega_2 = 2.16$, and $\omega_3 = 3.12$, and its Hankel singular values (for the infinite-frequency range) are $\gamma_1 = 203.40$, $\gamma_2 = 46.69$, and $\gamma_3 = 9.95$. The Hankel singular values for the frequency ranges $\Omega = [0, \omega]$ for $\omega \in [0, 4]$ rad/s were calculated using (4.124), and are plotted in Fig. 4.10. It is interesting to notice that each Hankel singular value reaches its maximal value (equal to the Hankel singular value of the infinite-frequency range) for a frequency range that includes the corresponding natural frequency.

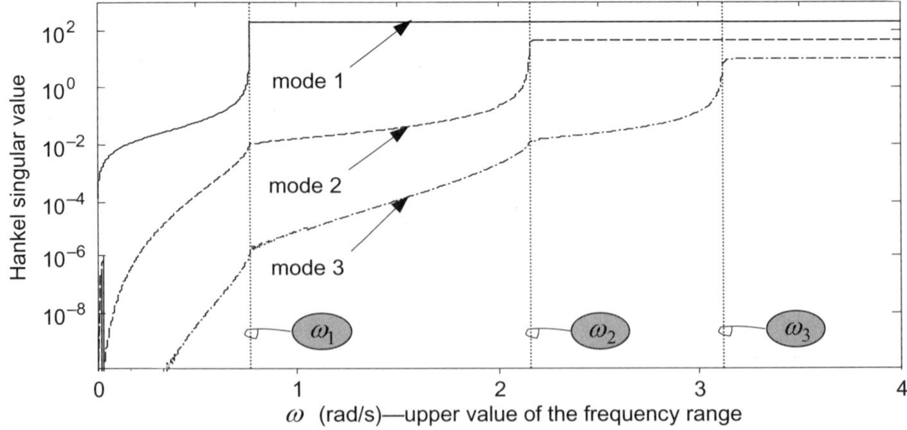

Figure 4.10. Hankel singular values versus frequency range $[0, \omega]$ rad/s of a discrete-time system: If frequency range includes the natural frequency the corresponding Hankel singular value reaches its maximum.

5

Norms

↳ *how to quantify structural dynamics*

Measure what is measurable,
and make measurable what is not so.
—*Galileo Galilei*

System norms serve as a measure of intensity of its response to standard excitations, such as unit impulse, or white noise of unit standard deviation. The standardized response allows comparing different systems. Three system norms, H_2, H_∞, and Hankel are used in this book. We show that for flexible structures the H_2 norm has an additive property: it is a root-mean-square (rms) sum of the norms of individual modes. We also show that the H_∞ and Hankel norms are determined from the corresponding modal norms, by selecting the largest one. All three norms of a mode with multiple inputs (or outputs) can be decomposed into the rms sum of norms of a mode with a single input (or output). Later in this book these two properties allow for the development of unique and efficient model reduction methods and actuator/sensor placement procedures.

5.1 Norms of the Continuous-Time Systems

Three system norms, H_2, H_∞, and Hankel are analyzed in this book. Their properties are derived and specified for structural applications.

5.1.1 The H_2 Norm

Let (A, B, C) be a system state-space representation of a linear system, and let $G(\omega) = C(j\omega I - A)^{-1}B$ be its transfer function. The H_2 norm of the system is defined as

$$\|G\|_2^2 = \frac{1}{2\pi}\int_{-\infty}^{\infty} \mathrm{tr}(G^*(\omega)G(\omega))\,d\omega. \tag{5.1}$$

Note that $\mathrm{tr}(G^*(\omega)G(\omega))$ is the sum of the squared magnitudes of all of the elements of $G(\omega)$, i.e., $\mathrm{tr}(G^*(\omega)G(\omega)) = \sum_{k,l}|g_{kl}(j\omega)|^2$. Thus, it can be interpreted as an average gain of the system, performed over all the elements of the matrix transfer function and over all frequencies.

Since the transfer function $G(\omega)$ is the Fourier transform of the system impulse response $g(t)$, from the Parseval theorem the above definition can be written as an average of the impulse response

$$\|G\|_2^2 = \|g(t)\|_2^2 = \int_0^{\infty} \mathrm{tr}(g^T(t)g(t))\,dt. \tag{5.2}$$

Again, $\mathrm{tr}(g^T(t)g(t))$ is the sum of squared magnitudes of impulse responses, that is, $\mathrm{tr}(g^T(t)g(t)) = \sum_{k,l} g_{kl}^2(t)$. Thus, it can be interpreted as an average impulse response of the system.

A system rms response to the white noise input is the third interpretation of the H_2 norm. Let u be a stationary random input with spectral density $S_u(\omega)$. A system response y is a stationary random process. Its spectral density $S_y(\omega)$ is obtained from the following equation, see, for example, [102],

$$S_y(\omega) = G(j\omega)S_u(\omega)G^*(j\omega).$$

For the unit variance white noise input we have $S_u(\omega) = I$; therefore, the output spectrum is $S_y(\omega) = G(j\omega)G^*(j\omega)$. The rms system response σ_y^2 we obtain as an average of the output spectra

$$\begin{aligned}\sigma_y^2 &= \frac{1}{2\pi}\int_0^{\infty} \mathrm{tr}(S_y(\omega))\,d\omega = \frac{1}{2\pi}\int_0^{\infty} \mathrm{tr}(G(j\omega)G^*(j\omega))\,d\omega \\ &= \frac{1}{2\pi}\int_0^{\infty} \mathrm{tr}(G^*(j\omega)G(j\omega))\,d\omega = \|G\|_2^2,\end{aligned}$$

showing equivalence of the H_2 norm and the rms system response to the white noise excitation.

A convenient way to determine the H_2 norm is to use the following formula:

$$\|G\|_2 = \sqrt{\mathrm{tr}(C^T C W_c)} = \sqrt{\mathrm{tr}(B B^T W_o)}, \tag{5.3}$$

where W_c and W_o are the controllability and observability grammians.

5.1.2 The H$_\infty$ Norm

The H_∞ norm is defined as

$$\|G\|_\infty = \sup_{u(t)\neq 0} \frac{\|y(t)\|_2}{\|u(t)\|_2} \tag{5.4}$$

or, alternatively, as

$$\|G\|_\infty = \max_\omega \sigma_{\max}(G(\omega)), \tag{5.5}$$

where $\sigma_{\max}(G(\omega))$ is the largest singular value of $G(\omega)$. The peak of the transfer function magnitude is the H_∞ norm of a single-input–single-output system $\|G\|_\infty = \max_\omega |G(\omega)|$.

This norm is particularly applicable to the system analysis and controller design since it is an induced norm, i.e., it can provide the bounds of the system output errors. Namely, let u and y be the system input and output, respectively, and G its transfer function, then from (5.4) we obtain

$$\|y\|_2 \leq \|G\|_\infty \|u\|_2 . \tag{5.6}$$

We can see from the above inequality and (5.5) that $\|G\|_\infty$ is the worst-case gain for sinusoidal inputs at any frequency, i.e., that it gives the bound of the output error.

There exists no general relationship between the H_2 and H_∞ norms (for flexible structures we derive this relationship later in this chapter). However, we would like to emphasize the difference between these two norms, see [129, p. 154]. In minimizing the H_2 norm one decreases the transfer function in average direction and average frequency (minimizes the sum of square of all singular values over all frequencies). In minimizing the H_∞ norm one decreases the transfer function in the worst direction and worst frequency (minimizes the largest singular value).

The H_∞ norm can be computed as a maximal value of ρ such that the solution S of the following algebraic Riccati equation is positive definite:

$$A^T S + SA + \rho^{-2} SBB^T S + C^T C = 0. \tag{5.7}$$

This is an iterative procedure where one starts with a large value of ρ and reduces it until negative eigenvalues of S appear.

5.1.3 The Hankel Norm

The Hankel norm of a system is a measure of the effect of its past input on its future output, or the amount of energy stored in and, subsequently, retrieved from the system [12, p. 103]. It is defined as

$$\|G\|_h = \sup \frac{\|y(t)\|_2}{\|u(t)\|_2} \qquad \text{where} \qquad \begin{cases} u(t)=0 & \text{for } t>0, \\ y(t)=0 & \text{for } t<0. \end{cases} \tag{5.8}$$

Comparing the definitions (5.4) of the H_∞ norm and (5.8) of the Hankel norm we see that the H_∞ norm is defined as the largest output for all possible inputs contained in the unit ball, while the Hankel norm is defined the largest *future* output for all the past inputs from the unit ball. From these definitions it follows that the Hankel norm never exceeds the H_∞ norm (since the set of outputs used to evaluate the Hankel norm is a subset of outputs used to evaluate the H_∞ norm); thus,

$$\|G\|_h \le \|G\|_\infty . \tag{5.9}$$

The Hankel norm can be determined from the controllability and observability grammians as follows:

$$\|G\|_h = \sqrt{\lambda_{\max}(W_c W_o)}, \tag{5.10}$$

where $\lambda_{\max}(.)$ denotes the largest eigenvalue, and W_c, W_o are the controllability and observability grammians, respectively. Thus, it follows from the definition of the Hankel singular value (4.8) that the Hankel norm of the system is the largest Hankel singular value of the system, $\gamma_{\max}$,

$$\|G\|_h = \gamma_{\max} . \tag{5.11}$$

Additionally, the Hankel and H_∞ norms are related to the Hankel singular values as follows; see [129, p. 156]:

$$\|G\|_h = \gamma_{\max} \le 2\sum_{i=1}^{N} \gamma_i . \|G\|_h = \gamma_{\max} \le 2\sum_{i=1}^{N} \gamma_i . \tag{5.12}$$

This estimation is rather rough since it says that $\gamma_{\max} \le 2\gamma_{\max}$. However, for structures, more precise estimation can be obtained, as shown later in this chapter.

5.2 Norms of the Discrete-Time Systems

In this section we present the H_2, H_∞, and Hankel norms for the discrete-time systems, and compare them with the norms of the continuous-time systems.

5.2.1 The H_2 Norm

The discrete-time H_2 norm is defined as an rms sum of integrals of the magnitudes of its transfer function, or as an rms sum of the impulse response

$$\|G_d\|_2 = \left(\frac{1}{2\pi} \int_0^{2\pi} \operatorname{tr}(G_d^*(e^{j\theta}) G_d(e^{j\theta}) d\theta \right)^{1/2} = \left(\sum_{i=0}^{\infty} g_d^2(i\Delta t) \right)^{1/2}. \quad (5.13)$$

In the above equation $\theta = \omega\Delta t$, and $g_d(i\Delta t)$ is the impulse response of the discrete-time system at $t = i\Delta t$.

Similarly to the continuous-time case we calculate the H_2 norm using the discrete-time grammians W_{dc} and W_{do},

$$\|G_d\|_2^2 = \operatorname{tr}(C^T C W_{dc}) = \operatorname{tr}(B_d B_d^T W_{do}). \quad (5.14)$$

A relationship between the discrete- and continuous-time H_2 norms is derived by introducing the relationships between discrete- and continuous-time grammians, as in (4.19) and (4.21) to the above equation. In this way one obtains

$$\|G\|_2 = \frac{1}{\sqrt{\Delta t}} \|G_d\|_2. \quad (5.15)$$

As we shall see, unlike the Hankel and H_∞ norm cases, the discrete-time H_2 norm does not converge to the continuous-time H_2 norm when the sampling time approaches zero. This can be explained by the system impulse responses. The continuous-time H_2 norm is obtained from the continuous-time unit impulse response

$$\|G\|_2^2 = \int_0^\infty g^2(\tau) d\tau,$$

which can be approximated as

$$\|G\|_2^2 \cong \sum_{i=0}^{\infty} g^2(i\Delta t)\Delta t. \quad (5.16)$$

The value of the applied impulse was 1. Note, however, that for the discrete-time system the impulse response is evaluated for the impulse value Δt. Indeed, for the discrete-time system the impulse amplitude was 1 and its duration was Δt. Thus the impulse value in this case, as a product of its amplitude and duration, is Δt. For this reason the relationship between the impulse responses of the continuous-time system and the equivalent discrete-time system is

$$g(i\Delta t) = \frac{g_d(i\Delta t)}{\Delta t}.$$

Introducing the latter equation to (5.16) we obtain

$$\|G\|_2^2 \cong \frac{1}{\Delta t}\sum_{i=0}^{\infty} g_d^2(i\Delta t) = \frac{1}{\Delta t}\|G_d\|_2^2,$$

which is identical with (5.15).

5.2.2 The H_∞ Norm

The H_∞ norm of the discrete-time system is defined as (see [15]),

$$\|G_d\|_\infty = \max_{\omega\Delta t} \sigma_{\max}\left(G_d(e^{j\omega\Delta t})\right). \tag{5.17}$$

Since for a small enough sampling time the discrete-time transfer function is approximately equal to the continuous-time transfer function, see [15],

$$G_d(e^{j\omega\Delta t}) \cong G(j\omega);$$

therefore, the discrete-time H_∞ norm is equal to the continuous-time H_∞ norm

$$\|G_d\|_\infty = \lim_{\Delta t \to 0} \|G\|_\infty \tag{5.18}$$

for the sampling time approaching zero.

5.2.3 The Hankel Norm

The Hankel norm of a discrete-time system is its largest Hankel singular value

$$\|G_d\|_h = \max_i \gamma_{di}, \tag{5.19}$$

where subscript d denotes a discrete-time system. In Chapter 4 we showed that the discrete-time Hankel singular values converge to the continuous-time Hankel singular values, see (4.23); therefore, the discrete-time Hankel norms converge to the continuous-time Hankel norms when the sampling time approaches zero,

$$\|G\|_h = \lim_{\Delta t \to 0} \|G_d\|_h \tag{5.20}$$

(absence of subscript "d" denotes a continuous-time system).

5.3 Norms of a Single Mode

For structures in the modal representation, each mode is independent, thus the norms of a single mode are independent as well (they depend on the mode properties, but not on other modes).

5.3.1 The H$_2$ Norm

Define $\Delta\omega_i$ as a half-power frequency at the ith resonance, $\Delta\omega_i = 2\zeta_i\omega_i$, see [18], [33]. This variable is a frequency segment at the ith resonance for which the value of the power spectrum is one-half of its maximal value. The determination of the half-power frequency is illustrated in Fig. 5.1. The half-power frequency is the width of the shaded area in this figure, obtained as a cross section of the resonance peak at the height of $h_i/\sqrt{2}$, where h_i is the height of the resonance peak.

Consider the ith natural mode and its state-space representation (A_{mi}, B_{mi}, C_{mi}), see (2.52). For this representation we obtain the following closed-form expression for the H$_2$ norm:

Property 5.1. *H$_2$ Norm of a Mode.* *Let $G_i(\omega) = C_{mi}(j\omega I - A_{mi})^{-1}B_{mi}$ be the transfer function of the ith mode. The H_2 norm of the ith mode is*

$$\|G_i\|_2 \cong \frac{\|B_{mi}\|_2\|C_{mi}\|_2}{2\sqrt{\zeta_i\omega_i}} = \frac{\|B_{mi}\|_2\|C_{mi}\|_2}{\sqrt{2\Delta\omega_i}} \cong \gamma_i\sqrt{2\Delta\omega_i}\,. \tag{5.21}$$

Proof. From the definition of the H$_2$ norm and (4.45) we obtain

$$\|G_i\|_2 \cong \sqrt{\operatorname{tr}(C_{mi}^T C_{mi} W_{ci})} \cong (\|B_{mi}\|_2\|C_{mi}\|_2)\big/(2\sqrt{\zeta_i\omega_i}). \qquad \blacksquare$$

We determine the norm of the second-order modal representation $(\omega_i, \zeta_i, b_i, c_i)$ by replacing B_{mi}, C_{mi} with b_i, c_i, respectively. Note also that $\|G_i\|_2$ is the modal cost of Skelton [124], Skelton and Hughes [126]. The Matlab function *norm_H2.m* in Appendix A.9 can be used to compute modal H_2 norms.

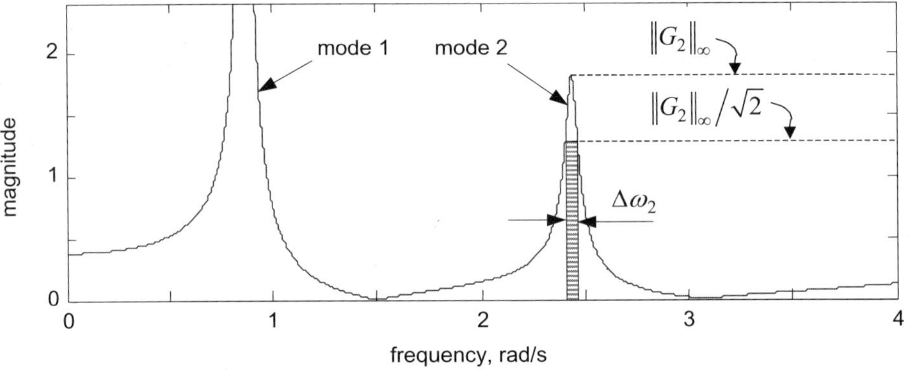

Figure 5.1. The determination of the half-power frequency, H_2 norm and H_∞ norm for the second mode.

Example 5.1. In this example we illustrate the determination of the H_2 norm for a simple system as in Fig. 1.1. For this system, the masses are $m_1 = 11$, $m_2 = 5$, and $m_3 = 10$, while the stiffness coefficients are $k_1 = 10$, $k_2 = 50$, $k_3 = 55$, and $k_4 = 10$. The damping matrix is proportional to the stiffness matrix $D = 0.01K$. The single input u is applied simultaneously to the three masses, such that $f_1 = u$, $f_2 = 2u$, $f_3 = -5u$, and the output is a linear combination of the mass displacements, $y = 2q_1 - 2q_2 + 3q_3$, where q_i is the displacement of the ith mass and f_i is the force applied to that mass.

The transfer function of the system and of each mode is shown in Fig. 5.2. We can see that each mode is dominant in the neighborhood of the mode natural frequency, thus the system transfer function coincides with the mode transfer function near this frequency. The shaded area shown in Fig. 5.3(a) is the H_2 norm of the mode. Note that this area is shown in the logarithmic scale for visualization purposes and that most of the actual area is included in the neighborhood of the peak; compare with the same plot in Fig. 5.1 in the linear coordinates. The system H_2 norm is shown as the shaded area in Fig. 5.3(b), which is approximately a sum of areas of each of the modes.

The H_2 norms of the modes determined from the transfer function are $\|G_1\|_2 = 1.9399$, $\|G_2\|_2 = 0.3152$, $\|G_3\|_2 = 0.4405$, and the system norm is $\|G\|_2 = 2.0141$. It is easy to check that these norms satisfy (5.25) since $\sqrt{2.0141^2 + 0.3152^2 + 0.4405^2} = 2.0141$.

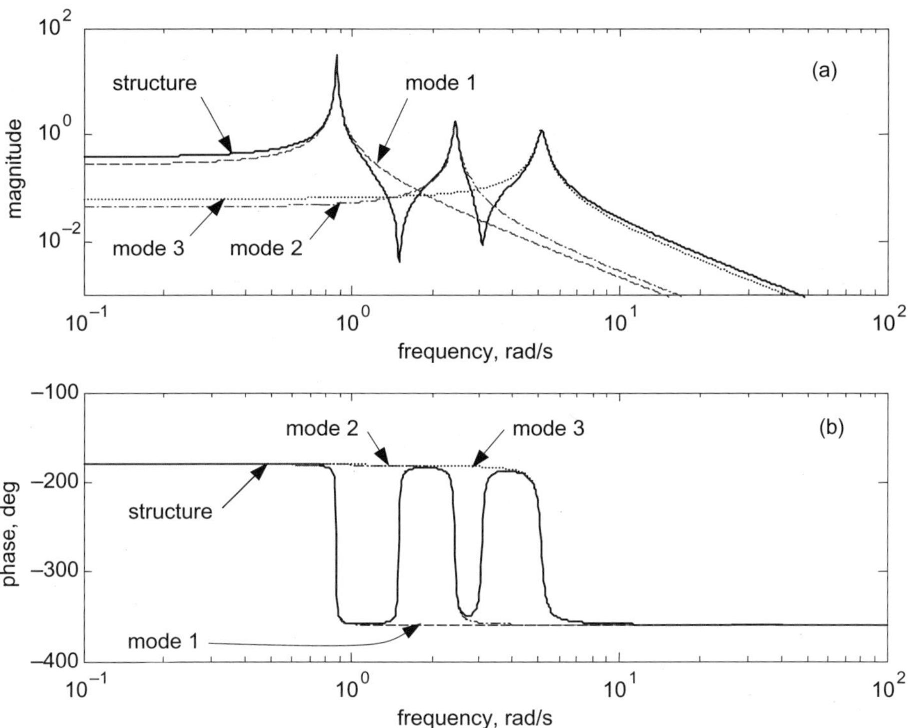

Figure 5.2. The transfer function of the structure (solid line) and of each mode: Mode 1 (dashed line), mode 2 (dash–dotted line), and mode 3 (dotted line).

5.3.2 The H$_\infty$ Norm

The H$_\infty$ norm of a natural mode can be expressed approximately in the closed-form as follows:

Property 5.2. H$_\infty$ *Norm of a Mode.* *Consider the ith mode* (A_{mi}, B_{mi}, C_{mi}) *or* $(\omega_i, \zeta_i, b_{mi}, c_{mi})$. *Its* H_∞ *norm is estimated as*

$$\|G_i\|_\infty \cong \frac{\|B_{mi}\|_2 \|C_{mi}\|_2}{2\zeta_i\omega_i} = \frac{\|b_{mi}\|_2 \|c_{mi}\|_2}{2\zeta_i\omega_i}. \tag{5.22}$$

Proof. In order to prove this, note that the largest amplitude of the mode is approximately at the *i*th natural frequency; thus,

$$\|G_i\|_\infty \cong \sigma_{\max}(G_i(\omega_i)) = \frac{\sigma_{\max}(C_{mi}B_{mi})}{2\zeta_i\omega_i} = \frac{\|B_{mi}\|_2 \|C_{mi}\|_2}{2\zeta_i\omega_i}. \qquad ▣$$

The modal H_∞ norms can be calculated using the Matlab function *norm_Hinf.m* given in Appendix A.10.

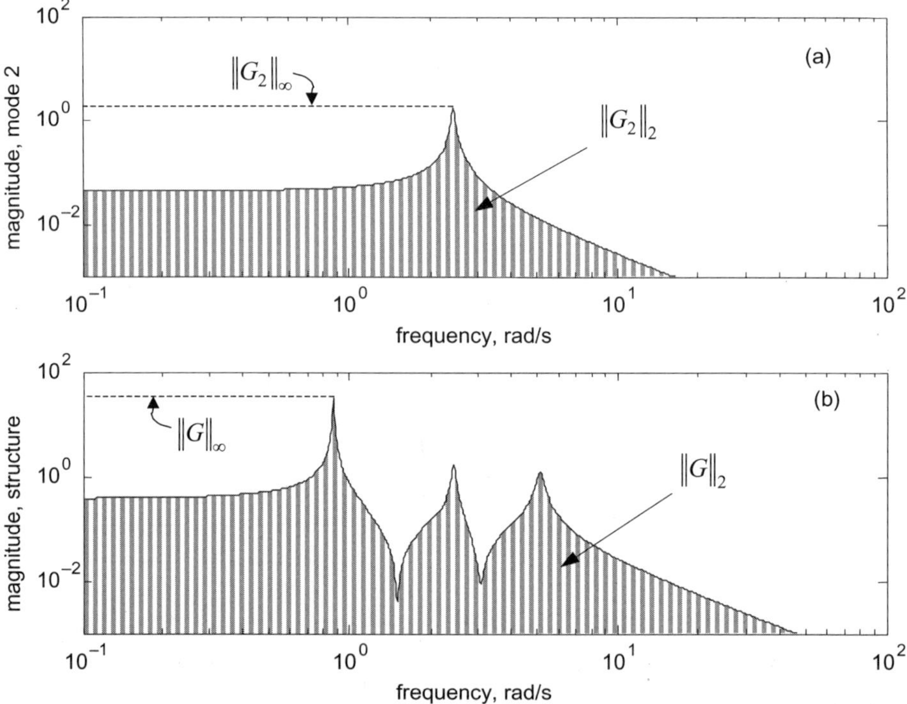

Figure 5.3. H_2 and H_∞ norms (a) of the second mode; and (b) of the system.

Example 5.2. In this example we illustrate the determination of the H_∞ norm of a simple structure, as in Example 5.1, and of its modes.

The H_∞ norm of the second mode is shown in Fig. 5.3(a) as the height of the second resonance peak. The H_∞ norm of the system is shown in Fig. 5.3(b) as the height of the highest (first in this case) resonance peak. The H_∞ norms of the modes, determined from the transfer function, are $\|G_1\|_\infty \cong 18.9229$, $\|G_2\|_\infty \cong 1.7454$, $\|G_3\|_\infty \cong 1.2176$, and the system norm is $\|G\|_\infty \cong \|G_1\|_\infty \cong 18.9619$.

5.3.3 The Hankel Norm

This norm is approximately evaluated from the following closed-form formula:

Property 5.3. *Hankel Norm of a Mode.* *Consider the ith mode in the state-space form* (A_{mi}, B_{mi}, C_{mi}), *or the corresponding second-order form* $(\omega_i, \zeta_i, b_{mi}, c_{mi})$. *Its Hankel norm is determined from*

$$\left\| G_i \right\|_h = \gamma_i \cong \frac{\left\| B_{mi} \right\|_2 \left\| C_{mi} \right\|_2}{4\zeta_i \omega_i} = \frac{\left\| b_{mi} \right\|_2 \left\| c_{mi} \right\|_2}{4\zeta_i \omega_i}. \tag{5.23}$$

The modal Hankel norms can be calculated using the Matlab function *norm_Hankel.m* given in Appendix A.11.

5.3.4 Norm Comparison

Comparing (5.21), (5.22), and (5.23) we obtain the approximate relationships between H_2, H_∞, and Hankel norms

$$\left\| G_i \right\|_\infty \cong 2\left\| G_i \right\|_h \cong \sqrt{\zeta_i \omega_i}\,\left\| G_i \right\|_2. \tag{5.24}$$

The above relationship is illustrated in Fig. 5.4, using (5.21), (5.22), and (5.23), assuming the same actuator and sensor locations.

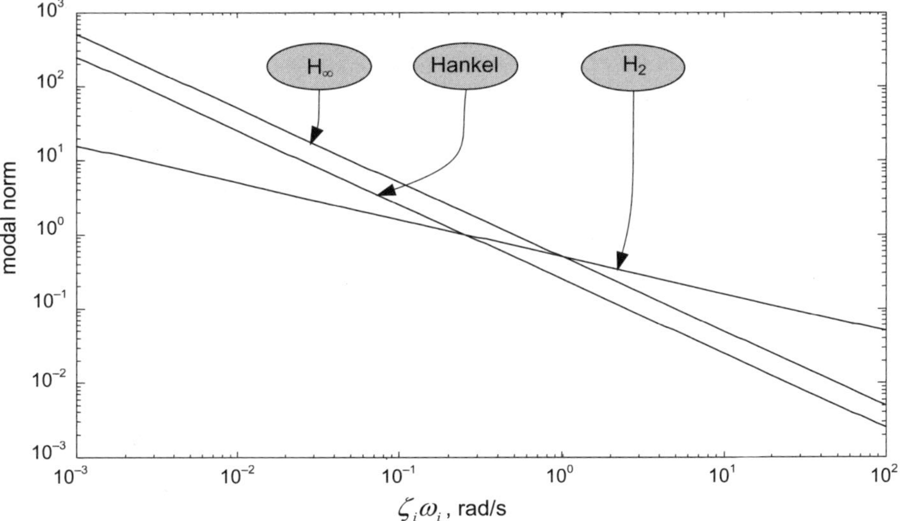

Figure 5.4. Modal norms versus $\zeta_i \omega_i$.

Example 5.3. The Matlab code for this example is in Appendix B. Consider a truss presented in Fig. 1.2. Vertical control forces are applied at nodes 9 and 10, and the output rates are measured in the horizontal direction at nodes 4 and 5. Determine the H_2 and H_∞ norms for each mode.

The norms are given in Fig. 5.5(a). From (5.24) it follows that the ratio of the H_2 and H_∞ norms is

$$\frac{\|G_i\|_2}{\|G_i\|_\infty} \cong \sqrt{\zeta_i \omega_i} = 0.707\sqrt{\Delta\omega_i};$$

hence, the relationship between the H_∞ and H_2 norms depends on the width of the resonance. For a wide resonant peak (large $\Delta\omega_i$) the H_2 norm of the ith mode is larger than the corresponding H_∞ norm. For a narrow resonant peak (small $\Delta\omega_i$) the H_∞ norm of the ith mode is larger than the corresponding H_2 norm. This is visible in Fig. 5.5(a), where neither norm is dominant.

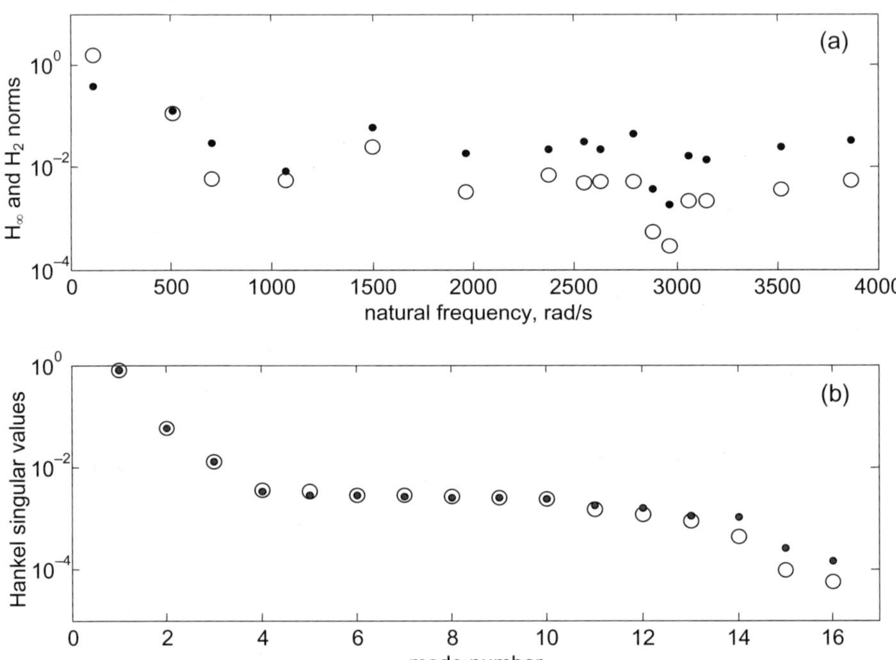

Figure 5.5. The 2D truss: (a) H_2 (○) and H_∞ (•) approximate norms; and (b) the exact (○) and the approximate (•) Hankel singular values.

Next, we obtained the exact Hankel singular values, and the approximate values from (5.10) and (5.23), respectively, and they are shown in Fig. 5.5(b), where we can see a good coincidence between the exact and approximate values.

5.4 Norms of a Structure

The H_2, H_∞, and Hankel norms of a structure are expressed in terms of the norms of its modes. The decomposition of the system norms in terms of its modal norms allows for the derivation of useful structural properties that are used in the dynamics and control algorithms presented in this book.

5.4.1 The H_2 Norm

For a structure the H_2 norm is obtained as follows:

Property 5.4. H_2 *Norm of a Structure.* *Let $G(\omega)=C_m(j\omega I-A_m)^{-1}B_m$ be the transfer function of a structure, and let (A_m,B_m,C_m) be its modal state-space representation. The system H_2 norm is, approximately, the rms sum of the modal norms*

$$\|G\|_2 \cong \sqrt{\sum_{i=1}^{n}\|G_i\|_2^2}, \tag{5.25}$$

where n is the number of modes, and $G_i=C_{mi}(j\omega I-A_{mi})^{-1}B_{mi}$.

Proof. Since the controllability grammian W_c in modal coordinates is diagonally dominant, its H_2 norm is as follows:

$$\|G\|_2^2=\mathrm{tr}(C_m^T C_m W_c)\cong\sum_{i=1}^{n}\mathrm{tr}(C_{mi}^T C_{mi} W_{ci})=\sum_{i=1}^{n}\|G_i\|_2^2 . \qquad \blacksquare$$

This property is illustrated in Fig. 5.6(a).

5.4.2 The H_∞ Norm

For a structure, the approximate H_∞ norm is proportional to its largest Hankel singular value $\gamma_{\max}$. The modal H_∞ norms can be calculated using the Matlab function *norm_Hinf.m* given in Appendix A.10.

Property 5.5. H_∞ *Norm of a Structure.* *Due to the almost independence of the modes, the system* H_∞ *norm is the largest of the mode norms, i.e.,*

$$\|G\|_\infty \cong \max_i \|G_i\|_\infty, \qquad i=1,\ldots,n. \tag{5.26}$$

This property is illustrated in Fig. 5.6(b), and it says that for a single-input–single-output system the largest modal peak response determines the worst-case response.

Example 5.4. Determine the H_∞ norm of a system and of a single almost-balanced mode using the Ricccati equation (5.7).

The norm is a smallest positive parameter ρ such that the solution S of this equation is positive definite. Due to the almost-independence of the modes the solution S of the Riccati equation is diagonally dominant, $S \cong \operatorname{diag}(s_1, s_2, ..., s_n)$, where, by inspection, one can find s_i as a solution of the following equation:

$$\text{(a)} \qquad s_i(A_{mi} + A_{mi}^T) + s_i^2 \rho_i^{-2} B_{mi} B_{mi}^T + C_{mi}^T C_{mi} \cong 0, \qquad i = 1, 2, ..., n,$$

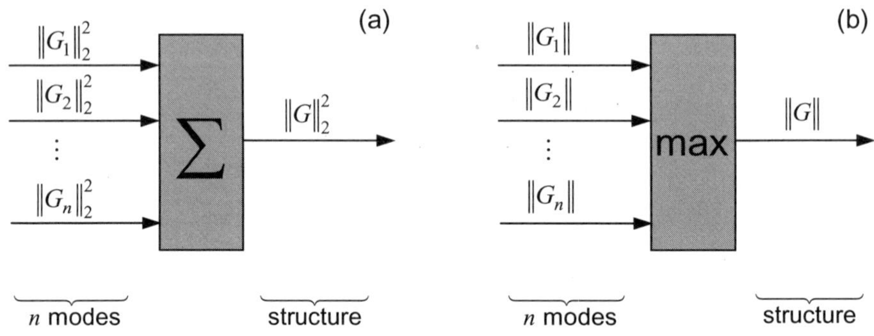

Figure 5.6. Combining modal norms into the norm of a structure for (a) H_2 norm; and for (b) H_∞ and Hankel norms.

and A_{mi} is given by (2.52), B_{mi} is a two-row block of B_m corresponding to block A_{mi} of A_m, and C_{mi} is a two-column block of C_m corresponding to block A_{mi} of A_m. For the almost-balanced mode the Lyapunov equations (4.5) are

$$\gamma_i(A_{mi} + A_{mi}^T) + B_{mi} B_{mi}^T \cong 0,$$
$$\gamma_i(A_{mi} + A_{mi}^T) + C_{mi}^T C_{mi} \cong 0.$$

Introducing them to (a) we obtain

$$s_i(A_{mi} + A_{mi}^T) - s_i^2 \rho_i^{-2} \gamma_i(A_{mi} + A_{mi}^T) - \gamma_i(A_{mi} + A_{mi}^T) \cong 0$$

or, for a stable system,

$$s_i^2 - \frac{\rho_i^2}{\gamma_i} s_i + \rho_i^2 \cong 0$$

with two solutions $s_i^{(1)}$ and $s_i^{(2)}$,

$$s_i^{(1)} = \frac{\rho_i^2(1 - \beta_i)}{2\gamma_i}, \qquad s_i^{(2)} = \frac{\rho_i^2(1 + \beta_i)}{2\gamma_i}, \qquad \text{and} \qquad \beta_i = \sqrt{1 - \frac{4\gamma_i^2}{\rho_i^2}}.$$

For $\rho_i = 2\gamma_i$ one obtains $s_i^{(1)} = s_i^{(2)} = 2\gamma_i = \rho_i$. Moreover, $\rho_i = 2\gamma_i$ is the smallest ρ_i for which a positive solution s_i exists. This is indicated in Fig. 5.7 by plots of $s_i^{(1)}$ (solid line) and $s_i^{(2)}$ (dashed line) versus ρ_i for $\gamma_i = 0.25$, 0.5, 1, 2, 3, and 4; circles "o" denote locations for which $\|G_i\|_\infty = \rho_{i\max} = 2\gamma_i$.

In order to obtain S positive definite, all s_i must be positive. Thus, the largest ρ_i from the set $\{\rho_1, \ \rho_2, \ \cdots \ \rho_n\}$ is the smallest one for which S is positive definite, which can easily be verified in Fig. 5.7. Thus,

$$\|G\|_\infty = \max_i \rho_i \cong 2\gamma_{\max}.$$

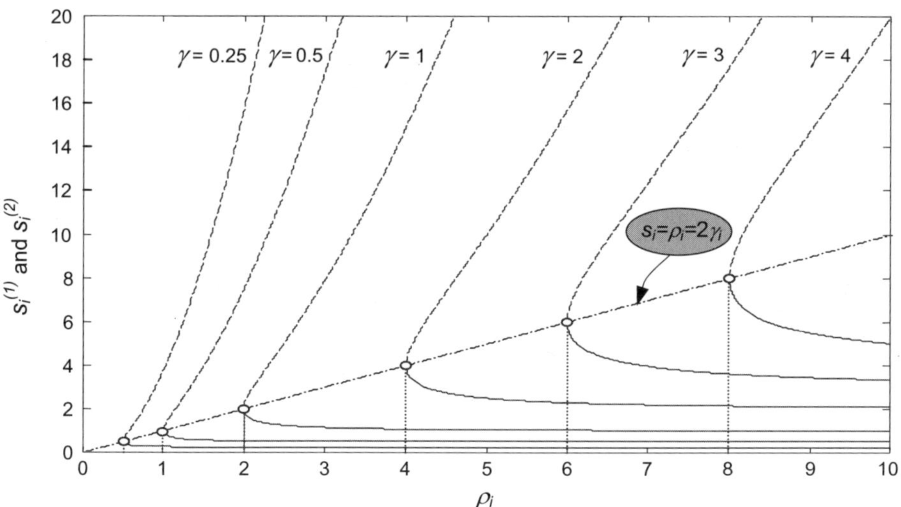

Figure 5.7 Solutions $s_i^{(1)}$ (solid lines) and $s_i^{(2)}$ (dashed lines): Note that $\rho_i = 2\gamma_i$ at locations marked "o".

5.4.3 The Hankel Norm

For a structure the Hankel norm is obtained as follows:

Property 5.6. *Hankel Norm of a Structure.* *The Hankel norm of the structure is the largest norm of its modes, and it is half of the H_∞ norm, i.e.,*

$$\|G\|_h \cong \max_i \|G_i\|_h = \gamma_{\max} = 0.5\|G\|_\infty, \tag{5.27}$$

where $\gamma_{\max}$ is the largest Hankel singular value of the system.

This property is illustrated in Fig. 5.6(b).

5.5 Norms of a Structure with a Filter

In structural testing or in controller design a structure is often equipped with a filter. The filter models disturbances or shapes the system performance. In the following we will analyze how the filter addition impacts the structural and modal norms. Consider a filter with a diagonal transfer function $F(\omega)$. The diagonal $F(\omega)$ of order s represents the input filter without cross-coupling between the inputs. Similarly, the diagonal $F(\omega)$ of order r represents the output filter without cross-coupling between the outputs. Denote by α_i the magnitude of the filter response at the ith natural frequency

$$\alpha_i = |F(\omega_i)| = \sqrt{F^*(\omega_i)F(\omega_i)}. \tag{5.28}$$

The filter is smooth if the slope of its transfer function is small when compared to the slope of the structure near the resonance, that is, at the half-power frequency

$$\left|\frac{\partial\sigma_{\max}(F)}{\partial\omega}\right| \ll \left|\frac{\partial\sigma_{\max}(G)}{\partial\omega}\right| \quad \text{for} \quad \omega = [\omega_i - 0.5\Delta\omega_i, \quad \omega_i + 0.5\Delta\omega_i], \tag{5.29}$$

for $i = 1, \ldots, n$. Above, $\sigma_{\max}(X)$ denotes the maximal singular value of X and $\Delta\omega_i$ denotes the half-power frequency at the ith resonance. The smoothness property is illustrated in Fig. 5.8.

5.5.1 The H_2 Norm

With the above assumptions the following property is valid:

Property 5.7. H_2 *Norm of a Structure with a Filter.* *The norm of a structure with a smooth filter is approximately an rms sum of scaled modal norms*

$$\|GF\|_2^2 \cong \sum_{i=1}^{n} \|G_i\alpha_i\|_2^2, \tag{5.30}$$

and the norm of the ith mode with a smooth filter is a scaled norm

$$\|G_iF\|_2 \cong \|G_i\alpha_i\|_2, \tag{5.31}$$

where the scaling factor α_i is given by (5.28).

Proof. Note that for the smooth filter the transfer function GF preserves the properties of a flexible structure given by Property 2.1; thus,

$$\|GF\|_2^2 = \frac{1}{2\pi}\int_{-\infty}^{\infty} \operatorname{tr}(F^*(\omega)G^*(\omega)G(\omega)F(\omega))\,d\omega$$
$$\cong \sum_{i=1}^{n} \frac{1}{2\pi}\int_{-\infty}^{\infty} \operatorname{tr}(F(\omega_i)F^*(\omega_i)G_i^*(\omega)G_i(\omega))\,d\omega$$
$$= \sum_{i=1}^{n} \frac{1}{2\pi}\int_{-\infty}^{\infty} \operatorname{tr}(|F(\omega_i)|^2\, G_i^*(\omega)G_i(\omega))\,d\omega$$
$$= \sum_{i=1}^{n} \frac{1}{2\pi}\int_{-\infty}^{\infty} \operatorname{tr}(\alpha_i^2 G_i^*(\omega)G_i(\omega))\,d\omega = \sum_{i=1}^{n}\|G_i\alpha_i\|_2^2.$$

In the above approximation we used (5.25), the trace commutative property $\operatorname{tr}(AB) = \operatorname{tr}(BA)$, and the following inequality:

$$\int_{-\infty}^{\infty} \operatorname{tr}(F^* G_j^* G_i F)\,d\omega \ll \int_{-\infty}^{\infty} \operatorname{tr}(F^* G_i^* G_i F)\,d\omega \qquad \text{for} \quad i \neq j. \qquad ▣$$

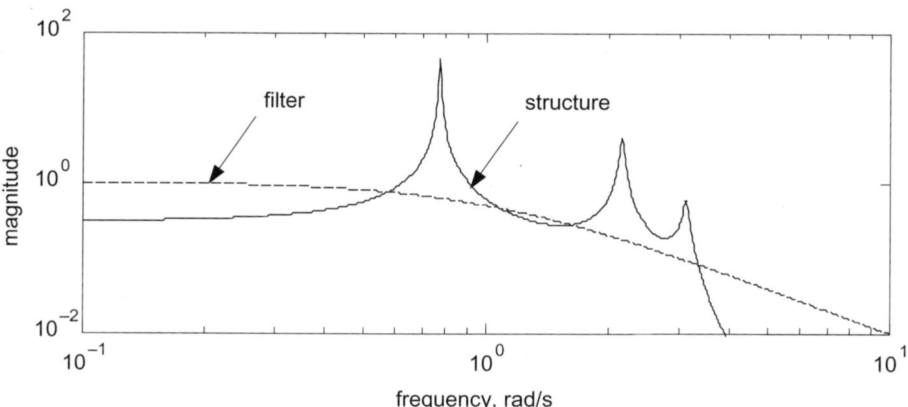

Figure 5.8. Comparing magnitudes of a transfer function of a structure (solid line) and a smooth filter (dashed line).

Property 5.7 says that a norm of a smooth filter in series with a flexible structure is approximately equal to the norm of a structure scaled by the filter gains at natural frequencies.

For a single-input–single-output system we obtain the following:

$$\|GF\|_2^2 \cong \sum_{i=1}^{n} \alpha_i^2 \|G_i\|_2^2,$$
$$\|G_iF\|_2 \cong \alpha_i \|G_i\|_2. \tag{5.32}$$

A similar result to Property 5.7 holds for a structure with a filter at the output.

5.5.2 The H_∞ Norm

Similarly to the H_2 norm, we consider a structure G with a smooth input filter F. The following property is valid:

Property 5.8. H_∞ *Norm of a Structure with a Filter.* *The H_∞ norm of a structure with a smooth filter is equal to the H_∞ norm of the structure with scaled modes*

$$\|GF\|_\infty \cong \max_i(\|G_i\alpha_i\|_\infty), \qquad i = 1,\dots n, \tag{5.33}$$

and the norm of the ith mode with a smooth filter is equivalent to the norm of the scaled mode

$$\|G_iF\|_\infty \cong \|G_i\alpha_i\|_\infty, \tag{5.34}$$

where the scaling factor α_i is defined in (5.28).

Proof. Note that for a smooth filter the transfer function GF preserves the properties of a flexible structure given by Property 2.1; thus,

$$\begin{aligned}\|GF\|_\infty &= \sup_\omega \sigma_{\max}(G(\omega)F(\omega)) \cong \max_i \sigma_{\max}(G(\omega_i)F(\omega_i))\\ &\cong \max_i \sigma_{\max}(G_i(\omega_i)\alpha_i) = \max_i(\|G_i\alpha_i\|_\infty).\end{aligned}$$

In the above approximation we took into consideration the fact that $\sigma_k(GF) = \sigma_k(G|F|)$ which can be proven as follows:

$$\begin{aligned}\sigma_k^2(GF) &= \lambda_k(F^*G^*GF) = \lambda_k(FF^*G^*G)\\ &= \lambda_k(|F|^2\,G^*G) = \lambda_k(|F|G^*G|F|) = \sigma_k^2(G|F|).\end{aligned}$$

▣

The above property says that a norm of a smooth filter in series with a flexible structure is approximately equal to the norm of a structure scaled by the filter gains at the natural frequencies.

For a single-input–single-output system the above formulas simplify to

$$\begin{aligned}\|GF\|_\infty &\cong \max_i(\alpha_i\|G_i\|_\infty),\\ \|G_iF\|_\infty &\cong \alpha_i\|G_i\|_\infty.\end{aligned} \tag{5.35}$$

A similar result to Property 5.8 holds for a structure with a filter at the output.

5.5.3 The Hankel Norm

Properties of the Hankel norm of a structure with a filter are similar to the H_∞ norm:

Property 5.9. ***Hankel Norm of a Structure with a Filter.*** *The Hankel norm of a structure with a smooth filter is equal to the Hankel norm of the structure with scaled modes*

$$\|GF\|_h \cong \max_i(\|G_i\alpha_i\|_h), \qquad i = 1, \ldots, n, \tag{5.36}$$

and the norm of the ith mode with a smooth filter is equivalent to the norm of the scaled mode

$$\|G_iF\|_h \cong \|G_i\alpha_i\|_h, \tag{5.37}$$

where the scaling factor α_i *is defined as* $\alpha_i = \|F(\omega_i)\|_2$.

For a single-input–single-output system the above formulas simplify to

$$\begin{aligned} \|GF\|_h &\cong \alpha_i \max_i(\|G_i\|_h), \\ \|G_iF\|_h &\cong \alpha_i \|G_i\|_h. \end{aligned} \tag{5.38}$$

A similar result to Property 5.9 holds for a structure with a filter at the output.

5.6 Norms of a Structure with Actuators and Sensors

Consider a flexible structure with s actuators (or s inputs) and n modes, so that the modal input matrix B consists of n block-rows of dimension $2 \times s$

$$B_m = \begin{bmatrix} B_{m1} \\ B_{m2} \\ \vdots \\ B_{mn} \end{bmatrix}, \tag{5.39}$$

and the ith block-row B_{mi} of B_m that corresponds to the ith mode has the form

$$B_{mi} = \begin{bmatrix} B_{mi1} & B_{mi2} & \cdots & B_{mis} \end{bmatrix}, \tag{5.40}$$

where B_{mik} corresponds to the kth actuator at the ith mode.

Similarly to the actuator properties we derive sensor properties. For r sensors of an n mode structure the output matrix is as follows:

$$C_m = \begin{bmatrix} C_{m1} & C_{m2} & \dots & C_{mn} \end{bmatrix}, \qquad \text{where} \qquad C_{mi} = \begin{bmatrix} C_{m1i} \\ C_{m2i} \\ \vdots \\ C_{mri} \end{bmatrix}, \tag{5.41}$$

where C_{mi} is the output matrix of the ith mode and C_{mji} is the 1×2 block of the jth output at the ith mode.

The question arises as to how the norm of a structure with a single actuator or sensor corresponds to the norm of the same structure with a set of multiple actuators or sensors. The answer is in the following properties of the H_2, H_∞, and Hankel norms.

5.6.1 The H_2 Norm

The H_2 norm has the following property:

Property 5.10. *Additive Property of the H_2 Norm for a Set of Actuators and for a Mode.* *The H_2 norm of the ith mode of a structure with a set of s actuators is the rms sum of H_2 norms of the mode with each single actuator from this set, i.e.,*

$$\|G_i\|_2 \cong \sqrt{\sum_{j=1}^{s} \|G_{ij}\|_2^2}, \qquad i = 1, \dots, n. \tag{5.42}$$

Proof. From (5.21) one obtains the norm of the ith mode with the jth actuator ($\|G_{ij}\|_2$) and the norm of the ith mode with all actuators ($\|G_i\|_2$),

$$\|G_{ij}\|_2 \cong \frac{\|B_{mij}\|_2 \|C_{mi}\|_2}{2\sqrt{\zeta_i \omega_i}}, \qquad \|G_i\|_2 \cong \frac{\|B_{mi}\|_2 \|C_{mi}\|_2}{2\sqrt{\zeta_i \omega_i}}.$$

But, from the definition of the norm and from (5.40), it follows that

$$\|B_{mi}\|_2 = \sqrt{\sum_{j=1}^{s} \|B_{mij}\|_2^2}, \tag{5.43}$$

introducing the above equation to the previous one, one obtains (5.42). ▣

This property is illustrated in Fig. 5.9(a).

In the following a similar property is derived for a whole structure:

Property 5.11. *Additive Property of the* H_2 *Norm for a Set of Actuators and for a Structure.* *The H_2 norm of a structure with a set of s actuators is the rms sum of norms of a structure with each single actuator from this set,*

$$\|G\|_2 \cong \sqrt{\sum_{j=1}^{s} \|G_j\|_2^2}. \tag{5.44}$$

Proof. From (5.25) and (5.42) one obtains

$$\|G\|_2 \cong \sqrt{\sum_{i=1}^{n} \|G_i\|_2^2} = \sqrt{\sum_{i=1}^{n}\sum_{j=1}^{s} \|G_{ij}\|_2^2} = \sqrt{\sum_{j=1}^{s}\left(\sum_{i=1}^{n} \|G_{ij}\|_2^2\right)} = \sqrt{\sum_{j=1}^{s} \|G_j\|_2^2}.$$ ▣

This property is illustrated in Fig. 5.10.

Similarly to the actuator properties we derive sensor properties. For r sensors of an n mode structure the output matrix is as in (5.41). For this output matrix the following property is obtained:

Property 5.12. *Additive Property of the* H_2 *Norm for a Set of Sensors and for a Mode.* *The H_2 norm of the ith mode of a structure with a set of r sensors is the rms sum of the H_2 norms of the mode with each single actuator from this set, i.e.,*

$$\|G_i\|_2 \cong \sqrt{\sum_{k=1}^{r} \|G_{ki}\|_2^2}, \qquad i = 1, \ldots, n. \tag{5.45}$$

Proof. Denote the norm of the ith mode with the kth sensor ($\|G_{ki}\|_2$) and the norm of the ith mode with all sensors ($\|G_i\|_2$). From (5.21) we have

$$\|G_{ik}\|_2 \cong \frac{\|B_{mi}\|_2 \|C_{mki}\|_2}{2\sqrt{\zeta_i \omega_i}}, \qquad \|G_i\|_2 \cong \frac{\|B_{mi}\|_2 \|C_{mi}\|_2}{2\sqrt{\zeta_i \omega_i}}. \tag{5.46}$$

From (5.41) it follows that

$$\left\|C_{mi}\right\|_2^2 = \sum_{k=1}^{r}\left\|C_{mki}\right\|_2^2. \tag{5.47}$$

Introducing the above equation to (5.46), we obtain (5.45). ▣

This property is illustrated in Fig. 5.9(b).

In the following a similar property is derived for a whole structure:

Property 5.13. *Additive Property of the H_2 Norm for a Set of Sensors and for a Structure.* *The H_2 norm of a structure with a set of r sensors is the rms sum of the H_2 norms of a structure with each single actuator from this set,*

$$\left\|G\right\|_2 \cong \sqrt{\sum_{j=1}^{r}\left\|G_j\right\|_2^2}. \tag{5.48}$$

Proof. Similar to the proof of Property 5.11. ▣

Equations (5.44) and (5.48) show that the H_2 norm of a mode with a set of actuators (sensors) is the rms sum of the H_2 norms of this mode with a single actuator (sensor). This is illustrated in Fig. 5.9(a),(b). The H_2 norm of a structure is also the rms sum of the H_2 norms of modes, as shown in (5.44) and (5.48), and this fact is illustrated in Fig. 5.6(a) and 5.10.

5.6.2 The H_∞ Norm

Consider a flexible structure with s actuators (or s inputs). Similarly to the H_2 norm, the question arises as to how the H_∞ norm of a structure with a single actuator corresponds to the H_∞ norm of the same structure with a set of s actuators. The answer is in the following property:

Property 5.14. *Additive Property of the H_∞ Norm for a Set of Actuators and for a Mode.* *The H_∞ norm of the ith mode of a structure with a set of s actuators is the rms sum of norms of the mode with each single actuator from this set, i.e.,*

$$\left\|G_i\right\|_\infty \cong \sqrt{\sum_{j=1}^{s}\left\|G_{ij}\right\|_\infty^2}, \qquad i = 1, \ldots, n. \tag{5.49}$$

Proof. From (5.22) one obtains the norm of the ith mode with the jth actuator ($\left\|G_{ij}\right\|_\infty$) and the norm of the ith mode with all actuators ($\left\|G_i\right\|_\infty$),

$$\|G_{ij}\|_\infty = \frac{\|B_{mij}\|_2 \|C_{mi}\|_2}{2\zeta_i \omega_i}, \qquad \|G_i\|_\infty \cong \frac{\|B_{mi}\|_2 \|C_{mi}\|_2}{2\zeta_i \omega_i}.$$

Introducing (5.43) to the above equation, we obtain (5.49). ▣

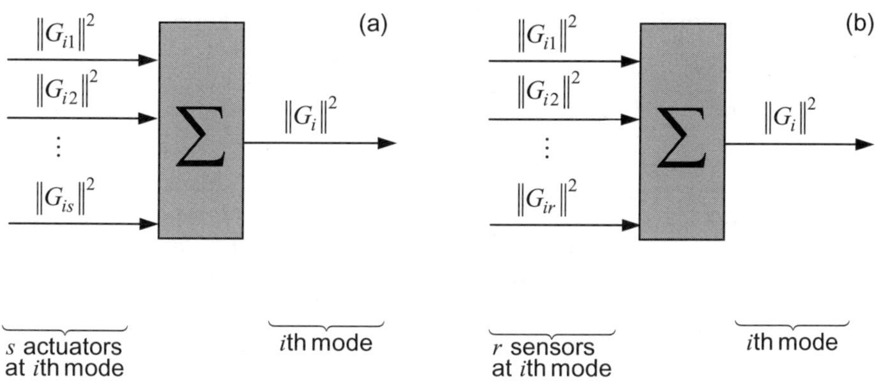

Figure 5.9. Combining (a) actuator norms and (b) sensor norms into the H_2, H_∞, and Hankel norms of a mode.

Similarly to the actuator properties one can derive sensor properties:

Property 5.15. ***Additive Property of the H_∞ Norm for a Set of Sensors and for a Mode.*** *The H_∞ norm of the ith mode of a structure with a set of r sensors is the rms sum of norms of the mode with each single actuator from this set, i.e.,*

$$\|G_i\|_\infty \cong \sqrt{\sum_{k=1}^{r} \|G_{ki}\|_\infty^2}, \qquad i = 1, \ldots, n. \tag{5.50}$$

Proof. Denote the norm of the ith mode with the kth sensor ($\|G_{ki}\|_\infty$) and the norm of the ith mode with all sensors ($\|G_i\|_\infty$). From (5.22) we have

$$\|G_{ik}\|_\infty = \frac{\|B_{mi}\|_2 \|C_{mki}\|_2}{2\zeta_i \omega_i}, \qquad \|G_i\|_\infty = \frac{\|B_{mi}\|_2 \|C_{mi}\|_2}{2\zeta_i \omega_i}.$$

Introducing (5.47) to the above equation, we obtain (5.50). ▣

Equations (5.49) and (5.50) show that the H_∞ norm of a mode with a set of actuators (sensors) is the rms sum of the H_∞ norms of this mode with a single actuator (sensor). This is illustrated in Fig. 5.9(a),(b). Note, however, that unlike the H_2 norm, this property does not hold for the whole structure. Instead, the maximum norm rule is applied; see (5.26) and Fig. 5.10.

5.6.3 The Hankel Norm

Since the Hankel norm is approximately one-half of the H_∞ norm $\|G_i\|_h \cong 0.5\|G_i\|_\infty$ and $\|G\|_h \cong 0.5\|G\|_\infty$; therefore, Properties 5.14 and 5.15 of the H_∞ norm apply to the Hankel norm as well, namely:

Property 5.16. ***Additive Property of the Hankel Norm for a Set of Actuators and for a Mode.*** *The Hankel singular value (Hankel norm) of the ith mode of a structure with a set of s actuators is the rms sum of the Hankel singular values of the mode with each single actuator from this set, i.e.,*

$$\gamma_i = \sqrt{\sum_{j=1}^{s} \gamma_{ij}^2}, \qquad i = 1, \ldots, n. \tag{5.51}$$

The sensor properties are similar:

Property 5.17. ***Additive Property of the Hankel Norm for a Set of Sensors and for a Mode.*** *The Hankel singular value (Hankel norm) of the ith mode of a structure with a set of r sensors is the rms sum of the Hankel singular values of the mode with each single sensor from this set, i.e.,*

$$\gamma_i = \sqrt{\sum_{k=1}^{r} \gamma_{ki}^2}, \qquad i = 1, \ldots, n. \tag{5.52}$$

Equations (5.42), (5.45), (5.49)–(5.52) show that the norms of a mode are the rms sums of norms of actuators or sensors for this mode. Additionally, equations (5.25)–(5.27) show that the norms of a structure can be obtained from norms of modes, either through the rms sum (H_2 norm), or through the selection of the largest modal norm (H_∞ and Hankel norms). This decomposition is very useful in the analysis of structural properties, as will be shown later, and is illustrated in Fig. 5.10 for actuators. A similar figure might be drawn for sensor norm decomposition.

Example 5.5. Using norms in structural damage detection problems. In this example we illustrate the application of modal and sensor norms to determine damage locations; in particular, using the H_2 norm we localize damaged elements of a structure, and assess the impact of the damage on the natural modes of the damaged structure.

Denote the norm of the *j*th sensor of a healthy structure by $\left\|G_{shj}\right\|_2$, and the norm of the *j*th sensor of a damaged structure by $\left\|G_{sdj}\right\|_2$. The *j*th sensor index of the structural damage is defined as a weighted difference between the *j*th sensor norms of a healthy and damaged structure, i.e.,

$$\sigma_{sj} = \frac{\left|\left\|G_{shj}\right\|_2^2 - \left\|G_{sdj}\right\|_2^2\right|}{\left\|G_{shj}\right\|_2^2}.$$

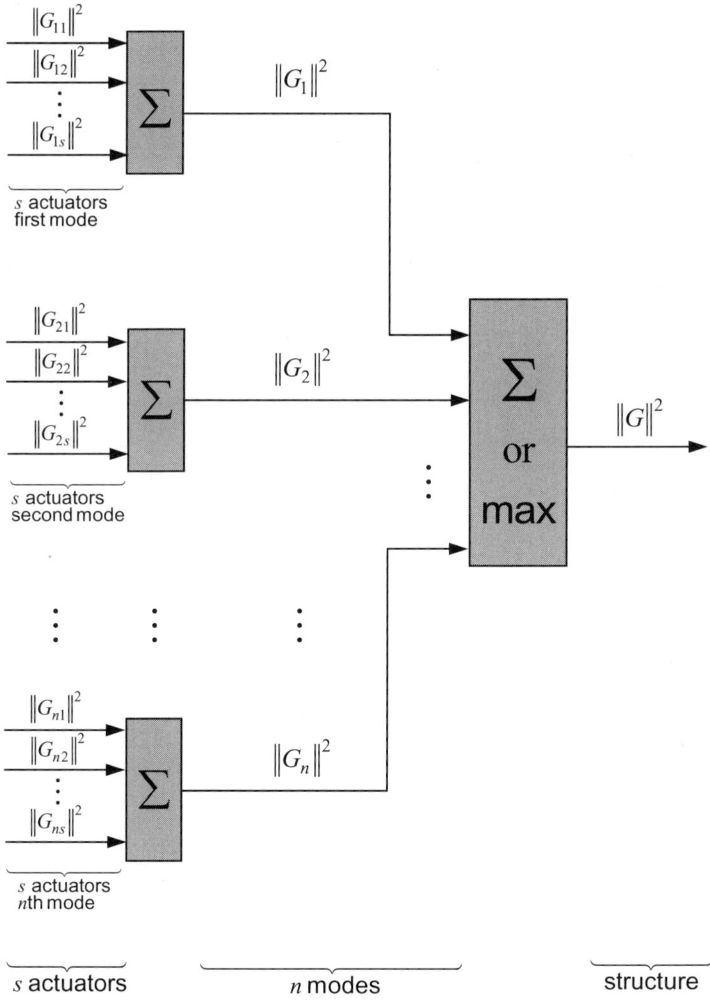

Figure 5.10. Combining modal norms and actuator norms into norms of a structure (H_2, H_∞, and Hankel).

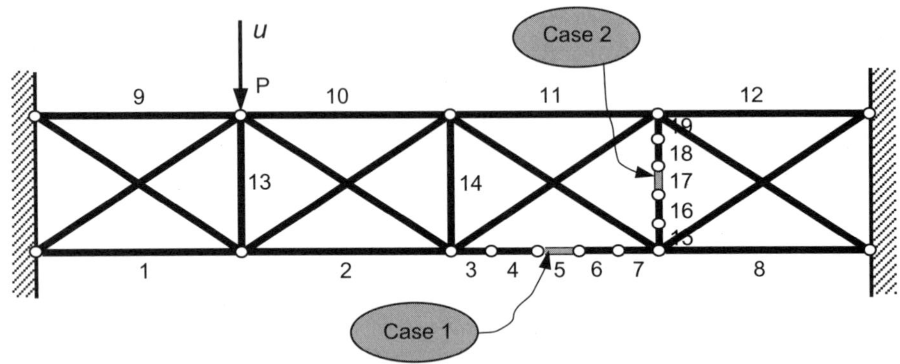

Figure 5.11. The beam structure: Healthy elements are black, damaged elements are gray, and numbers refer to the sensors.

The sensor index reflects the impact of the structural damage on the *j*th sensor.

Similarly, denote the norm of the *i*th mode of a healthy structure by $\|G_{mhi}\|_2$, and the norm of the *i*th mode of a damaged structure by $\|G_{mdi}\|_2$. The *i*th mode index of the structural damage is defined as a weighted difference between the *i*th mode norm of a healthy and damaged structure, i.e.,

$$\sigma_{mi} = \frac{\left|\|G_{mhi}\|_2^2 - \|G_{mdi}\|_2^2\right|}{\|G_{mhi}\|_2^2}.$$

The *i*th mode index reflects the impact of the structural damage on the *i*th mode.

The question arises as to how to measure the sensor and modal norms. It follows from (5.21) that the norm is determined from the system natural frequencies, modal damping ratios, and the modal input and output gains. The gains, on the other hand, are determined from the natural mode shapes at the actuator and sensor locations. Thus, the measurements of natural frequencies, modal damping, and the modal displacements at the actuator and sensor locations of healthy and damaged structures allows for the localization of structural damage.

We analyze a beam structure with fixed ends as in Fig. 5.11. The cross-section area of the steel beams is 1 cm^2. We consider two cases of damage. The first damage is modeled as a 20% reduction of the stiffness of beam No. 5, and the second case is modeled as a 20% reduction of the stiffness of beam No. 17. The structure is more densely divided near the damage locations to reflect more accurately the stress concentration. Nineteen strain-gauge sensors are placed at beams 1 to 19. A vertical force at node *P* excites the structure.

For the first case the sensor and modal indices are shown in Fig. 5.12(a),(b). The sensor indices in Fig. 5.12(a) indicate that sensor No. 5, located at the damaged beam, indicates the largest changes. The modal indices in Fig. 5.12(b) show that the first mode is heavily affected by the damage.

The sensor and modal indices for the second case are shown in Fig. 5.13(a),(b). Figure 5.13(a) shows the largest sensor index at location No. 17 of the damaged beam. The modal indices in Fig. 5.13(b) show that the tenth and second modes are mostly affected by the damage.

5.7 Norms of a Generalized Structure

Consider a structure as in Fig. 3.10, with inputs w and u and outputs z and y. Let G_{wz} be the transfer matrix from w to z, let G_{wy} be the transfer matrix from w to y, let G_{uz} be the transfer matrix from u to z, and let G_{uy} be the transfer matrix from u to y. Let G_{wzi}, G_{uyi}, G_{wyi}, and G_{uzi} be the transfer functions of the ith mode. The following multiplicative properties of modal norms hold:

Property 5.18. *Modal Norms of a General Plant.* *The following norm relationships hold:*

$$\|G_{wzi}\|\|G_{uyi}\| \cong \|G_{wyi}\|\|G_{uzi}\|, \quad \text{for} \quad i = 1, \ldots n, \tag{5.53}$$

where $\|.\|$ *denotes either* H_2, H_∞, *or Hankel norms.*

Proof. We denote by B_{mw} and B_{mu} the modal input matrices of w and u, respectively, and let C_{mz} and C_{my} be the modal output matrices of z and y, respectively, and let B_{mwi}, B_{mui}, C_{mzi}, and C_{myi} be their ith blocks related to the ith mode. The H_∞ norms are approximately determined from (5.22) as

$$\|G_{wzi}\|_\infty \cong \frac{\|B_{mwi}\|_2 \|C_{mzi}\|_2}{2\zeta_i\omega_i}, \qquad \|G_{uyi}\|_\infty \cong \frac{\|B_{mui}\|_2 \|C_{myi}\|_2}{2\zeta_i\omega_i},$$

$$\|G_{wyi}\|_\infty \cong \frac{\|B_{mwi}\|_2 \|C_{myi}\|_2}{2\zeta_i\omega_i}, \qquad \|G_{uzi}\|_\infty \cong \frac{\|B_{mui}\|_2 \|C_{mzi}\|_2}{2\zeta_i\omega_i}.$$

Introducing the above equations to (5.53) the approximate equality is proven by inspection. We prove similarly the H_2 and Hankel norm properties, using (5.21) and (5.23) instead of (5.22). ▣

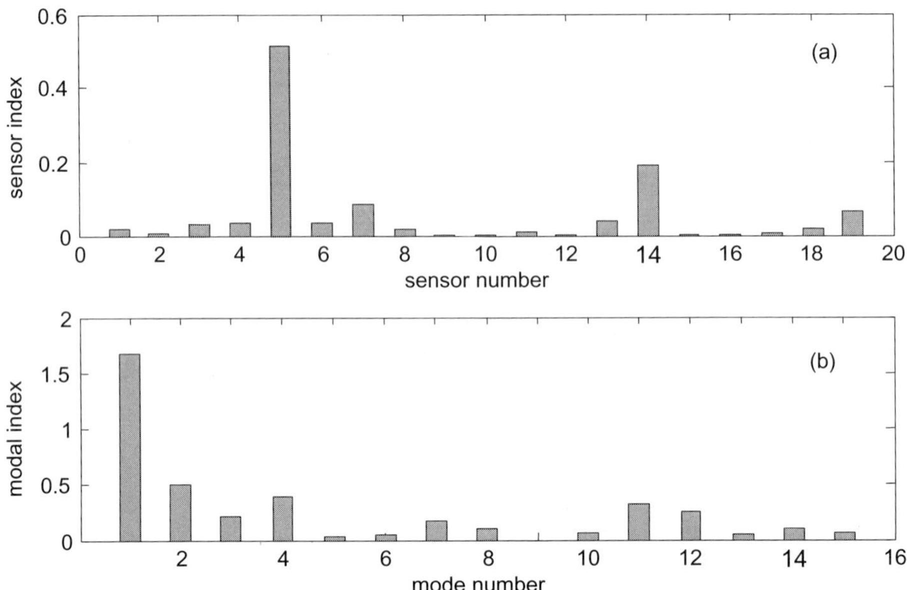

Figure 5.12. Sensor and modal indices for the beam structure, damage case 1: Sensor index for the damaged element No. 5 is high; the modal index shows that the first mode is predominantly impacted.

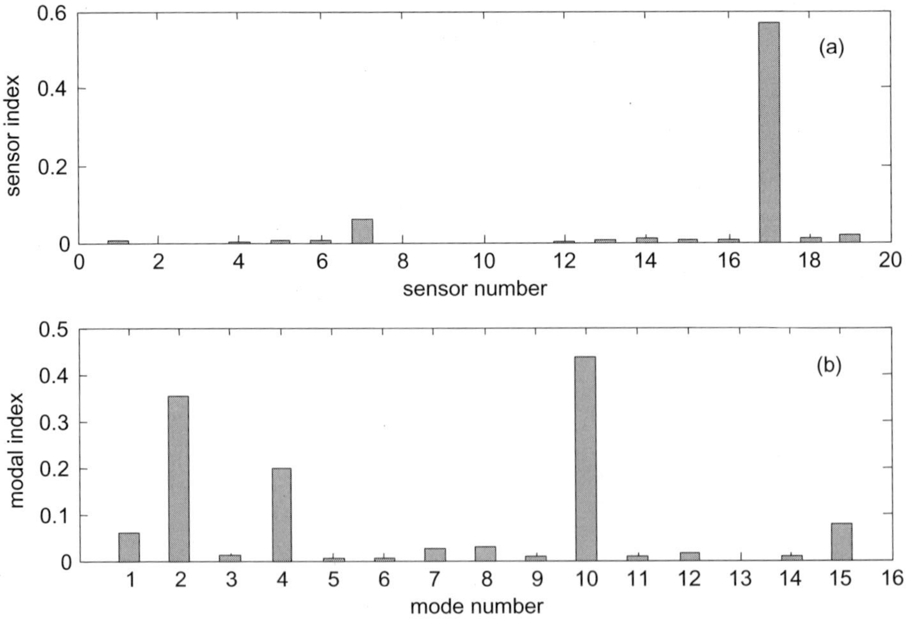

Figure 5.13. Sensor and modal indices for the beam structure, damage case 2: Sensor index for the damaged element No. 17 is high; the modal index shows that modes Nos. 10, 2, and 4 are predominantly impacted.

Property 5.18 shows that for each mode the product of norms of the performance loop (i.e., from the disturbance to the performance) and the control loop (i.e., from the actuators to the sensors) is approximately equal to the product of the norms of the cross-couplings: between the disturbance and sensors, and between the actuators and performance. The physical meaning of this property lies in the fact that by increasing the actuator–sensor connectivity, one increases automatically the cross-connectivity for the *i*th mode: actuator-to-performance and disturbance-to-sensors. This shows that sensors not only respond to the actuator input, but also to disturbances, and actuators not only impact the sensors, but also the performance.

This property is important for the closed-loop design. For the plant as in Fig. 3.11 one obtains

$$z = G_{wz} w + G_{uz} u \qquad \text{and} \qquad y = G_{wy} w + G_{uy} u.$$

The closed-loop transfer matrix G_{cl} from w to z, with the controller K such that $u = Ky$, is as follows:

$$G_{cl} = G_{wz} + G_{uz} K (I - G_{uy} K)^{-1} G_{wy}. \tag{5.54}$$

From the second part of the right-hand side of the above equation it follows that the controller impacts the closed-loop performance not only through the action from u to y, but also through the cross-actions from u to z, and from w to y. Therefore, if the transfer matrices G_{wy} or G_{uz} are zero, the controller has no impact whatsoever on the performance z. Thus the controller design task consists of simultaneous gain improvement between u and y, w and y, and u and z. However, Property 5.18 shows that the improvement in G_{uy} automatically leads to the improvement of G_{wy} and G_{uz}. Thus, the task of actuator and sensor location simplifies to the manipulation of G_{uy} alone, as it will be shown later in this book.

5.8 Norms of the Discrete-Time Structures

The norms of discrete-time structures are obtained in a similar way to the norms of the continuous-time structures. First of all, the system matrix A in discrete-time modal coordinates is block-diagonal, similar to the continuous-time case. For a diagonal A the structural norms are determined from the norms of structural modes, as described previously in this chapter. However, the norms of modes in discrete time are not exactly the same as the norms of modes for the continuous-time case. Later in this section they are obtained in closed-form and compared to the continuous-time norms.

5.8.1 The H_2 Norm

The H_2 norm of a structure is the rms sum of the H_2 norms of its modes. The H_2 norm of the ith mode, on the other hand, is obtained (similar to the continuous case) as follows:

$$\|G_{di}\|_2 = \sqrt{\operatorname{tr}(B_{di}^T W_{doi} B_{di})} \tag{5.55}$$

or, alternatively,

$$\|G_{di}\|_2 = \sqrt{\operatorname{tr}(C_{di} W_{dci} C_{di}^T)}. \tag{5.56}$$

Using the first equation, (3.50), and (4.19) we obtain

$$\|G_{di}\|_2^2 = \frac{1}{\Delta t}\operatorname{tr}(B_{mi}^T S_i^T W_{oi} S_i B_{mi}) = \frac{w_{oi}}{\Delta t}\operatorname{tr}(B_{mi}^T S_i^T S_i B_{mi}),$$

where w_{oi} is the continuous-time grammian given by (4.45). Note also that

$$S_i^T S_i = \frac{2(1-\cos(\omega_i \Delta t)}{\omega_i^2} I_2.$$

Thus,

$$\begin{aligned}\|G_{di}\|_2^2 &= \frac{2w_{oi}(1-\cos(\omega_i\Delta t))}{\Delta t\omega_i^2}\operatorname{tr}(B_{mi}^T B_{mi}) = w_{oi}\|B_{mi}\|_2^2 \frac{2(1-\cos(\omega_i\Delta t))}{\Delta t\omega_i^2} \\ &= \frac{\|C_{mi}\|_2^2\|B_{mi}\|_2^2}{4\zeta_i\omega_i}\Delta t \; \frac{2(1-\cos(\omega_i\Delta t))}{\Delta t^2\omega_i^2} = \|G_i\|_2^2 \Delta t\, k_i^2,\end{aligned}$$

where $\|G_i\|_2$ is the H_2 norm of the mode in continuous time. Therefore,

$$\|G_{di}\|_2 = k_i\sqrt{\Delta t}\,\|G_i\|_2. \tag{5.57}$$

For fast sampling $k_i \to 1$; thus,

$$\lim_{\Delta t\to 0}\frac{\|G_{di}\|_2}{\sqrt{\Delta t}} = \|G_i\|_2. \tag{5.58}$$

The above equation indicates that the discrete-time H_2 norm does not converge to continuous time. This is a consequence of nonconvergence of the discrete-time controllability and observability grammians, and to the continuous-time grammians; see [109], [98], and Chapter 4 of this book.

5.8.2 The H$_\infty$ Norm

The H$_\infty$ norm of a discrete-time system is defined as the peak magnitude over the segment $0 \le \omega\Delta t \le \pi$, i.e.,

$$\|G_d\|_\infty = \sup_{\omega\Delta t} \sigma_{\max}\left(G_d(e^{j\omega\Delta t})\right). \tag{5.59}$$

The H$_\infty$ norm of the ith mode is approximately equal to the magnitude of its transfer function at its resonant frequency ω_i, thus,

$$\|G_{di}\|_\infty = \sigma_{\max}\left(G_{di}(e^{j\omega_i\Delta t})\right) = \lambda_{\max}^{1/2}\left(G_{di}(e^{j\omega_i\Delta t})G_{di}^*(e^{j\omega_i\Delta t})\right), \tag{5.60}$$

where G_{di} is the discrete-time transfer function of the ith mode, ω_i is its natural frequency, and $\lambda_{\max}$ denotes its largest eigenvalue.

In order to obtain its H$_\infty$ norm we use the discrete-time transfer function at $\omega = \omega_i$ of the ith mode as in (3.55). First, note that $z = e^{j\omega_i\Delta t} = \cos(\omega_i\Delta t) + j\sin(\omega_i\Delta t)$ and that, for small ζ_i, one can use the approximation $e^{-\zeta_i\omega_i\Delta t} \cong 1 - \zeta_i\omega_i\Delta t$. Now using (3.55) we obtain

$$(zI - A)^{-1}\Big|_{z=e^{j\omega_i\Delta t}} = \frac{1}{2\zeta_i\omega_i\Delta t}\begin{bmatrix} j & 1 \\ -1 & j \end{bmatrix}.$$

For B_{dmi} as in (3.50) and $B_{mi} = \begin{bmatrix} 0 \\ b_{oi} \end{bmatrix}$, the modal transfer function at its resonance frequency is therefore as follows:

$$G_{dmi}(\omega_i) = \frac{C_{mi}}{2\zeta_i\omega_i^2\Delta t}\begin{bmatrix} 1 - \cos(\omega_i\Delta t) - j\sin(\omega_i\Delta t) \\ \sin(\omega_i\Delta t) + j(1 - \cos(\omega_i\Delta t)) \end{bmatrix} b_{oi}.$$

Introducing the above to (5.60) we obtain

$$\|G_{di}\|_\infty = \lambda_{\max}^{1/2}\left(G_{di}(e^{j\omega_i\Delta t})G_{di}^*(e^{j\omega_i\Delta t})\right) = \frac{\|C_{mi}\|_2 \|b_{oi}\|_2}{\zeta_i\omega_i^2\Delta t}\left(1 - \cos(\omega_i\Delta t)\right),$$

which can be presented in the following form:

$$\|G_{di}\|_\infty = k_i \|G_i\|_\infty, \tag{5.61}$$

where $\|G_i\|_\infty$ is the H$_\infty$ norm of the continuous-time mode and k_i is the coefficient given by (4.84). Comparing the above equation with (5.63) we see that relationships between the continuous- and discrete-time norms are similar for the H_∞ and Hankel cases, and since $k_i = 1$ for $\Delta t \to 0$, the discrete-time H_∞ norm converges to the continuous-time norm.

5.8.3 The Hankel Norm

The Hankel norm of the discrete-time system is defined (similarly to the continuous-time case) as a "geometric mean" of the discrete-time controllability and observability grammians, i.e.,

$$\|G_d\|_h = \sqrt{\lambda_{\max}(W_{dc}W_{do})} \tag{5.62}$$

(subscript d is added to emphasize the discrete-time system). The grammians in modal coordinates are diagonally dominant, therefore for a single (ith) mode we obtain, from (4.81),

$$\|G_{di}\|_h = k_i \|G_i\|_h, \tag{5.63}$$

where $k_i = \sqrt{2(1-\cos\omega_i\Delta t)}/\omega_i\Delta t$ (see (4.84)) and $\|G_{di}\|_h$ is the Hankel norm of the ith mode in discrete time, while $\|G_i\|_h$ is the same norm as the ith mode in continuous time. For fast sampling, i.e., when $\Delta t \to 0$, one obtains $k_i = 1$, which means that the discrete Hankel norm converges to the continuous Hankel norm.

5.8.4 Norm Comparison

From (5.63), (5.61), and (5.57) we obtain the following relationship between the norms of a single mode of a discrete-time system:

$$\|G_{di}\|_\infty \cong 2\|G_{di}\|_h \cong \sqrt{\zeta_i\omega_i\Delta t}\,\|G_{di}\|_2. \tag{5.64}$$

Example 5.6. Consider a beam in Fig. 1.4, divided into $n = 15$ elements, and its first 15 modes. Let a vertical force be applied at node 6, and let its velocity be measured at node 6 in the vertical direction. Determine the H_2 and Hankel norms for their continuous- and discrete-time models.

The norms are plotted in Fig. 5.14. The beam's largest natural frequency is 6221 rad/s. We choose a sampling time of 0.0003 s. The Nyquist frequency for this sampling time is $\pi/\Delta t = 10{,}472$ rad/s, so that the largest natural frequency is quite close to the Nyquist frequency. The plots of the norms are shown in Fig. 5.14, for the continuous-time model in solid line and for the discrete-time model in dotted line. The Hankel norms for the continuous- and discrete-time models are almost identical, except for some discrepancy at higher modes, with natural frequencies close to the Nyquist frequency. The H_2 norms of the continuous- and discrete-time systems are separated by a distance of $1/\sqrt{\Delta t} = 57.74$, as predicted in (5.58).

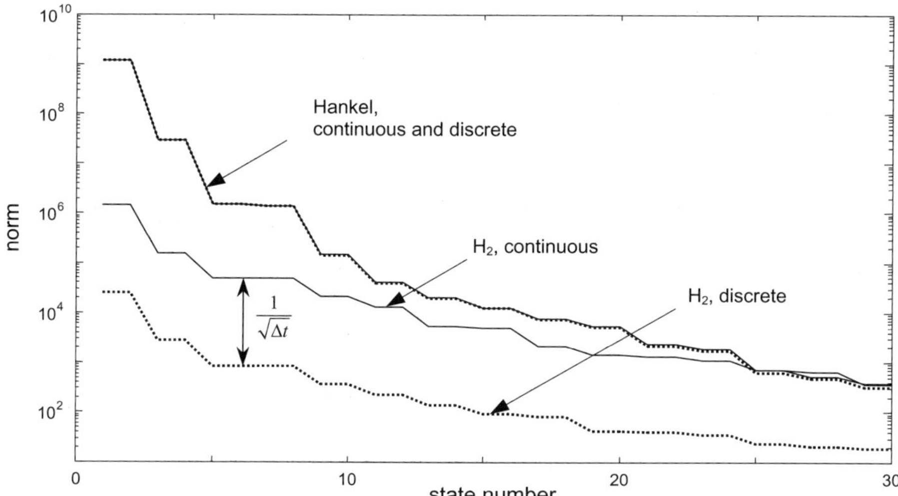

Figure 5.14. Hankel and H_2 norms of a beam: Continuous-time model—solid line; discrete-time model—dotted line. The Hankel norms almost coincide, while the H_2 norms are distanced by $1/\Delta t$.

6 Model Reduction

 *how to obtain small but accurate models*

Make everything as simple as possible,
but not simpler.
—*Albert Einstein*

Model reduction is a part of dynamic analysis, testing planning, and the control design of structures. Typically, a model with a large number of degrees of freedom, such as one developed for static analysis, causes numerical difficulties in dynamic analysis, to say nothing of the high computational cost. In system identification, on the other hand, the order of the identified system is determined by the reduction of the initially oversized model that includes a noise model. Finally, in structural control design the complexity and performance of a model-based controller depends on the order of the structural model. In all cases the reduction is a crucial part of the analysis and design. Thus, the reduced-order system solves the above problems if it acquires the essential properties of the full-order model.

Many reduction techniques have been developed. Some of them, as in [78], [138], and [139], give optimal results, but they are complex and computationally expensive. Some methods, comparatively simple, give results close to the optimal one. The latter include balanced and modal truncation, see [109], [69], [81], [124], [125], [128], [51], [53], and [64]. In this chapter we discuss the reduction in modal coordinates.

6.1 Reduction Through Truncation

In this chapter we consider a structural model in modal coordinates, namely, modal models 1, 2, and 3, as in (2.52), (2.53), and (2.54)). The states of the model are ordered as follows:

$$x = \begin{Bmatrix} x_1 \\ x_2 \\ \vdots \\ x_n \end{Bmatrix} \leftarrow \text{modal norm indicator}, \tag{6.1}$$

where x_i is the state corresponding to the ith mode. It consists of two states; see (2.55), (2.56), and (2.57):

$$x_i = \begin{Bmatrix} x_{i1} \\ x_{i2} \end{Bmatrix}. \tag{6.2}$$

Let $\|G_i\|$ denote either H_2, H_∞, or Hankel norms of the ith mode, and order the states in the state vector (6.1) in the descending norm order. Now, the norm of the first mode is the largest one, and the norm of the last mode is the smallest, which is marked in (6.1) with the norm value indicator located to the right of the equation. In the indicator the largest norm is marked in black and the smallest norm in white.

We obtain a reduced-order model by evaluating the modal states and truncating the least important. Since the modes with the smallest norm are the last ones in the state vector, a reduced-order model is obtained here by truncating the last states in the modal vector. How many of them? This will be determined later in this section by evaluating the reduction errors. Let (A, B, C) be the modal representation (the subscript m is dropped for simplicity of notation) corresponding to the modal state vector x as in (6.1). Let x be partitioned as follows:

$$x = \begin{Bmatrix} x_r \\ x_t \end{Bmatrix}, \tag{6.3}$$

where x_r is the vector of the retained states and x_t is a vector of truncated states. If there are $k < n$ retained modes, x_r is a vector of $2k$ states, and x_t is a vector of $2(n-k)$ states. Let the state triple (A, B, C) be partitioned accordingly,

$$A = \begin{bmatrix} A_r & 0 \\ 0 & A_t \end{bmatrix}, \qquad B = \begin{bmatrix} B_r \\ B_t \end{bmatrix}, \qquad C = \begin{bmatrix} C_r & C_t \end{bmatrix}. \tag{6.4}$$

We obtain the reduced model by deleting the last $2(n-k)$ rows of A, B, and the last $2(n-k)$ columns of A, C. Formally, this operation can be written as follows:

$$A_r = LAL^T, \qquad B_r = LB, \qquad C_r = CL^T, \tag{6.5}$$

where $L = \begin{bmatrix} I_{2k} & 0 \end{bmatrix}$.

Modal reduction by truncation of stable models always produces a stable reduced model, since the poles of the reduced model are a subset of the poles of the full-order model.

The problem is to order the states so that the retained states x_r will be the best reproduction of the full system response. The choice depends on the definition of the reduction index.

6.2 Reduction Errors

We use H_2, H_∞, and Hankel norms to evaluate the reduction errors. The first approach, based on the H_2 norm, is connected to the Skelton reduction method, see [125]. The second method, based on the H_∞ and Hankel norms, is connected with the Moore reduction method; see [109].

6.2.1 H_2 Model Reduction

The H_2 reduction error is defined as

$$e_2 = \|G - G_r\|_2, \tag{6.6}$$

where G is the transfer function of the full model and G_r is the transfer function of the reduced model. Note that in modal coordinates the transfer function is a sum of its modes (see Property 2.1); therefore,

$$G = \sum_{i=1}^{n} G_i \qquad \text{and} \qquad G_r = \sum_{i=1}^{k} G_i;$$

thus, $G - G_r = G_t$, where G_t is the transfer function of the truncated part. Thus,

$$e_2 = \|G_t\|_2. \tag{6.7}$$

Also, the squares of the mode norm are additive, see Property 5.4, therefore the norm of the reduced system with k modes is the root-mean-square (rms) sum of the mode norms

$$\|G_t\|_2^2 = \sum_{i=k+1}^{n} \|G_i\|_2^2. \tag{6.8}$$

Thus, the reduction error is

$$e_2 = \left(\sum_{i=k+1}^{n} \|G_i\|_2^2 \right)^{1/2}. \tag{6.9}$$

It is clear from the above equation that we obtain the optimal reduction if the truncated mode norms $\|G_i\|_2$ for $i = k+1, \ldots, n$, are the smallest ones. Therefore, we rearrange the modal state vector, starting from the mode with the largest H_2 norm and ending with the mode with the smallest norm. Truncation of the last $n-k$ modes will give, in this case, the optimal reduced model of order k.

6.2.2 H_∞ and Hankel Model Reduction

It can be seen from (5.27) that the H_∞ norm is approximately twice the Hankel norm; hence, the reduction using one of these norms is identical with the reduction using the other norm. Thus, we consider here the H_∞ reduction only.

The H_∞ reduction error is defined as

$$e_\infty = \|G - G_r\|_\infty. \tag{6.10}$$

It was shown by Glover [66] that the upper limit of the H_∞ reduction error is as follows:

$$e_\infty = \|G - G_r\|_\infty \le \sum_{i=k+1}^{n} \|G_i\|_\infty. \tag{6.11}$$

However, for the flexible structures in the modal coordinates the error can be estimated less conservatively. Recall that $G - G_r = G_t$ where G_t is the transfer function of the truncated part; therefore,

$$e_\infty = \|G - G_r\|_\infty = \|G_t\|_\infty \cong \|G_{k+1}\|_\infty, \tag{6.12}$$

i.e., the error is equal to the H_∞ norm of the largest truncated mode. It is clear that we obtain the near-optimal reduction if the H_∞ norms of the truncated modes are the smallest ones. Similar results were obtained for the Hankel norm.

Example 6.1. Consider a reduction of a simple system as in Example 2.9 using H_∞ and H_2 norms.

For this system we obtain the H_∞ modal norms from (5.21), namely, $\|G_1\|_\infty \cong 6.7586$ (mode of the natural frequency 1.3256 rad/s), $\|G_2\|_\infty \cong 4.9556$ (mode of the natural frequency 2.4493 rad/s), and $\|G_3\|_\infty \cong 2.6526$ (mode of the natural frequency 3.200 rad/s). The H_2 mode norms (see (5.20)) are as follows:

$\|G_1\|_2 \cong 3.2299$, $\|G_2\|_2 \cong 3.3951$, and $\|G_3\|_2 \cong 0.5937$. The reduction errors after the truncation of the last mode (of frequency 3.200 rad/s) are $e_\infty = 2.6526$ and $e_2 = 0.5937$, while the system norms are $\|G\|_\infty = 6.7586$ and $\|G\|_2 \cong 4.7235$.

Example 6.2. Reduce the model of a 2D truss as in Figure 1.2 in the modal coordinates using the H_∞ norm. Determine the reduction error.

The approximate norms of the modes are shown in Fig. 6.1. From this figure we obtain the system norm (the largest of the mode norms) as $\|G\|_\infty = 1.6185$. Using modal norm values we decided that in the reduced-order model we reject all modes of the H_∞ norm less than 0.01. The area of the H_∞ norm less than 0.01 lies in Fig. 6.1, below the dashed line, and the modes with the H_∞ norm in this area are deleted. Consequently, the reduced model consists of three modes. The transfer function of the full and reduced models (from the second input to the second output) is shown in Fig. 6.2(a), and the corresponding impulse response is shown in Fig. 6.2(b). The reduction error is obtained as $(\|G\|_\infty - \|G_r\|_\infty)/\|G\|_\infty \cong \|G_4\|_\infty / \|G\|_\infty = 0.0040$.

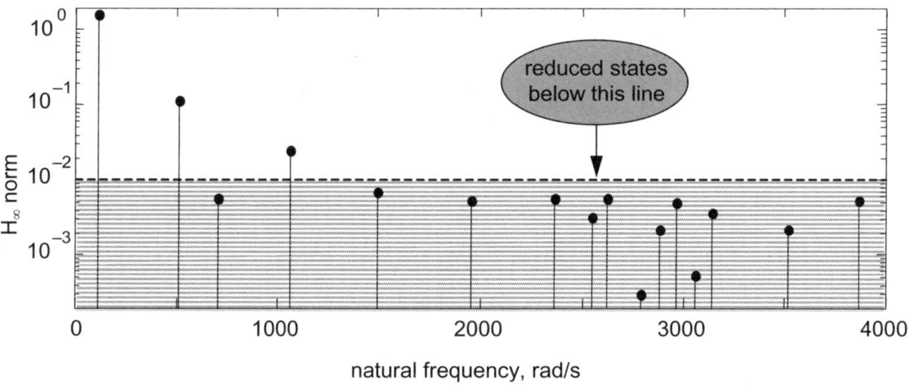

Figure 6.1. H_∞ norms of the 2D truss modes.

6.3 Reduction in the Finite-Time and -Frequency Intervals

We introduced the time- and frequency-limited grammians in Chapter 4. They are used in model reduction such that the response of the reduced system fits the response of the full system in the prescribed time and/or frequency intervals. This approach is useful, for example, in the model reduction of unstable plants (using time-limited grammians) or in filter design (using band-limited grammians).

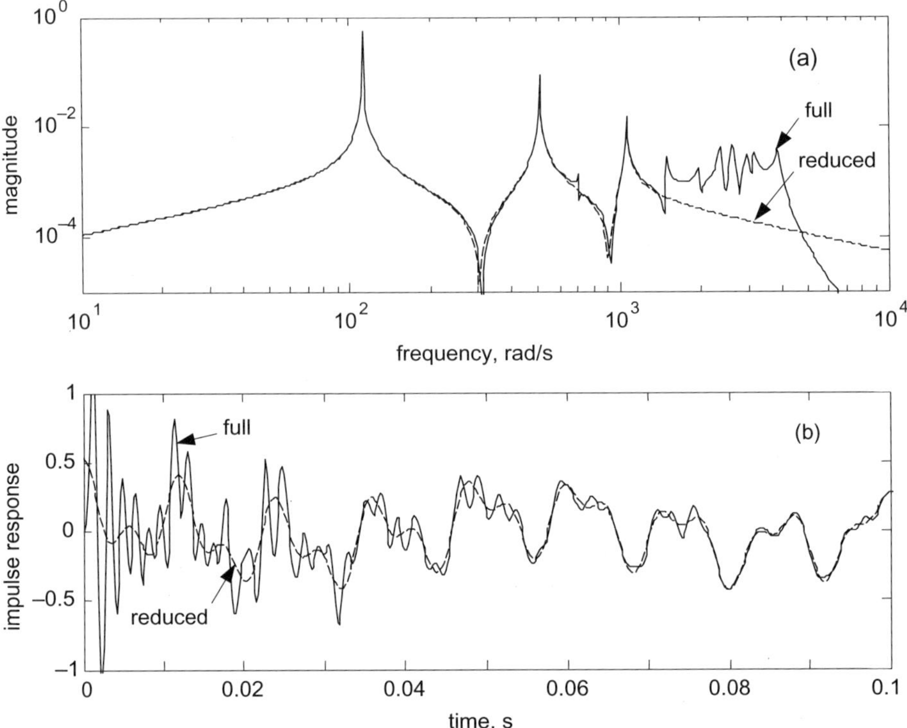

Figure 6.2. (a) Magnitude of the transfer function; and (b) impulse responses of the full (solid line) and reduced (dashed line) truss models show that predominantly high-frequency modes were deleted.

6.3.1 Reduction in the Finite-Time Interval

In the finite-time interval, the reduction is similar to the infinite-time interval. The difference occurs in the use of time-limited grammians instead of the infinite-time grammians. Note that in modal coordinates the reduced model of a stable system is stable, since the poles of the reduced model are the subset of the full model.

The computational procedure is summarized as follows:

1. For given (A,B,C) determine grammians W_c and W_o from (4.5).
2. Compute $W_c(t_i)$, $W_o(t_i)$, $S(t_i)$ for $i = 1, 2$ from (4.89) and (4.87).
3. Determine $W_c(T)$ and $W_o(T)$ from (4.88).
4. Apply the reduction procedure to obtain (A_r, B_r, C_r) for $W_c(T)$ and $W_o(T)$.

The Matlab function *modal_time_fr.m* given in Appendix A.7 determines the modal representation 1, the corresponding time-limited grammians $W_c(T)$ and $W_o(T)$, and the Hankel singular values.

The following example illustrates the application of the time-limited grammians in the model reduction:

Example 6.3. Consider a simple system with masses $m_1 = 11$, $m_2 = 5$, and $m_3 = 10$, stiffnesses $k_1 = k_4 = 10$, $k_2 = 50$, $k_3 = 55$, and damping proportional to the stiffness, $d_i = 0.01k_i$, $i = 1, 2, 3, 4$. A single input u is applied to all masses, such that $f_1 = u$, $f_2 = 2u$, and $f_3 = -5u$, and f_i is a force applied to the ith mass. The output is a linear combination of the three mass displacements, $y = 2q_1 - 2q_2 + 3q_3$, where q_i is a displacement of the ith mass. The system poles are

$$\lambda_{1,2} = -0.0038 \pm j0.8738,$$
$$\lambda_{3,4} = -0.0297 \pm j2.4374,$$
$$\lambda_{5,6} = -0.1313 \pm j5.1217.$$

Consider two cases of model reduction from 6 to 4 state variables. Case 1, the reduction over the interval $T_1 = [0, 8]$ and Case 2, the reduction over the interval $T_2 = [10, 18]$.

In Case 1, the first and third pair of poles are retained in the reduced model. The impulse responses of the full and reduced models are compared in Fig. 6.3(a), where the area outside the interval T_1 is shaded.

Case 2 is obtained from Case 1 by shifting the interval T_1 by 10 s, i.e., $T_2 = T_1 + 10$. In this case the first two pairs of poles are retained. The impulse responses of the reduced and full systems are presented in Fig. 6.3(b), where the area outside the interval T_2 is shaded. Comparison of Fig. 6.3(a) and Fig. 6.3(b) shows that the third mode is less visible for $t > 10$ s, thus it was eliminated in Case 2.

Example 6.4. Re-examine the above example with negative damping, $D = -0.006K$, which makes the system unstable.

The system poles are

$$\lambda_{1,2} = 0.0023 \pm j0.8738,$$
$$\lambda_{3,4} = 0.0178 \pm j2.4375,$$
$$\lambda_{5,6} = 0.0787 \pm j5.1228.$$

We determine the system grammians over the time interval [0, 8] s. The reduced model has four states (or two modes), and the impulse responses of the full and reduced models within the interval [0, 8] s are shown in Fig. 6.4, nonshaded area, showing a good coincidence between the full and reduced models.

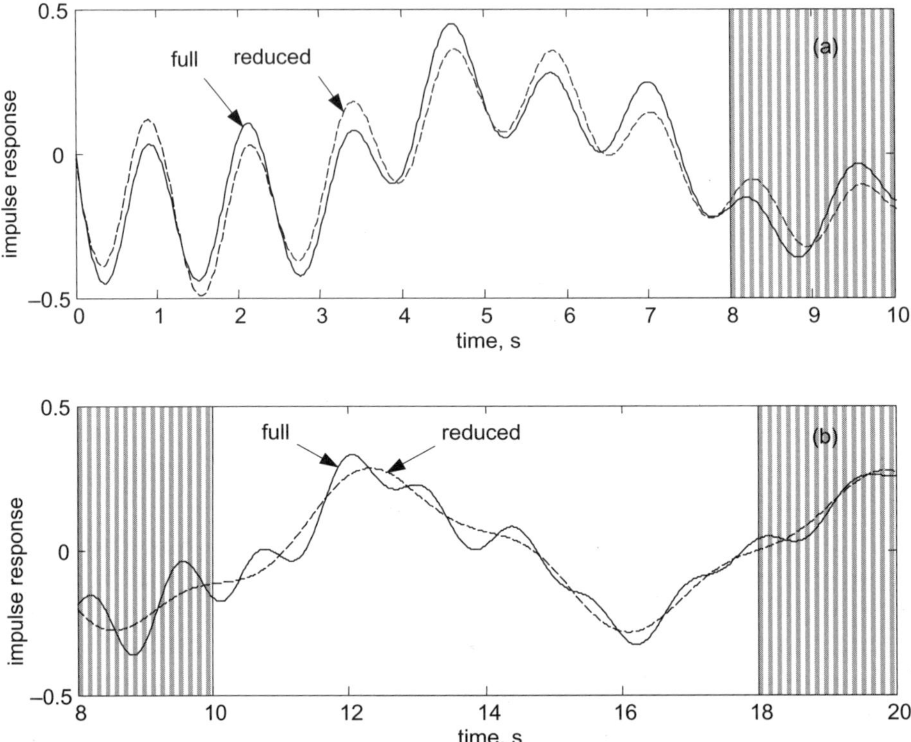

Figure 6.3. Impulse responses of the full- and reduced-order systems show that different modes were retained within each time interval: (a) Reduction in the time interval $T_1 = [0,\ 8]$ s; and (b) reduction in the time interval $T_2 = [10,\ 18]$ s.

6.3.2 Reduction in the Finite-Frequency Interval

One approach to reduce a model in frequency domain is to impose frequency weighting on input and/or output; see [32], [140]. Presented here is a model reduction in a finite-frequency interval, where we use the band-limited grammians rather than weighting, and this approach is formally the same as the standard problem presented earlier. Hankel singular values are determined from the band-limited grammians, and the states with small singular values are truncated.

The computational procedure is summarized as follows:

1. Determine the stationary grammians W_c and W_o from (4.5) for a given (A,B,C).
2. Determine $W_c(\omega_1)$, $W_c(\omega_2)$ and $W_o(\omega_1)$, $W_o(\omega_2)$ from (4.100), and $S(\omega)$ from (4.101).
3. Determine $W_c(\Omega)$ and $W_o(\Omega)$ from (4.104).
4. Apply the reduction procedure to obtain the reduced state-space triple (A_r, B_r, C_r) using grammians $W_c(\Omega)$ and $W_o(\Omega)$.

The Matlab function *modal_time_fr.m* given in Appendix A.7 determines the modal representation 1, the corresponding frequency-limited grammians $W_c(\Omega)$ and $W_o(\Omega)$, and Hankel singular values.

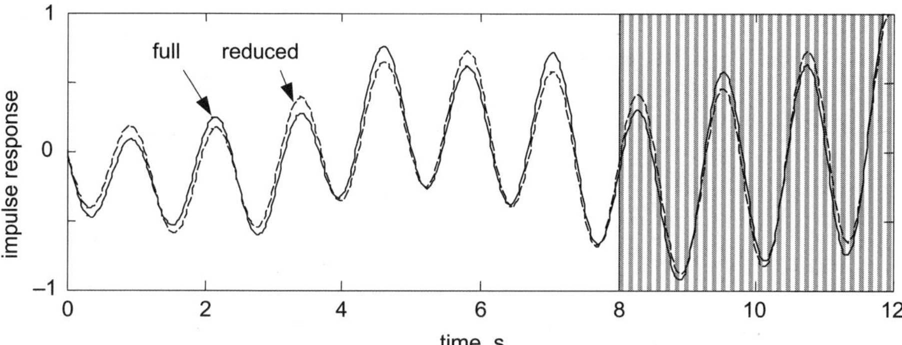

Figure 6.4. Impulse responses of the full- and reduced-order unstable system: The reduced model is obtained for the time interval [0, 8] s.

Example 6.5. Consider Example 6.3 in two cases: Case 1, in the frequency interval $\Omega_1 = [0,\ 3.2]$ rad/s, and Case 2, in the frequency interval $\Omega_2 = [1.5,\ 3.2]$ rad/s.

In the first case we obtained the reduced model with four states (two modes), whereas in the second case we obtained the reduced model with two states (one mode). Figures 6.5(a) and (b) show a good fit of magnitude of the transfer function in the frequency bands Ω_1 and Ω_2 (nonshaded areas). This indicates that the reduction in the finite-frequency intervals can serve as a filter design tool. For example, in Case 2 the output signal is filtered such that the resulting output of the reduced model is best fitted to the output of the original system within the interval Ω_2.

6.3.3 Reduction in the Finite-Time and -Frequency Intervals

The reduction technique is similar to that above. The computational procedure can be set up alternatively, either by first applying frequency and then time transformation of grammians, or by first applying time and then frequency transformation. Since both procedures are similar, only the first one is presented as follows:

1. Determine stationary grammians W_c and W_o from (4.5) for a given (A,B,C).
2. Determine $W_c(\omega_i)$ and $W_o(\omega_i)$, $i = 1, 2$, from (4.100).
3. Determine $W_c(t_i,\omega_j)$ and $W_o(t_i,\omega_j)$, $i, j = 1, 2$, from (4.118).
4. Determine $W_c(T,\omega_j)$ and $W_o(T,\omega_j)$, $i = 1, 2$, from (4.116) and (4.117).
5. Determine $W_c(T,\Omega)$ and $W_o(T,\Omega)$ from (4.115).

6. Apply the reduction procedure to obtain the reduced state-space triple (A_r, B_r, C_r) using grammians $W_c(T,\Omega)$ and $W_o(T,\Omega)$.

The Matlab function *modal_time_fr.m* given in Appendix A.7 determines the modal representation 1, the corresponding frequency-limited grammians $W_c(T,\Omega)$ and $W_o(T,\Omega)$, and Hankel singular values.

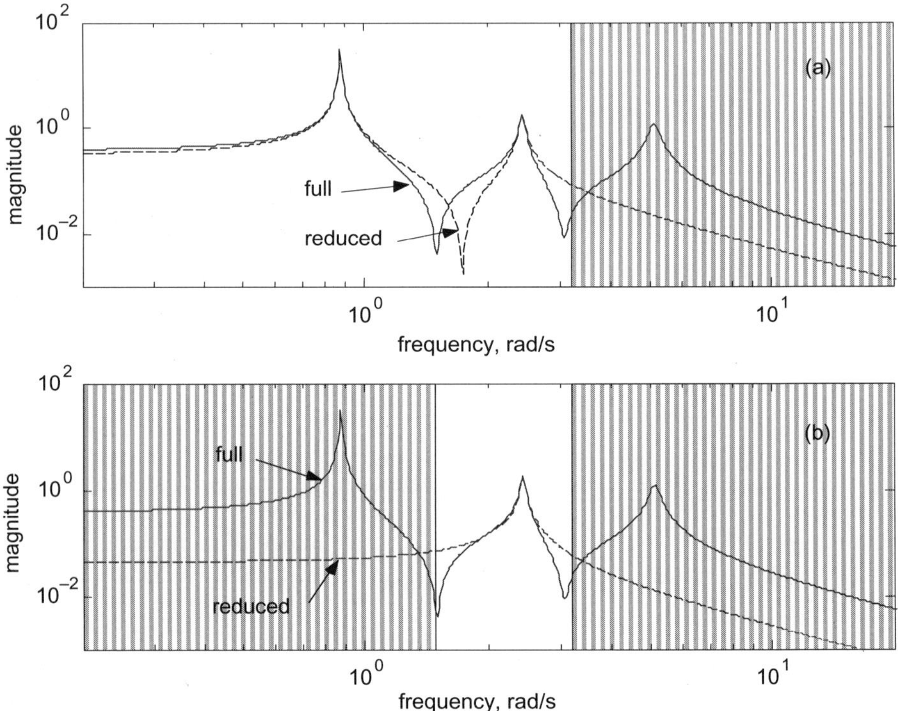

Figure 6.5. Magnitudes of transfer functions of the full and reduced systems: (a) Reduction in the frequency interval $\Omega_1 = [0,\ 3.2]$ rad/s; and (b) reduction in the frequency interval $\Omega_2 = [1.5,\ 3.2]$ rad/s. In both cases the reduced-order transfer function fits the full-order transfer function within the prescribed frequency interval.

Example 6.6. Consider Example 6.3 in the time interval T = [0, 8] s and in the frequency interval Ω = [0, 3.2] rad/s.

Figures 6.6(a),(b), in nonshaded areas, show the impulse responses and magnitudes of the transfer functions of the full and reduced systems, within intervals T and Ω. This example is a combination of Examples 6.3 and 6.5. However, the results are different. In Example 6.3 the second pole was deleted for reduction within the interval T = [0, 8] s. In this example the third pole is deleted as a result of the additional restriction on the frequency band.

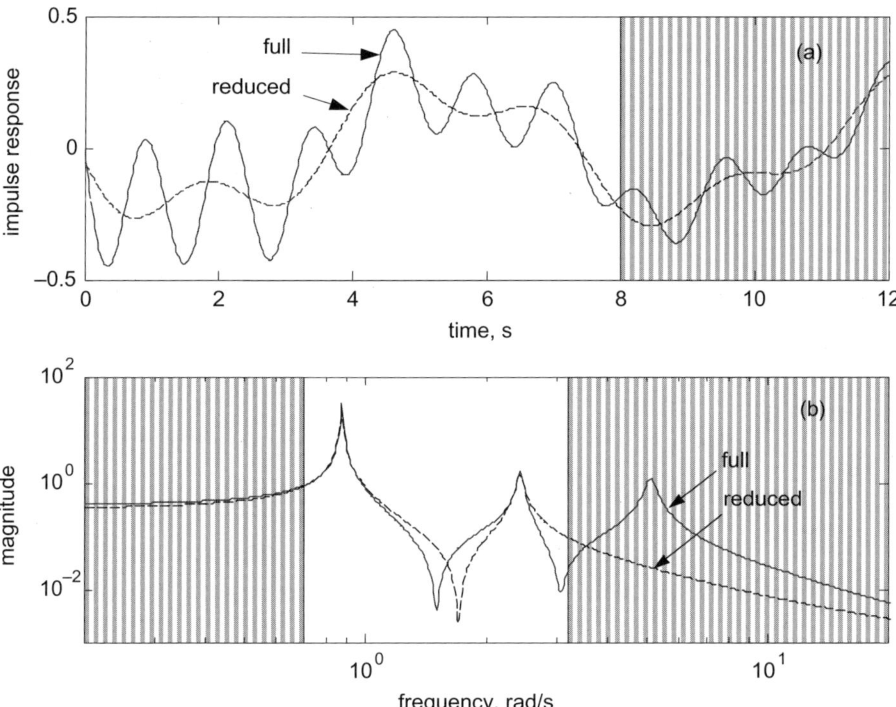

Figure 6.6. Full and reduced system in time interval $T = [0, 8]$ s and frequency interval $\Omega = [0, 3.2]$ rad/s: (a) Impulse responses; and (b) magnitudes of transfer functions. The reduced-order system fits the time response within interval T and the frequency response within interval Ω.

Example 6.7. The Matlab code for this example is in Appendix B. Apply model reduction for the Advanced Supersonic Transport. The Advanced Supersonic Transport was a project on a supersonic passenger plane. Its control system was analyzed by Colgren [19]. Its model is a linear unstable system of eight order with four inputs and eight outputs. In order to make our presentation concise the model presented in this book is restricted to a single output. The system triple (A,B,C) is as follows:

$$A = \begin{bmatrix} -0.0127 & -0.0136 & -0.036 & 0 & 0 & 0 & 0 & 0 \\ -0.0969 & -0.401 & 0 & 0.961 & 19.59 & -0.1185 & -9.2 & -0.1326 \\ 0 & 0 & 0 & 1 & 0 & 0 & 0 & 0 \\ -0.229 & 1.726 & 0 & -0.722 & -12.021 & -0.342 & 1.8422 & 0.881 \\ 0 & 0 & 0 & 0 & 0 & 1 & 0 & 0 \\ 0 & 0.1204 & 0 & 0.0496 & -44 & -1.2741 & -4.0301 & -0.508 \\ 0 & 0 & 0 & 0 & 0 & 0 & 0 & 1 \\ 0 & 0.1473 & 0 & 0.301 & -7.4901 & -0.1257 & -21.7 & -0.803 \end{bmatrix},$$

$$B^T = \begin{bmatrix} 0 & -0.0215 & 0 & -1.097 & 0 & -0.640 & 0 & -1.882 \\ 0.0194 & 0 & 0 & 0 & 0 & 0 & 0 & 0 \\ 0 & -0.004 & 0 & 0.366 & 0 & 0.1625 & 0 & 0.472 \\ 0 & -1.786 & 0 & -0.0569 & 0 & -0.037 & 0 & -0.0145 \end{bmatrix},$$

$$C = \begin{bmatrix} 0 & 1 & 0 & 0 & 0 & 0 & 0 & 0 \end{bmatrix}.$$

The system poles are

$$\begin{aligned} \lambda_1 &= 6.6873, \\ \lambda_2 &= -1.7756, \\ \lambda_{3,4} &= -0.0150 \pm j0.0886, \\ \lambda_{5,6} &= -0.3122 \pm j4.4485, \\ \lambda_{7,8} &= -0.7257 \pm j6.7018, \end{aligned}$$

so that there is one unstable pole. Colgren reduced the model by removing the unstable pole from the model and applying the reduction procedure to the stable part of the model. After reduction the unstable pole was returned to the reduced model. Here, apply the finite-time reduction to the full aircraft model, without removal of the unstable pole.

We choose the time interval $T = [0.0, 3.5]$ s and perform the model reduction in modal coordinates from eight to four states within this interval. We obtain the following reduced model (A_r, B_r, C_r):

$$A_r = \begin{bmatrix} 0.6687 & 0 & 0 & 0 \\ 0 & -0.3122 & -4.4485 & 0 \\ 0 & 4.4485 & -0.3122 & 0 \\ 0 & 0 & 0 & -1.7756 \end{bmatrix},$$

$$B_r = \begin{bmatrix} -0.4762 & -0.0093 & 0.1192 & -2.4733 \\ -0.0601 & 0.0000 & 0.0140 & 0.0592 \\ 2.0225 & -0.0001 & -0.5083 & -0.0409 \\ 1.0491 & -0.0009 & -0.3840 & -1.6410 \end{bmatrix},$$

$$C_r = \begin{bmatrix} 0.3966 & -0.0630 & -0.4215 & 0.4764 \end{bmatrix},$$

with the poles λ_1, λ_2 , λ_5, and λ_6 preserved in the reduced model. The step responses of the full and reduced models due to the first input are shown in Fig. 6.7. The step responses of the reduced model overlap the responses of the full model for all four inputs.

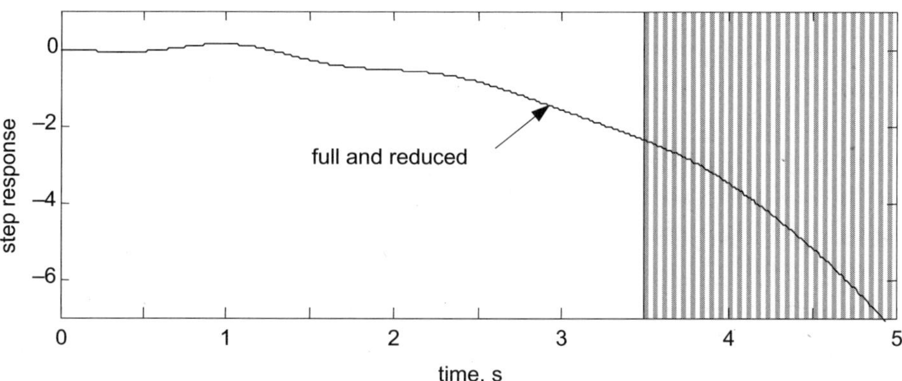

Figure 6.7. Step responses of the full and reduced models of the Advanced Supersonic Transport show that the response of the reduced unstable system coincides with the response of the full system.

6.4 Structures with Rigid-Body Modes

Structures with rigid-body modes have poles at zero, therefore they behave like systems with integrators. The corresponding H_2, H_∞, and Hankel norms for systems with poles at zero do not exist as their values tend to infinity. However, the infinite values of the norms of some modes should not be an obstacle in the reduction process. These values indicate that the corresponding states should be retained in the reduced model, regardless of the norms of other modes. The reduction problem can be solved by determining the inverses of grammians, as in [58]. Here we use two simple approaches for the reduction of systems with integrators.

In the first approach, the system is represented in modal coordinates by the following system triple (see Section 4.3):

$$A = \begin{bmatrix} 0_m & 0 \\ 0 & A_o \end{bmatrix}, \qquad B = \begin{bmatrix} B_r \\ B_o \end{bmatrix}, \qquad C = \begin{bmatrix} C_r & C_o \end{bmatrix}, \tag{6.13}$$

where 0_m is an $m \times m$ zero matrix. The triple (A_o, B_o, C_o) has no poles at zero. It is itself in modal coordinates. The vector of the corresponding modal H_∞ norms is denoted h_o. This vector is arranged in descending order, and the remaining infinite norms are added

$$h = \{\text{inf},\ h_o\} \tag{6.14}$$

to obtain the vector of H_∞ norms of the (A, B, C) representation, where $\text{inf} = \{\infty, \infty, \ldots, \infty\}$ contains m values at infinity. The system is reduced by truncation, as described at the beginning of this chapter.

The second approach is based on the approximate evaluation of the H_∞ norms. From (5.22) we find

$$\|G_i\|_\infty \cong \frac{\|B_{mi}\|_2 \|C_{mi}\|_2}{2\zeta_i\omega_i},$$

and $\omega_i = 0$ for the poles at zero; thus, $\|G_i\|_\infty \to \infty$. For nonzero poles we determine the finite norms from the above equation, and order in a descending order. The corresponding state-space representation is reduced by truncation.

Example 6.8. Consider a simple system from Fig. 1.1 with the following parameters: $m_1 = m_2 = 1$, $m_3 = 2$, $k_1 = k_4 = 0$, $k_2 = 0.3$, and $k_3 = 1$. Damping is proportional to the stiffnesses, $d_i = 0.03k_i$, i = 1, 2, 3, the input force is applied at mass m_2, and the output rate is measured at the same location. This system has two poles at zero. Find its H_∞ norms and reduce the system.

The modal representation of the system without rigid-body modes is as follows; see (6.13):

$$A_o = \left[\begin{array}{cc|cc} -0.0264 & 1.3260 & 0 & 0 \\ -1.3260 & -0.0264 & 0 & 0 \\ \hline 0 & 0 & -0.0051 & 0.5840 \\ 0 & 0 & -0.5840 & -0.0051 \end{array}\right],$$

$$B_o = \left[\begin{array}{c} -0.3556 \\ -1.1608 \\ \hline -0.0072 \\ 0.1642 \end{array}\right],$$

$$C_o = \left[\begin{array}{cc|cc} -0.161 & -0.5836 & -0.0043 & -0.0821 \end{array}\right],$$

$$B_r = \begin{bmatrix} 6.0883\times 10^7 \\ 35.8240 \end{bmatrix},$$

$$C_r = \begin{bmatrix} 4.1062\times 10^{-9} & 0 \end{bmatrix},$$

with the following H_∞ norms $\|G_1\|_\infty \cong 13.9202$ and $\|G_2\|_\infty \cong 1.3247$. Therefore, the vector of the norms of the modes of the (A_o, B_o, C_o) representation is

$$h_o = \{13.9202,\ 13.9202,\ 1.3247,\ 1.3247\},$$

and the vector of the H_∞ norms of the full system (A, B, C) is

$$h = \{\infty, \infty, 13.9202, 13.9202, 1.3247, 1.3247\}.$$

By deleting the last two states in the state-space representation, related to the smallest norms, one obtains the reduced-order model as follows:

$$A_r = \left[\begin{array}{cc:cc} 0 & 0 & 0 & 0 \\ 0 & 0 & 0 & 0 \\ \hdashline 0 & 0 & -0.0264 & -1.3260 \\ 0 & 0 & 1.3260 & -0.0264 \end{array}\right],$$

$$B_r = \left[\begin{array}{c} 6.0883 \times 10^7 \\ 35.8240 \\ \hdashline -0.3556 \\ -1.1608 \end{array}\right],$$

$$C_r = \left[\begin{array}{cc:cc} 4.1062 \times 10^{-9} & 0 & -0.161 & -0.5840 \end{array}\right].$$

The plots of the impulse response and the magnitude of the transfer function of the full and reduced models are shown in Fig. 6.8(a),(b). The plots show that the error of the reduction is small. In fact, for the impulse response, the error was less than 1%, namely,

$$\varepsilon = \frac{\|y - y_r\|_2}{\|y\|_2} = 0.0021.$$

In the above, y denotes the impulse response of the full model and y_r denotes the impulse response of the reduced model.

Example 6.9. Consider the Deep Space Network antenna azimuth model that has a pole at zero. The identified state-space representation of the open-loop model has $n = 36$ states, including states with a pole at zero. Reduce this model in modal coordinates, by determining the Hankel norms (or Hankel singular values) for states with nonzero poles.

The plot of the Hankel singular values is shown in Fig. 6.9. By deleting the states with Hankel singular values smaller than 0.003 we obtain the 18-state model. The reduced model preserves properties of the full model, as is shown by the magnitude and phase of the transfer function in Fig. 6.10(a),(b). The state-space representation of the reduced antenna model is given in Appendix C.3.

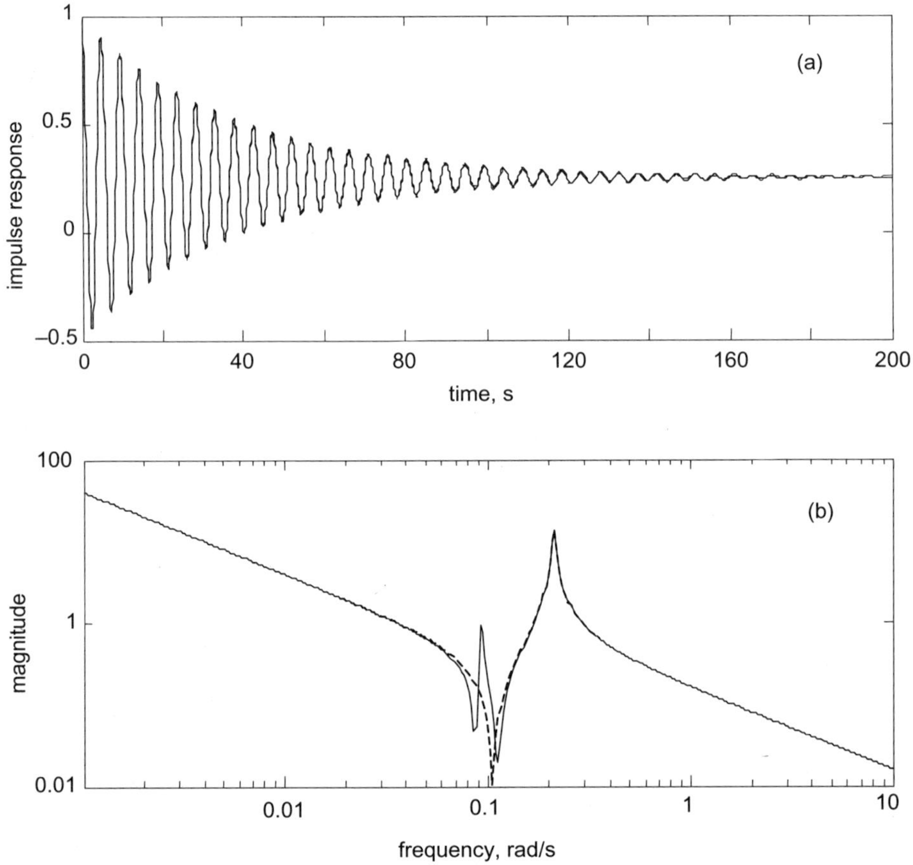

Figure 6.8. Full model (solid line) and reduced model (dashed line) of a simple structure with poles at zero: (a) Impulse responses; and (b) magnitudes of the transfer function. The figure shows good coincidence between the responses of the reduced and full models.

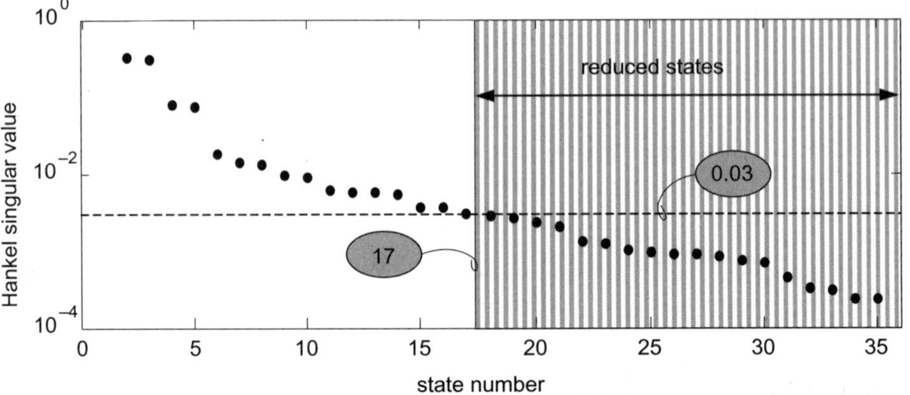

Figure 6.9. Hankel singular values of the DSS26 antenna rate-loop model: 17 states out of 35 states are retained in the reduced model.

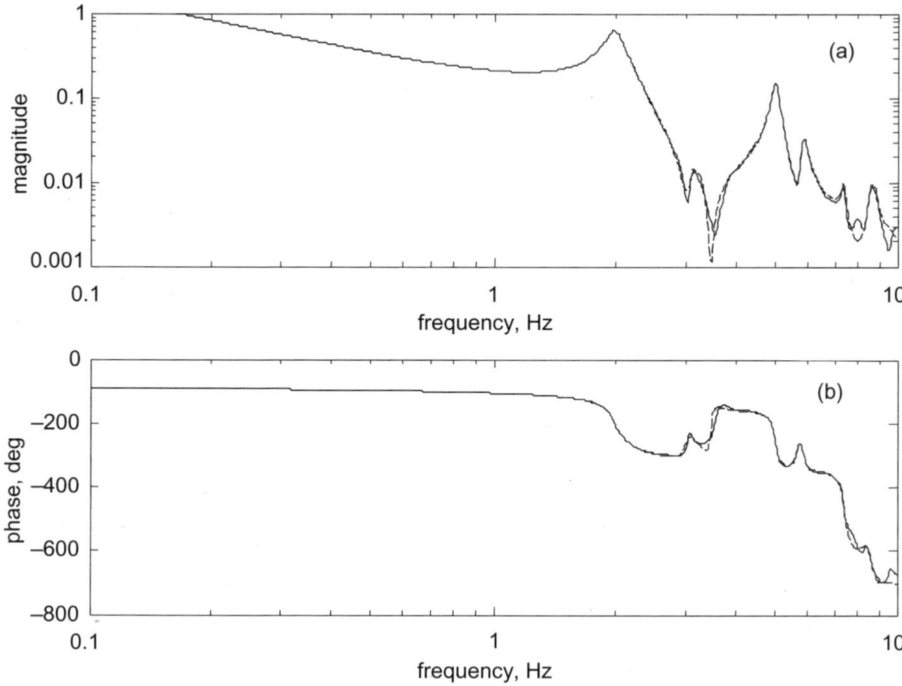

Figure 6.10. Transfer function of the full- (solid line) and reduced-order (dashed line) models of the DSS26 antenna shows good coincidence in (a) magnitude; and (b) phase.

6.5 Structures with Actuators and Sensors

A flexible structure in testing, or in a closed-loop configuration, is equipped with sensors and actuators. Does the presence of sensors and actuators impact the process of model reduction? This question is answered for four important cases: sensors and actuators in cascade connection with a structure, accelerometers as sensors, the proof-mass actuators, and inertial actuators attached to a structure.

6.5.1 Actuators and Sensors in a Cascade Connection

We consider actuator dynamics only. In particular, the reconstruction of the norms of modes and of a structure from the norms of the actuator-structure norms is discussed. The problem of sensors in a cascade connection with a structure is similar to the actuator case.

Properties of the actuators in a series connection with structures can be derived from the properties of a smooth filter in series connection with a structure, which was discussed in Chapter 5. Let G_i and G_{si} be a transfer function of the ith mode with and without actuators, respectively. Additionally, let G and G_s be a transfer function of the structure with and without actuators, respectively. As a corollary of Property 5.7, the norms of modes and a structure for a smooth actuator transfer function are determined approximately as follows:

Property 6.1. ***The H_2 Norms of a Mode and a Structure with an Actuator.***

$$\begin{aligned} \|G_i\|_2 &\cong \alpha_i \|G_{si}\|_2, \\ \|G\|_2 &\cong \sum_{i=1}^{n} \sqrt{\alpha_i^2 \|G_{si}\|_2^2}. \end{aligned} \tag{6.15}$$

where

$$\alpha_i = \|G_a(\omega_i)\|_2, \tag{6.16}$$

with ω_i being the ith resonance frequency and G_a the transfer function of the actuator.

Similarly, based on Property 5.8, we have the following property of the H_∞ and Hankel norms of a mode and a structure with a smooth actuator.

Property 6.2. ***The H_∞ Norms of a Mode and a Structure with an Actuator.***

$$\begin{aligned} \|G_i\|_\infty &\cong \alpha_i \|G_{si}\|_\infty, \\ \|G\|_\infty &\cong \max_i(\alpha_i \|G_{si}\|_\infty). \end{aligned} \tag{6.17}$$

Property 6.3. ***The Hankel Norms of a Mode and a Structure with an Actuator.***

$$\begin{aligned} \|G_i\|_h &\cong \alpha_i \|G_{si}\|_h, \\ \|G\|_h &\cong \max_i(\alpha_i \|G_{si}\|_h). \end{aligned} \tag{6.18}$$

Example 6.10. Consider the 3D truss from Fig. 1.3, with the longitudinal (x-direction) input at node 21 and the longitudinal rate output at node 14. The actuator located at node 21 has the following smooth transfer function:

$$G_a(s) = \frac{0.1}{(1+0.01s)^2}.$$

The truss modal damping is identical for each mode, 0.5% ($\zeta_i = 0.005$), $i = 1,\ldots,72$. Compare the exact and approximate H_∞ norms of the modes of the structure with the actuator.

From the definition (5.5) we obtain the exact H_∞ norms of the modes of the structure with the actuator; they are marked by circles in Fig. 6.11. We obtain the approximate H_∞ norms of the modes of the structure with the actuator from Property 5.2 (using the Matlab function *norm_Hinf.m* from Appendix A.10), and plot as dots in the same figure. The exact and approximate norms overlap each other in this figure, showing that the approximation error is negligible.

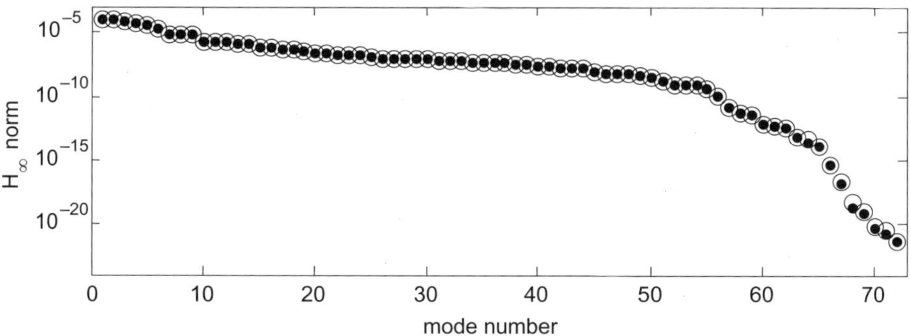

Figure 6.11. The H_∞ norms of the modes of the 3D truss with an actuator: Exact (○) and approximate (•) values are almost the same.

6.5.2 Structure with Accelerometers

Accelerometers as structural sensors were described in Section 3.2. The state-space equations of a structure with accelerometers include the feed-through term, D, in the state output equation, i.e., the output for the accelerometer sensors is in the form of $y = Cx + Du$, see (3.12), rather than $y = Cx$, as in (2.1). The difficulty with this equation follows from the fact that the grammians do not depend on matrix D. Thus the grammian-based model reduction does not reflect the presence of the accelerometers. However, this problem can be solved by using the series connection of a structure and sensors, and Properties 5.7–5.9.

We consider a structure with the accelerometers as a structure with rate sensors cascaded with differentiating devices. Thus, we determine the norms of a structure equipped with accelerometers as the scaled norms of a structure with rate sensors. For simplicity of notation consider a structure with a single accelerometer. Denote (A_r, B_r, C_r) and $G_r = C_r(sI - A_r)^{-1}B_r$ as the state-space triple and as the transfer function, respectively, of the structure with a rate sensor. The transfer function G_a of the structure with an accelerometer is therefore

$$G_a = j\omega G_r. \qquad (6.19)$$

According to (6.16) the scaling factor is $\alpha_i = \|j\omega_i\| = \omega_i$, thus from (6.15)–(6.18) the following property holds:

Property 6.4. ***Norms of Modes with Accelerometers.*** *The norms of modes with accelerometers are related to modes with the rate output as follows:*

$$\|G_{ai}\| \cong \omega_i \|G_{ri}\|, \qquad i = 1, \ldots, n, \qquad (6.20)$$

where ω_i is the ith natural frequency and $\|.\|$ denotes either H_2, H_∞, or Hankel norms.

The above equations show that the norm of the *i*th mode with an accelerometer sensor is obtained as a product of the norm of the *i*th mode with a rate sensor and the *i*th natural frequency.

Example 6.11. Consider the truss from the previous example. The longitudinal input force is applied to node 21 and the longitudinal acceleration is measured at node 14. Determine the H_∞ norms of the modes for the structure with the accelerometer.

The exact norms are marked by circles in Fig. 6.12. We obtained from (6.20) the approximate H_∞ norms of the modes of the structure with the accelerometers and plotted as dots in Fig. 6.12. The exact and approximate norms overlap each other in this figure, showing that the approximation error is negligible.

6.5.3 Structure with Proof-Mass Actuators

Proof-mass actuators are widely used in structural dynamics testing. In this subsection we study the relationship between the norms of a structure with a proof-mass actuator and the norms of the structure alone (i.e., with an ideal actuator) and analyze the influence of the proof-mass actuator on model dynamics and reduction.

Let us consider a structure with proof-mass actuator, shown in Fig. 3.5, position (a). Let M_s, D_s, and K_s be the mass, damping, and stiffness matrices of the structure, respectively, and let B_s be the matrix of the actuator location,

$$B_s = \begin{bmatrix} 0 & 0 & \ldots & 0 & 1 & 0 & \ldots & 0 \end{bmatrix}^T$$

with a nonzero term at the actuator location *na*. Denote $G_s(\omega) = -\omega^2 M_s + j\omega D_s + K_s$ and $j = \sqrt{-1}$, then the dynamic stiffness of a structure at the actuator location is defined as

$$k_s = \frac{1}{B_s^T G_s^{-1} B_s}. \tag{6.21}$$

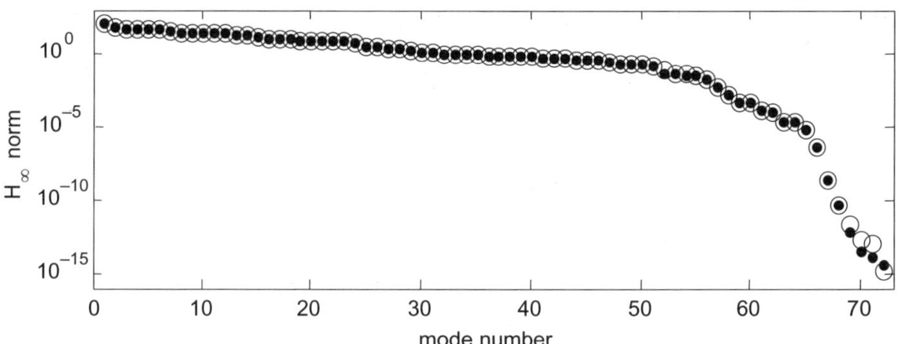

Figure 6.12. The H$_\infty$ norms of the modes of the 3D truss with an accelerometer: Exact (○) and approximate (•) values coincide for almost all modes.

The dynamic stiffness is the inverse of the frequency response function at the actuator location. At zero frequency, it is reduced to the stiffness constant at the actuator location. Denote by q_a, *m*, *k*, *d*, the displacement, mass, stiffness, and damping of the actuator. Denoting

$$\rho = \frac{\omega_o}{\omega}, \qquad \omega_o = \sqrt{\frac{k}{m}}, \qquad \beta = \frac{k}{k_s}, \tag{6.22}$$

we obtain the following relationship between the proof-mass actuator force (f_o) and the ideal actuator force (f), see (3.29),

$$f_o = \alpha_c f, \qquad \alpha_c = \frac{1}{1 + \beta - \rho^2}. \tag{6.23}$$

It follows from the above equation that the actuator force, f_o, approximately reproduces the ideal force f if $\alpha_c \cong 1$. This is obtained if

$$\beta \ll 1 \qquad \text{and} \qquad \rho \ll 1. \tag{6.24}$$

The above conditions are satisfied when the actuator stiffness is small (compared with the structural stiffness), and the actuator mass is large enough, such that the actuator natural frequency is smaller than the structural principal frequency. Hence, the above conditions can be replaced with the following ones:

$$k \ll k_s \qquad \text{and} \qquad \omega \gg \omega_o. \tag{6.25}$$

If these conditions are satisfied, we obtain $f \cong f_o$ and, consequently, the transfer function of the system with the proof-mass actuator is approximately equal to the transfer function of the system without the proof-mass actuator. Based on these considerations the following norm properties are derived:

Property 6.5. ***Norms of a Mode with Proof-Mass Actuators.*** *Norms of the ith structural mode with a proof-mass actuator* (G_{ci}) *and of the ith structural mode alone* (G_{si}) *are related as follows:*

$$\|G_{si}\| = \frac{\|G_{ci}\|}{\alpha_{ci}}, \qquad i = 1, \ldots, n, \tag{6.26}$$

where $\|.\|$ *denotes either* H_2, H_∞, *or Hankel norms, where*

$$\alpha_{ci} = \frac{1}{1 + \beta_i - \rho_i^2} \tag{6.27}$$

and

$$\rho_i = \frac{\omega_o}{\omega_i} \qquad \text{and} \qquad \beta_i = \frac{k}{k_{si}}, \tag{6.28}$$

while

$$k_{si} = k_s(\omega_i) = \frac{1}{B_s^T G_s^{-1}(\omega_i) B_s}. \tag{6.29}$$

The variable k_{si} is the ith modal stiffness of the structure.

Proof. The force f_o acting on the structure is related to the actuator force f as in (6.23). Hence, replacing f_o with f in the structural model gives (6.26). ▣

In addition to conditions (6.25), consider the following ones:

$$\omega_o \ll \omega_1 \qquad \text{and} \qquad k \ll \min_i k_{si}, \tag{6.30}$$

where ω_1 is the fundamental (lowest) frequency of the structure. These conditions say that the actuator natural frequency should be significantly lower than the fundamental frequency of the structure, and that the actuator stiffness should be much smaller than the dynamic stiffness of the structure at any frequency of interest. If the aforementioned conditions are satisfied, we obtain $\alpha_{ci} \cong 1$ for $i = 1,\ldots,n$; thus, the norms of the structure with the proof-mass actuator are equal to the norms of the structure without the proof-mass actuator. Also, the controllability and observability properties of the system are preserved. In particular, the presence of the proof-mass actuator will not affect the model order reduction. Note also that for many cases, whenever the first condition of (6.25) is satisfied, the second condition (6.30) is satisfied too.

Example 6.12. Consider the 3D truss as in Example 3.4 with and without the proof-mass actuator. Let the force input act at node 21 in the y-direction, and let the rate without output be measured at node 14 in the y-direction. Determine the Hankel singular values of the truss for the ideal actuator (force applied directly at node 21) and of the truss with a proof-mass actuator. The mass of the proof-mass actuator is $m = 0.1\ \mathrm{Ns}^2/\mathrm{cm}$, and its stiffness is $k = 1\ \mathrm{N/cm}$. Its natural frequency is $\omega_o = 3.1623\ \mathrm{rad/s}$, much lower than the truss fundamental frequency.

For the ideal force applied in the y-direction of node 21 the Hankel singular values are shown as dots in Fig. 6.13. Next, a proof-mass actuator was attached to node 21 to generate the input force. Circles in Fig. 6.13 denote Hankel singular values of the truss with the proof-mass actuator. Observe that the Hankel singular values are the same for the truss with and without the proof-mass actuator, except for the first Hankel singular value, related to the proof-mass actuator itself.

6.5.4 Structure with Inertial Actuators

In the inertial actuator, force is proportional to the square of the excitation frequency. It consists of mass m and a spring with stiffness k, and they are attached to a structure at node n_b, Fig. 3.5, position (b). The force acts on mass m exclusively. It is assumed that the stiffness of the actuator is much smaller than the dynamic stiffness of the structure (often it is zero).

This configuration is shown in Fig. 3.5(b). The force acting on mass m is proportional to the squared frequency

$$f = \kappa\omega^2, \tag{6.31}$$

where κ is a constant. The relationship between transfer functions of a structure with (G_c) and without (G_s) in an inertial actuator is as follows:

$$G_c = \alpha_c G_s, \qquad \alpha_c = \frac{\kappa \omega_o}{1 + \beta - \rho^2}, \tag{6.32}$$

which was derived in Subsection 3.3.2.

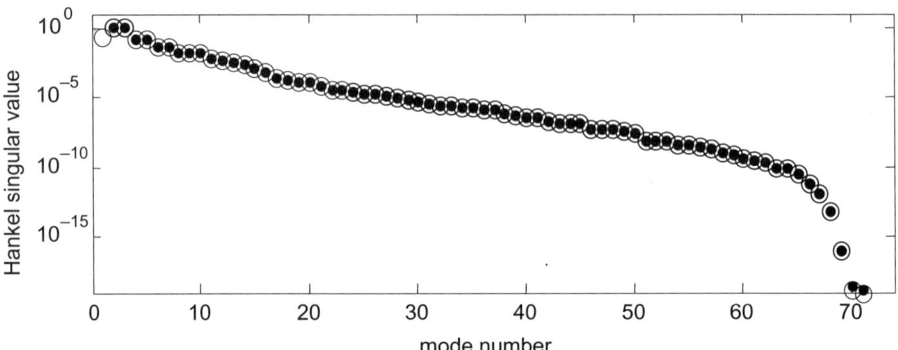

Figure 6.13. 3D truss with and without the proof-mass actuator: Hankel singular values with (•) and without (○) the proof-mass actuator; they are identical except for the additional Hankel singular value of the proof-mass actuator itself.

Also, the relationship between the actuator force (f_o) and the force acting on the structure (f) was derived in Subsection 3.3.2,

$$f_o = \alpha_c f, \qquad \alpha_c = \frac{\kappa \omega^2}{1 + \beta - \rho^2}. \tag{6.33}$$

The above result shows that the structural transfer function with the inertial actuator is proportional to the structural transfer function without the actuator.

Property 6.6. *Norms of Modes with Inertial Actuators.* The norms of the ith structural mode (G_{si}), and of the ith structural mode with an inertial actuator (G_{ci}), are related as in (6.26); however, the factor α_{ci} is now

$$\alpha_{ci} = \frac{\kappa \omega_o^2}{1 + \beta_i - \rho_i^2}. \tag{6.34}$$

Proof. Similar to Property 6.5. ▣

With the conditions in (6.30) satisfied, one obtains $\alpha_{ci} = \kappa \omega_o^2$ for $i = 1, \ldots, n$; thus, the norms of the structure with the inertial actuator are proportional to the norms of the structure without the actuator. This scaling does not influence the results of model reduction, since the procedure is based on ratios of norms rather than their absolute values.

7 Actuator and Sensor Placement

how to set up a test procedure and control strategy

Experimentalists think that it is a mathematical theorem
while the mathematicians believe it to be an experimental fact.
—*Gabriel Lippman*

A typical actuator and sensor location problem for structural dynamics testing can be described as a structural test plan. The plan is based on the available information on the structure itself, on disturbances acting on the structure, and on the required structural performance. The preliminary information on structural properties is typically obtained from a structural finite-element model. The disturbance information includes disturbance location and disturbance spectral contents. The structure performance is commonly evaluated through the displacements or accelerations of selected structural locations. The actuator and sensor placement problem was investigated by many researchers, see, for example, [1], [7], [24], [47], [55], [86], [89], [90], [96], [97], [101], [103], [105], [106], [127], and a review article [131].

It is not possible to duplicate the dynamics of a real structure during testing. This happens, not only due to physical restrictions or a limited knowledge of disturbances, but also because the test actuators cannot often be located at the actual location of disturbances, and sensors cannot be placed at locations where the performance is evaluated. Thus, to conduct the test close to the conditions of a structure in a real environment one uses the available (or candidate) locations of actuators and sensors and formulates the selection criteria and selection mechanisms.

The control design problem of a structure can be defined in a similar manner. Namely, actuators are placed at the allowable locations, and they are not necessarily collocated with the locations where disturbances are applied; sensors are placed at the sensor allowable locations, which are generally outside the locations where performance is evaluated.

For simple test articles, an experienced test engineer can determine the appropriate sensor or actuator locations in an ad hoc manner. However, for the first-time testing of large and complex structures the placement of sensors and actuators is neither an obvious nor a simple task. In practice heuristic means are combined with engineering judgment and simplified analysis to determine actuator and sensor locations. In most cases the locations vary during tests (in a trial and error approach) to obtain acceptable data to identify target modes.

For a small number of sensors or actuators a typical solution to the location problem is found through a search procedure. For large numbers of locations the search for the number of possible combinations is overwhelming, time-consuming, and gives not necessarily the optimal solution. The approach proposed here consists of the determination of the norm of each sensor (or actuator) for selected modes, and then grading them according to their participation in the system norm. This is a computationally fast (i.e., nonsearch) procedure with a clear physical interpretation.

7.1 Problem Statement

Given a larger set of sensors and actuators, the placement problem consists of determining the locations of a smaller subset of sensors or actuators such that the H_2, H_∞, or Hankel norms of the subset are as close as possible to the norms of the original set. In this chapter we solve this placement problem in modal coordinates using the previously derived structural properties. We propose a comparatively simple methodology of choice of a small subset of sensors and/or actuators from a large set of possible locations.

Let $\mathcal{R}$ and $\mathcal{S}$ be the sets of the candidate sensor and actuator locations, respectively. These are chosen in advance as allowable locations of actuators of population S, and as allowable locations of sensors of population R. The placement of s actuators within the given $\mathcal{S}$ actuator candidate locations, and the placement of r sensors within the given $\mathcal{R}$ sensor candidate locations is considered. Of course, the number of candidate locations is larger than the number of final locations, i.e., $R > r$ and $S > s$.

7.2 Additive Property of Modal Norms

The properties of modal norms that are used in the actuator and sensor placement procedures are discussed in this section.

7.2.1 The H_2 Norm

Consider a flexible structure in modal representation. The H_2 norm of the ith mode is given by (5.21), i.e.,

$$\left\|G_i\right\|_2 \cong \frac{\left\|B_{mi}\right\|_2 \left\|C_{mi}\right\|_2}{2\sqrt{\zeta_i \omega_i}}, \tag{7.1}$$

where B_{mi} and C_{mi} are the input and output matrices of the ith mode. For S inputs and R outputs, these matrices are

$$B_{mi} = \begin{bmatrix} B_{mi1} & B_{mi2} & \cdots & B_{miS} \end{bmatrix}, \qquad C_{mi} = \begin{bmatrix} C_{m1i} \\ C_{m2i} \\ \vdots \\ C_{mRi} \end{bmatrix}, \tag{7.2}$$

and B_{mij} is the 2×1 block of the jth input, while C_{mji} is the 1×2 block of the jth output. From Properties 5.10 and 5.12 we obtain the following additive properties of the H_2 norm:

$$\left\|G_i\right\|_2^2 \cong \sum_{j=1}^{R} \left\|G_{ij}\right\|_2^2 \qquad \text{or} \qquad \left\|G_i\right\|_2^2 \cong \sum_{k=1}^{S} \left\|G_{ik}\right\|_2^2, \tag{7.3}$$

where

$$\left\|G_{ij}\right\|_2 = \frac{\left\|B_{mij}\right\|_2 \left\|C_{mi}\right\|_2}{2\sqrt{\zeta_i \omega_i}}, \qquad \left\|G_{ik}\right\|_2 = \frac{\left\|B_{mi}\right\|_2 \left\|C_{mki}\right\|_2}{2\sqrt{\zeta_i \omega_i}}, \tag{7.4}$$

are the H_2 norms of the ith mode with the jth actuator only, or the ith mode with the kth sensor only. Equation (7.3) shows that the H_2 norm of a mode with a set of actuators (sensors) is the root-mean-square (rms) sum of the H_2 norms of this mode with a single actuator (sensor).

7.2.2 The H_∞ and Hankel Norms

A similar relationship can be obtained for the H_∞ norm. From (5.22) one obtains

$$\left\|G_i\right\|_\infty \cong \frac{\left\|B_{mi}\right\|_2 \left\|C_{mi}\right\|_2}{2\zeta_i \omega_i}, \tag{7.5}$$

and from Properties 5.14 and 5.15 the additive property of the H_∞ norm has the following form:

$$\left\|G_i\right\|_\infty^2 \cong \sum_{j=1}^{S}\left\|G_{ij}\right\|_\infty^2 \qquad \text{or} \qquad \left\|G_i\right\|_\infty^2 \cong \sum_{k=1}^{R}\left\|G_{ik}\right\|_\infty^2, \tag{7.6}$$

where

$$\left\|G_{ij}\right\|_\infty = \frac{\left\|B_{mij}\right\|_2 \left\|C_{mi}\right\|_2}{2\zeta_i\omega_i}, \qquad \left\|G_{ik}\right\|_\infty = \frac{\left\|B_{mi}\right\|_2 \left\|C_{mki}\right\|_2}{2\zeta_i\omega_i}, \tag{7.7}$$

are the H_∞ norms of the *i*th mode with the *j*th actuator only, or the *i*th mode with the *k*th sensor only. Equation (7.6) shows that the H_∞ norm of a mode with a set of actuators (sensors) is the rms sum of the H_∞ norms of this mode with a single actuator (sensor).

Hankel norm properties are similar to the H_∞ norm properties and follow from Properties 5.16 and 7.17.

7.3 Placement Indices and Matrices

Actuator and sensor placement are solved independently, and both procedures are similar.

7.3.1 H_2 Placement Indices and Matrices

Denote by G the transfer function of the system with all S candidate actuators. The placement index σ_{2ki} that evaluates the *k*th actuator at the *i*th mode in terms of the H_2 norm is defined with respect to all the modes and all admissible actuators

$$\sigma_{2ki} = w_{ki}\frac{\left\|G_{ki}\right\|_2}{\left\|G\right\|_2}, \qquad k = 1,\ldots S, \quad i = 1,\ldots,n, \tag{7.8}$$

where $w_{ki} \geq 0$ is the weight assigned to the *k*th actuator and the *i*th mode, n is the number of modes, and G_{ki} is the transfer function of the *i*th mode and *k*th actuator, as given in the first equation of (7.4). The Matlab function *norm_H2.m* given in Appendix A.9 determines modal H_2 norms. The weight reflects the importance of the mode and the actuator in applications, and reflects the dimensions of the inputs. In applications it is convenient to represent the H_2 placement indices as a placement matrix in the following form:

$$\Sigma_2 = \begin{bmatrix} \sigma_{211} & \sigma_{212} & \cdots & \sigma_{21k} & \cdots & \sigma_{21S} \\ \sigma_{221} & \sigma_{222} & \cdots & \sigma_{22k} & \cdots & \sigma_{22S} \\ \cdots & \cdots & \cdots & \cdots & \cdots & \cdots \\ \sigma_{2i1} & \sigma_{2i2} & \cdots & \sigma_{2ik} & \cdots & \sigma_{2iS} \\ \cdots & \cdots & \cdots & \cdots & \cdots & \cdots \\ \sigma_{2n1} & \sigma_{2n2} & \cdots & \sigma_{2nk} & \cdots & \sigma_{2nS} \end{bmatrix} \begin{matrix} \\ \\ \\ \leftarrow i\text{th mode.} \\ \\ \\ \end{matrix} \tag{7.9}$$

$\uparrow$

kth actuator

The kth column of the above matrix consists of indexes of the kth actuator for every mode, and the ith row is a set of the indexes of the ith mode for all actuators.

Similarly to actuators, the placement index σ_{ki} evaluates the kth sensor at the ith mode

$$\sigma_{2ki} = w_{ki} \frac{\|G_{ki}\|_2}{\|G\|_2}, \qquad k = 1, \ldots, R, \qquad i = 1, \ldots, n, \tag{7.10}$$

where $w_{ki} \geq 0$ is the weight assigned to the kth sensor and ith mode, n is a number of modes, and G_{ki} is the transfer function of the ith mode and kth sensor, as given in the second equation of (7.4). We define the sensor placement matrix as follows:

$$\Sigma_2 = \begin{bmatrix} \sigma_{211} & \sigma_{212} & \cdots & \sigma_{21k} & \cdots & \sigma_{21R} \\ \sigma_{221} & \sigma_{222} & \cdots & \sigma_{22k} & \cdots & \sigma_{22R} \\ \cdots & \cdots & \cdots & \cdots & \cdots & \cdots \\ \sigma_{2i1} & \sigma_{2i2} & \cdots & \sigma_{2ik} & \cdots & \sigma_{2iR} \\ \cdots & \cdots & \cdots & \cdots & \cdots & \cdots \\ \sigma_{2n1} & \sigma_{2n2} & \cdots & \sigma_{2nk} & \cdots & \sigma_{2nR} \end{bmatrix} \begin{matrix} \\ \\ \\ \leftarrow i\text{th mode,} \\ \\ \\ \end{matrix} \tag{7.11}$$

$\uparrow$

kth sensor

where the kth column consists of indexes of the kth sensor for every mode, and the ith row is a set of the indexes of the ith mode for all sensors.

7.3.2 H_∞ and Hankel Placement Indices and Matrices

Similarly to the H_2 index, the placement index $\sigma_{\infty ki}$ evaluates the kth actuator at the ith mode in terms of the H_∞ norm. It is defined in relation to all the modes and all admissible actuators, i.e.,

$$\sigma_{\infty ki} = w_{ki} \frac{\|G_{ki}\|_\infty}{\|G\|_\infty}, \qquad k = 1,\ldots,S, \quad i = 1,\ldots,n, \tag{7.12}$$

where $w_{ki} \geq 0$ is the weight assigned to the kth actuator and ith mode.

Using the above indices we introduce the H_∞ placement matrix, similar to the H_2 matrix introduced earlier, namely,

$$\Sigma_\infty = \begin{bmatrix} \sigma_{\infty 11} & \sigma_{\infty 12} & \cdots & \sigma_{\infty 1k} & \cdots & \sigma_{\infty 1S} \\ \sigma_{\infty 21} & \sigma_{\infty 22} & \cdots & \sigma_{\infty 2k} & \cdots & \sigma_{\infty 2S} \\ \cdots & \cdots & \cdots & \cdots & \cdots & \cdots \\ \sigma_{\infty i1} & \sigma_{\infty i2} & \cdots & \sigma_{\infty ik} & \cdots & \sigma_{\infty iS} \\ \cdots & \cdots & \cdots & \cdots & \cdots & \cdots \\ \sigma_{\infty n1} & \sigma_{\infty n2} & \cdots & \sigma_{\infty nk} & \cdots & \sigma_{\infty nS} \end{bmatrix} \begin{matrix} \\ \\ \\ \leftarrow i\text{th mode.} \\ \\ \\ \end{matrix} \tag{7.13}$$

$$\uparrow$$
$$k\text{th actuator}$$

The Hankel placement index and matrix is one-half of the H_∞ placement index and Σ_∞ matrix, respectively. The Matlab functions *norm_Hinf.m* and *norm_Hankel.m* given in Appendix A.10 and A.11 determine the modal H_∞ and Hankel norms.

In the sensor placement procedure the placement index $\sigma_{\infty ki}$ evaluates the kth sensor at the ith mode in terms of the H_∞ norm

$$\sigma_{\infty ki} = w_{ki} \frac{\|G_{ki}\|_\infty}{\|G\|_\infty}, \qquad k = 1,\ldots,R, \quad i = 1,\ldots,n, \tag{7.14}$$

where $w_{ki} \geq 0$ is the weight assigned to the kth sensor and ith mode.

The H_∞ norm placement matrix is similar to the 2-norm matrix, i.e.,

$$\Sigma_\infty = \begin{bmatrix} \sigma_{\infty 11} & \sigma_{\infty 12} & \cdots & \sigma_{\infty 1k} & \cdots & \sigma_{\infty 1R} \\ \sigma_{\infty 21} & \sigma_{\infty 22} & \cdots & \sigma_{\infty 2k} & \cdots & \sigma_{\infty 2R} \\ \cdots & \cdots & \cdots & \cdots & \cdots & \cdots \\ \sigma_{\infty i1} & \sigma_{\infty i2} & \cdots & \sigma_{\infty ik} & \cdots & \sigma_{\infty iR} \\ \cdots & \cdots & \cdots & \cdots & \cdots & \cdots \\ \sigma_{\infty n1} & \sigma_{\infty n2} & \cdots & \sigma_{\infty nk} & \cdots & \sigma_{\infty nR} \end{bmatrix} \begin{matrix} \\ \\ \\ \leftarrow i\text{th mode.} \\ \\ \\ \end{matrix} \qquad (7.15)$$

$\uparrow$

kth sensor

The Hankel placement index and matrix is one-half of the H_∞ placement index and Σ_∞ matrix, respectively.

The placement matrix is a quick visual tool for previewing the importance of each sensor (or actuator) and each mode. Indeed, each column represents the sensor (or actuator) importance of every mode, and each row represents the mode importance for every sensor (or actuator).

7.3.3 Actuator/Sensor Indices and Modal Indices

The placement matrix gives an insight into the placement properties of each actuator, since the placement index of the kth actuator is determined as the rms sum of the kth column of Σ. (For convenience in further discussion we denote by Σ the placement matrix either of the two- or the infinity-norm.) The vector of the actuator placement indices is defined as $\sigma_a = \begin{bmatrix} \sigma_{a1} & \sigma_{a2} & \cdots & \sigma_{aS} \end{bmatrix}^T$, and its kth entry is the placement index of the kth actuator. In the case of the H_2 norm, it is the rms sum of the kth actuator indexes over all modes,

$$\sigma_{ak} = \sqrt{\sum_{i=1}^{n} \sigma_{ik}^2}, \qquad k = 1, \ldots, S, \tag{7.16}$$

and in the case of the H_∞ and Hankel norms it is the largest index over all modes

$$\sigma_{ak} = \max_i(\sigma_{ik}), \qquad i = 1, \ldots, n, \quad k = 1, \ldots, S, \tag{7.17}$$

Similarly, we define the vector of the sensor placement indices as $\sigma_s = \begin{bmatrix} \sigma_{s1} & \sigma_{s2} & \cdots & \sigma_{sR} \end{bmatrix}^T$, and its kth entry is the placement index of the kth sensor. In the case of the H_2 norm, it is the rms sum of the kth sensor indexes over all modes,

$$\sigma_{sk} = \sqrt{\sum_{i=1}^{n} \sigma_{ik}^2}, \qquad k = 1, \ldots, R, \tag{7.18}$$

and in the case of the H_∞ and Hankel norms it is the largest index over all modes

$$\sigma_{sk} = \max_i(\sigma_{ik}), \qquad i = 1, \ldots, n, \quad k = 1, \ldots, R, \tag{7.19}$$

We define the vector of the mode indices as $\sigma_m = \begin{bmatrix} \sigma_{m1} & \sigma_{m2} & \ldots & \sigma_{mn} \end{bmatrix}^T$, and its ith entry is the index of the ith mode. This entry is an rms sum of the ith mode indices over all actuators

$$\sigma_{mi} = \sqrt{\sum_{k=1}^{S} \sigma_{ik}^2}, \qquad i = 1, \ldots, n, \tag{7.20}$$

or an rms sum of the ith mode indices over all sensors

$$\sigma_{mi} = \sqrt{\sum_{k=1}^{R} \sigma_{ik}^2}, \qquad i = 1, \ldots, n, \tag{7.21}$$

The actuator placement index, σ_{ak}, is a nonnegative contribution of the kth actuator at all modes to the H_2 or H_∞ norms of the structure. The sensor placement index, σ_{sk}, is a nonnegative contribution of the kth sensor at all modes to the H_2 or H_∞ norms of the structure. The mode index, σ_{mi}, is a nonnegative contribution of the ith mode for all actuators (or all sensors) to the H_2 or H_∞ norms of the structure. We illustrate the determination of the H_∞ actuator and modal indices for the pinned beam in Fig. 7.1. Six actuators are located on the beam and four modes are considered. The second mode index is the rms sum of indices of all actuators for this mode, and the third actuator index is the largest index of this actuator over four modes.

From the above properties it follows that the index σ_{ak} (σ_{sk}) characterizes the importance of the kth actuator (sensor), thus it serves as the actuator (sensor) placement index. Namely, the actuators (sensors) with small index σ_{ak} (σ_{sk}) can be removed as the least significant ones. Note also that the mode index σ_{mi} can be used as a reduction index. Indeed, it characterizes the significance of the ith mode for the given locations of sensors and actuators. The norms of the least significant modes (those with the small index σ_{mi}) should either be enhanced by the reconfiguration of the actuators or sensors, or be eliminated.

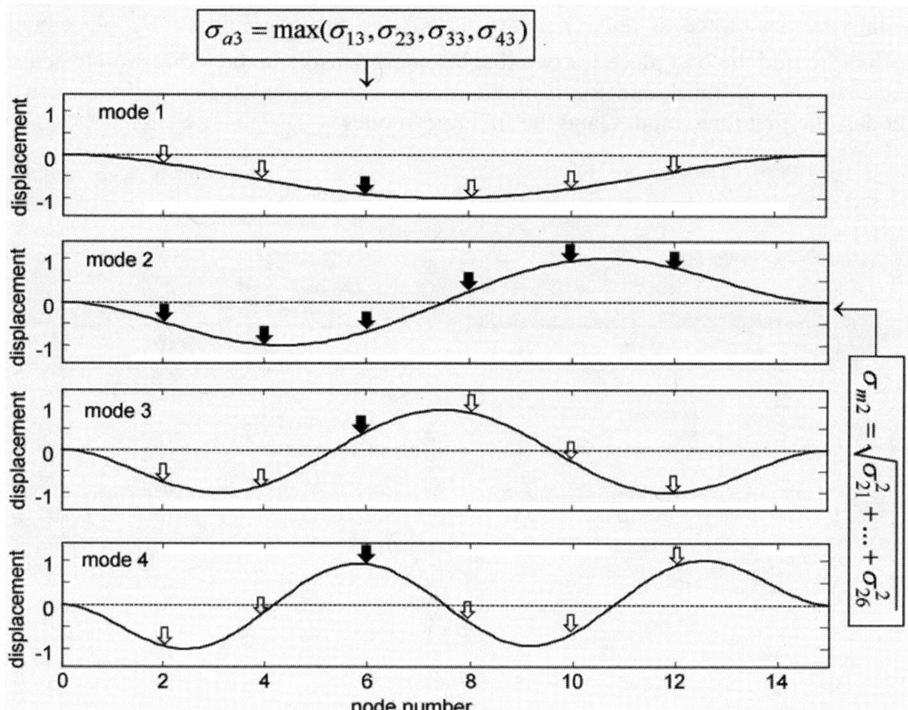

Figure 7.1. Determination of the H_∞ actuator and modal indices of a pinned beam (⇩—actuator location; and ⬇—actuators used for the calculation of the indices): The mode index is the rms sum of indices of all six actuators for this mode, while the actuator index is the largest of the actuator indices over four modes.

Example 7.1. Consider the 2D truss from Fig. 1.2. It is excited in the *y*-direction by an actuator located at node 4. Accelerometers serve as sensors. The task is to find four accelerometer locations within all 16 possible locations, that is, within all but 1 and 6 nodes, in the *x*- and *y*-directions. Assume the unit weights for all modes, and chose the 2-norm indices for the analysis.

We calculated the placement indices σ_{si}, $i = 1,...,16$, of each accelerometer location and show them in Fig. 7.2 for lower (2–5) nodes of the truss, and in Fig. 7.3 for upper (7–10) nodes of the truss. The left column of these figures represents the H_2 index σ_{si} for the *x*-direction accelerometers, while the right column represents the index for the *y*-direction accelerometers. The largest value indices are for nodes 5, 10, 4, and 9, all in the *y*-direction. Note that the chosen locations are the nodes at the tip in the same direction, and that a single accelerometer would probably do the same job as the four put together. This problem is addressed in the following section.

Example 7.2. Placing two sensors on a beam for the best sensing of up to four modes. The Matlab code for this example for $n = 15$ elements is in Appendix B. Consider a beam as in Subsection 1.1.4 for $n = 100$ elements, and shown in Fig. 7.4

with a vertical force at node $n_a = 40$. Using the presented above H_∞ placement technique find the best place for two displacement sensors in the y-direction to sense the first, second, third, and fourth mode, and to sense simultaneously the first two modes, the first three modes, and the first four modes.

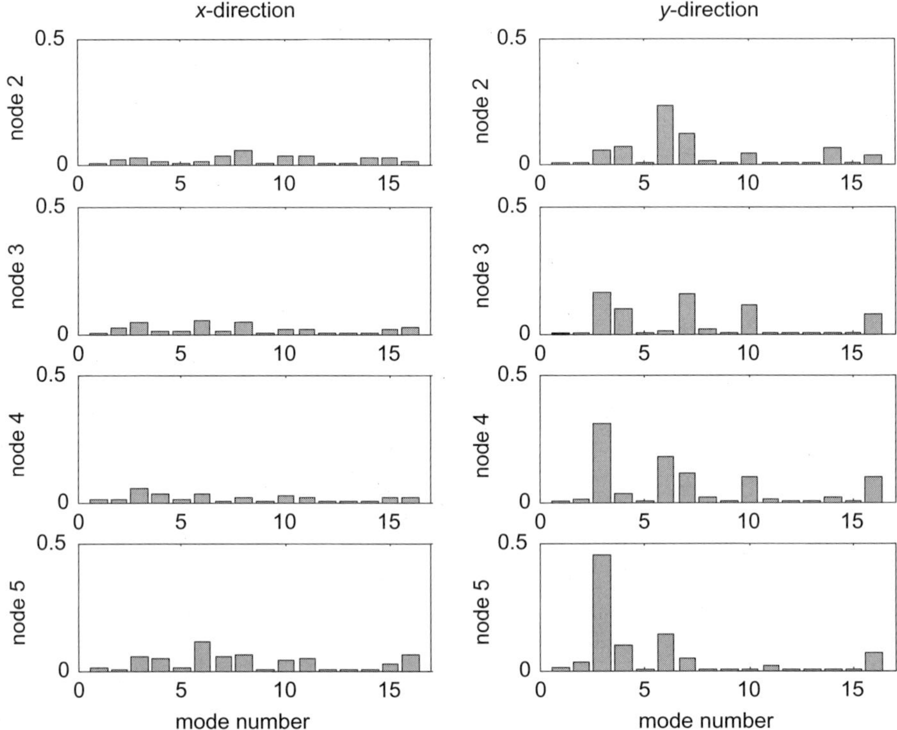

Figure 7.2. The 2D truss sensor indices for nodes 2–5.

Each node of a beam has three degrees of freedom $\{x, y, \theta\}$: horizontal displacement x, vertical displacement y, and rotation in the figure plane θ. Denote a unit vector $e_i = [0, 0, \ldots, 1, \ldots, 0]$ that has all zeros except 1 at the ith location, then the displacement output matrix for sensors located at the ith node is

$$C_{qij} = e_{3i-1}.$$

The input matrix is $B_o = e_{3n_a-1}^T = e_{119}^T$.

We obtain the H_∞ norm $\|G_{ki}\|_\infty$ for the kth mode ($k = 1,2,3,4$) and ith sensor location from (7.7) using B_o and C_{qi} as above. From these norms we obtain the sensor placement indices for each mode from (7.14), using weight such that $\max_i(\sigma_{\infty ki}) = 1$.

The plots of $\sigma_{\infty ki}$ are shown in Fig. 7.5(a),(b),(c),(d). The plot of the sensor placement indices for the first mode in Fig. 7.5(a) shows the maximum at node 50, and indicates that the sensors shall be placed at this node. The plot of the sensor placement indices for the second mode in Fig. 7.5(b) shows two maxima, at nodes 29 and 71, and indicates these two locations as the best for sensing the second mode. The plot of the sensor placement indices for the third mode in Fig. 7.5(c) shows two maxima, at nodes 21 and 79, and indicates that these two locations are the best for sensing the third mode. Finally, the plot of the sensor placement indices for the fourth mode in Fig. 7.5(d) shows two maxima, at nodes 16 and 84, and indicates that these two locations are the best for sensing the fourth mode.

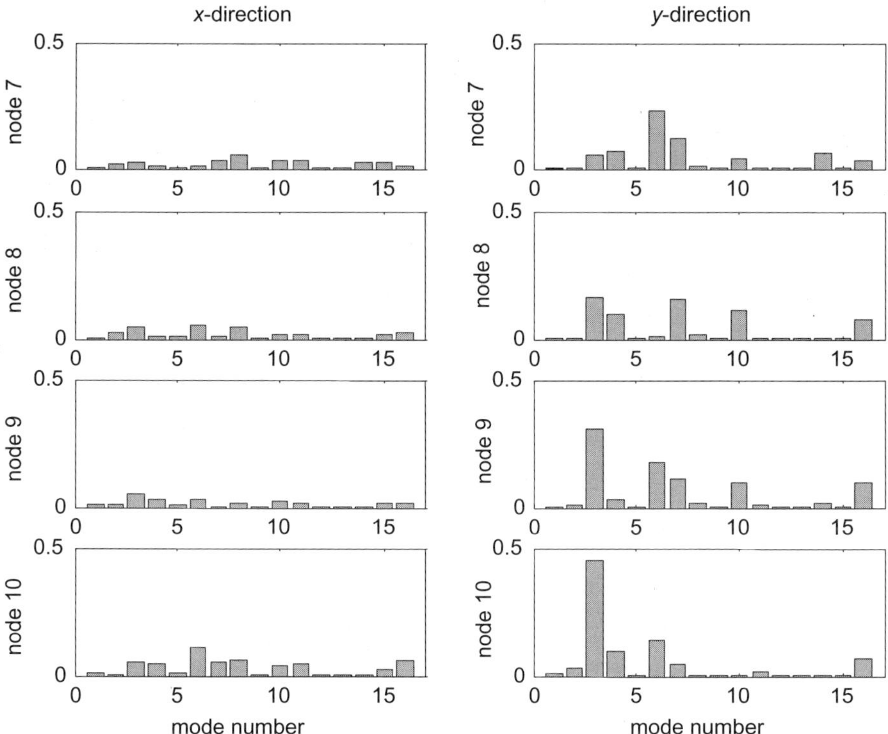

Figure 7.3. The 2D truss sensor indices for nodes 7–10.

Next, we determine the indices for the first two modes, using (7.19), namely,

$$\sigma_{\infty 12i} = \max(\sigma_{\infty 1i}, \sigma_{\infty 2i}).$$

The plot of this index is shown in Fig. 7.6(b). It shows that the index reaches its maximum at three locations: 29, 50, and 71. These locations are the best for sensing the first and second modes. Clearly, location 50 serves for the first mode sensing, while locations 29 and 71 serve for the second mode sensing.

Next, the indices for the first three modes are determined, using (7.19),

$$\sigma_{\infty 123i} = \max(\sigma_{\infty 1i}, \sigma_{\infty 2i}, \sigma_{\infty 3i}).$$

The plot of this index is shown in Fig. 7.6(c). It shows that the index reaches its maximum at five locations: 21, 29, 50, 71, and 79. These locations are the best for sensing the first, second, and third modes. Obviously, location 50 serves for the first mode sensing, locations 29 and 71 serve for the second mode sensing, and locations 21 and 79 serve for the third mode sensing.

Finally, the indices for the first four modes are determined, using (7.19),

$$\sigma_{\infty 1234i} = \max(\sigma_{\infty 1i}, \sigma_{\infty 2i}, \sigma_{\infty 3i}, \sigma_{\infty 4i}).$$

The plot of this index is shown in Fig. 7.6(d). It shows that the index reaches its maximum at seven locations: 16, 21, 29, 50, 71, 79, and 84. These locations are the best for sensing the first, second, third, and fourth modes. Location 50 serves for the first mode sensing, locations 29 and 71 serve for the second mode sensing, locations 21 and 79 serve for the third mode sensing, and locations 16 and 84 serve for the fourth mode sensing.

So far in this example we used the H_∞ norms and indices. It would be interesting to compare the sensor placement using the H_2 norms and indices. First, the H_2 norm $\|G_{ki}\|_2$ for the *k*th mode (*k*=1,2,3,4) and *i*th sensor location is obtained from (7.4) using B_o and C_{qi} as above.

We determine the indices for the first two modes using (7.18), namely,

$$\sigma_{2,12i} = \sqrt{\sigma_{2,1i}^2 + \sigma_{2,2i}^2}.$$

The plot of this index is shown in Fig. 7.7(b). It shows that the index reaches its maximum at two locations: 33 and 67.

Next, we determine the indices for the first three modes using (7.18),

$$\sigma_{2,123i} = \sqrt{\sigma_{2,1i}^2 + \sigma_{2,2i}^2 + \sigma_{2,3i}^2}.$$

The plot of this index is shown in Fig. 7.7(c). It shows that the index reaches its maximum at two locations: 25 and 75.

Finally, we determine the indices for the first four modes using (7.18),

$$\sigma_{2,1234i} = \sqrt{\sigma_{2,1i}^2 + \sigma_{2,2i}^2 + \sigma_{2,3i}^2 + \sigma_{2,4i}^2}.$$

The plot of this index is shown in Fig. 7.7(d). It shows that the index reaches its maximum at two locations: 20 and 80.

A comparison of the H_∞ and H_2 indices in Figs. 7.6 and 7.7 shows that the H_2 index determines different sensor locations than the H_∞ index, and that it changes more dramatically with the change of sensor location, while the H_∞ index becomes more flat (the result of selection of maximal values), thus the first one can be considered a more sensitive measure of the sensor (or actuator) location. Due to the flattening action of the H_∞ norm they indicate slightly different sensor locations.

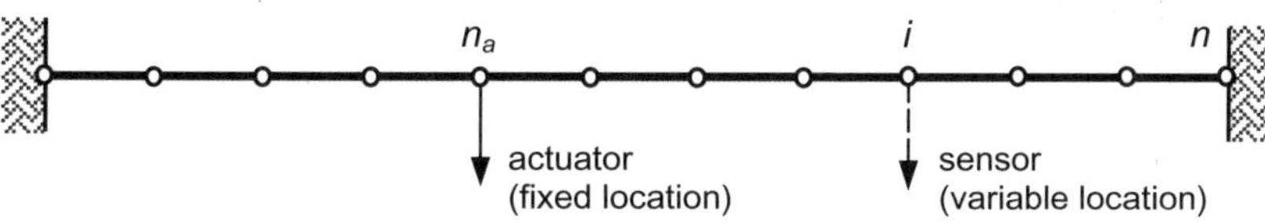

Figure 7.4. A beam with a fixed actuator and a moving sensor.

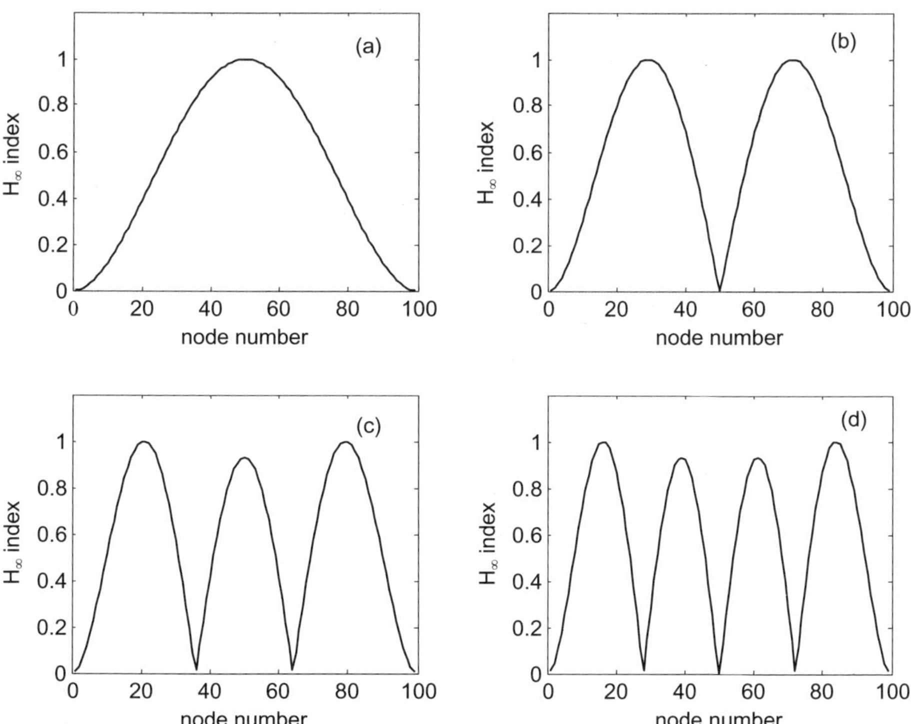

Figure 7.5. Sensor placement H_∞ indices as a function of sensor locations: (a) For the first beam mode; (b) for the second beam mode; (c) for the third beam mode; and (d) for the fourth beam mode.

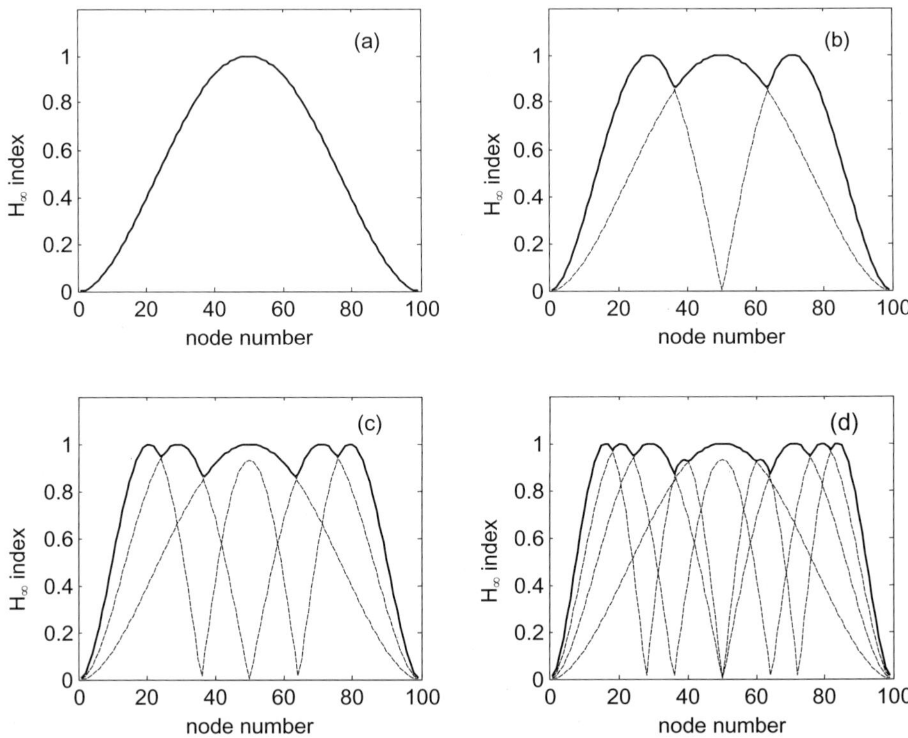

Figure 7.6. Sensor placement H_∞ indices as a function of sensor locations: (a) For the first mode; (b) for the first two modes; (c) for the first three modes; and (d) for the first four modes.

7.4 Placement for Large Structures

In the case of the placement of a very large number of sensors, the maximization of the performance index alone may be either a sufficient or satisfactory criterion. Suppose that a specific sensor location gives a high-performance index. Inevitably, locations close to it will have a high-performance index as well. But the locations in the neighborhood of the original sensor are not necessarily the best choice, since the sensors at these locations can be replaced by the appropriate gain adjustment of the original sensor. We want to find sensor locations that cannot be compensated for by original sensor gain adjustment. These locations we determine using an additional criterion, which is based on the correlation of each sensor modal norm. We define a vector of the ith sensor norms, which is composed of the squares of the modal norms

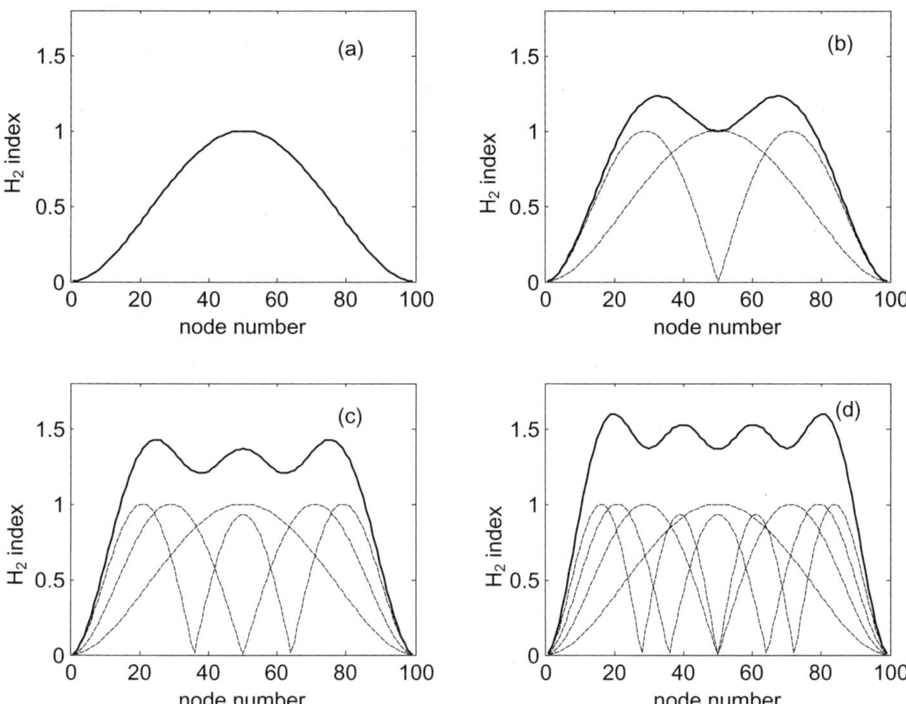

Figure 7.7. Sensor placement H_2 indices as a function of sensor locations: (a) For the first mode; (b) for the first two modes; (c) for the first three modes; and (d) for the first four modes.

$$g_i = \begin{bmatrix} \|G_{i1}\|^2 \\ \|G_{i2}\|^2 \\ \vdots \\ \|G_{in}\|^2 \end{bmatrix}, \tag{7.22}$$

where G_{ik} denotes the transfer function of the kth mode at the ith sensor. The norm $\|.\|$ denotes the H_2, H_∞, or Hankel norms. Next, we define the correlation coefficient r_{ik} as follows:

$$r_{ik} = \frac{g_i^T g_k}{\|g_i\|_2 \|g_k\|_2}, \qquad i = 1,...,r, \quad k = i+1,...,R. \tag{7.23}$$

Denote a small positive number, ε, say $\varepsilon = 0.01 - 0.20$. Denote the membership index $I(k)$, $k = 1,\ldots, R,$ where R is the number of sensors. We define this index as follows:

$$I(k)=\begin{cases}0 & \text{if } r_{ik}>1-\varepsilon \quad \text{for } \sigma_k \le \sigma_i \text{ and for } k>i, \\ 1 & \text{elsewhere},\end{cases} \tag{7.24}$$

for $k > i$. If $I(k) = 1$, the kth sensor is accepted, and if $I(k) = 0$, the kth sensor is rejected (in this case the two locations i and k are either highly correlated, or the ith location has a higher performance σ_i).

Based on the above analysis we establish the placement strategy. For technical and economic reasons the number of sensors significantly exceeds the number of actuators. Therefore, the actuator selection comes first, as a less flexible procedure.

7.4.1 Actuator Placement Strategy

1. Place sensors at all accessible degrees of freedom.
2. Based on engineering experience, technical requirements, and physical constraints select possible actuator locations. In this way, S candidate actuator locations are selected.
3. For each mode (k) and each selected actuator location (i), determine the actuator placement index $\sigma_k(i)$.
4. For each mode select the s_1 most important actuator locations (those with the largest $\sigma_k(i)$). The resulting number of actuators s_2 for all the modes under consideration (i.e., $s_2 \le n \times s_1$) is much smaller than the number of candidate locations S, i.e., $s_2 \ll S$.
5. Check the correlation indices for the remaining s_2 actuators. Reject all but one actuator with a correlation index higher than $1-\varepsilon$ (i.e., those with the zero membership index). The resulting number of actuators is now $s_3 < s_2$, typically $s_3 \ll s_2$.
6. If the already small number s_3 is still too large, the actuator importance index and the modal importance index are recalculated. The actuator number is further reduced to the required one by reviewing the indices.

7.4.2 Sensor Placement Strategy

1. Actuator locations are already determined.
2. Select the areas where the sensors can be placed, obtaining the R candidate sensor locations.
3. Determine the sensor placement indices $\sigma_k(i)$ for all the candidate sensor locations ($i = 1, \ldots, R$), and for all the modes of interest ($k = 1, \ldots, n$).
4. For each mode, select r_1 for the most important sensor locations. The resulting number of sensors r_2 for all the modes considered (i.e., $r_2 \le n \times r_1$) is much smaller than the number of candidate locations, i.e., $r_2 \ll R$.

5. For the given small positive number ε check the correlation indices for the remaining r_2 sensors. Reject the sensors with correlation indices higher than $1-\varepsilon$ (i.e., those with the zero membership indices). The resulting number of sensors is $r_3 < r_2$, typically $r_3 \ll r_2$.

Example 7.3. Reconsider the 2D truss accelerometer location as in Example 7.1. Using $\varepsilon = 0.15$, determine the membership index I for each location.

The plot of the index is in Fig. 7.8. This indicates four accelerometer locations, namely, at nodes 2, 5, and 8 in the y-direction, and at node 7 in the x-direction. These are the locations that are not heavily correlated, and have the best detection of modes 6, 3, 7, and 8, respectively.

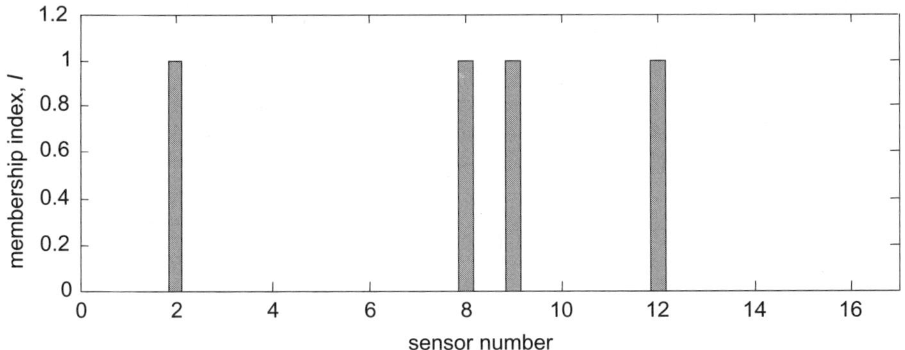

Figure 7.8. The 2D truss placement index I indicates four accelerometer locations.

Example 7.4. International Space Station Structure. This example is based on [115]. The Z1 module of the International Space Station structure, see Fig.1.9, is shaped with a basic truss frame and numerous appendages and attachments such as control moment gyros and a cable tray. The total mass of the structure is 30,000 lb.

The finite-element model of the structure consists of 11,804 degrees of freedom with 56 modes below the frequency of 70 Hz. The natural frequencies are listed in Table 7.1. The task is to identify all modes below 70 Hz by generating dynamic test data, with accelerometers used as sensors. This nontrivial undertaking requires extensive pretest analysis and careful planning of the actuator and sensor locations, especially if one does not have the freedom to repeat the test and modify the sensor/actuator location for retesting.

Actuator Placement. The first part of the analysis involves the selection of four actuator locations. The initial selection procedure combines engineering judgment, practical experience, and physical constraints including the following criteria:

- All target modes should be excited with relatively equal amplitudes.
- The structure is excited in three axes.

We examined the structure drawings and the finite-element model in order to select 2256 actuator candidate locations out of the 11,804 translational degrees of freedom. The selection was based on accessibility of the locations, strength of the structural parts, modal masses, and local flexibility. It was assumed at this stage of analysis that accelerometers were located at all degrees of freedom. We determined the Hankel norms of each actuator and used them to evaluate the actuator importance indices. For each of 56 modes the six most important actuators were selected, obtaining 268 actuator locations (it is less than 6×56, because some locations were the same for two or more modes). Next, we calculated the correlation coefficients of the Hankel norm vectors (see (7.23)) for each actuator location. Those highly correlated were discarded and the one with the highest placement index, out of all the highly correlated actuators, was kept.

Table 7.1. Natural frequencies (Hz) of the International Space Station structure.

Modes 1–8	Modes 9–16	Modes 17–24	Modes 25–32	Modes 33–40	Modes 41–48	Modes 49–56
9.34	28.93	35.07	40.71	49.78	60.91	65.91
16.07	29.44	35.16	41.18	50.98	61.53	66.79
19.21	30.19	36.43	42.10	51.39	62.92	67.05
21.14	30.42	37.21	42.46	54.82	63.25	67.26
22.67	31.21	37.61	43.34	57.02	63.46	67.49
23.81	32.25	38.30	44.83	57.61	64.22	67.63
25.24	33.88	39.79	46.42	58.42	64.70	69.17
26.33	34.71	40.37	47.34	59.24	65.23	69.67

In this process the number of actuators was reduced to 52 locations. The next step of the selection process involved the re-evaluation of the importance indices of each actuator and their comparison with the threshold value. In this step the number of actuator locations was reduced to seven. The final step involved evaluation of the actual location of these actuators using the finite-element model simulations, along with determination of accessibility, structural strength, and the importance index. The final four actuators were located at the nodal points, shown in Fig. 1.9 as white circles. These four locations are essentially near the four corners of the structure.

Sensor Placement. The sensor selection criteria includes the following:

- Establishing the maximum allowable number of sensors. In our case it was 400.
- Determination of the sensor placement indices for each mode. Sensors with the highest indices were selected.
- Using the correlation procedure to select uncorrelated sensors by evaluating the membership index.

The excitation level of each mode by the four selected actuators is represented by the Hankel norms and is shown in Fig. 7.9(a). We see that some modes are weakly excited, providing a weaker measurement signal; thus, they are more difficult to identify. Figure 7.9(b) presents an overview of the sensor importance index for each sensor as the sum of the indices for all modes. Sensors at the degrees of freedom with a larger amplitude of modal vibrations have higher indices. By looking at the sensor importance indices for a particular mode we can roughly evaluate the participation of each mode at a particular sensor location. The highly participating modes have a high index at this location. The set of illustrations presented in Fig. 7.10 shows the placement indices of each sensor for the first 10 modes. The first mode (Fig. 7.10(a)) is a global (or system) mode with indices for all sensors almost identical. The second mode (Fig. 7.10(b)) is a global mode of more complex configuration. The third, fourth, fifth, and seventh modes (Figs. 7.10(c),(d),(e),(g)) show more dominant responses from the cable tray attachment. The sixth mode is dominated by the local motion at locations 1000–2000, which correspond to the attachments and cross-beams near the circular dish on the side of the structure. The eighth and ninth modes (Figs. 7.10(h),(i)) are local modes of the control moment gyros—see the four columns sticking up at the end. The last one (Fig. 7.10(j)) shows a highly dominant mode of a beam sticking out of the structure.

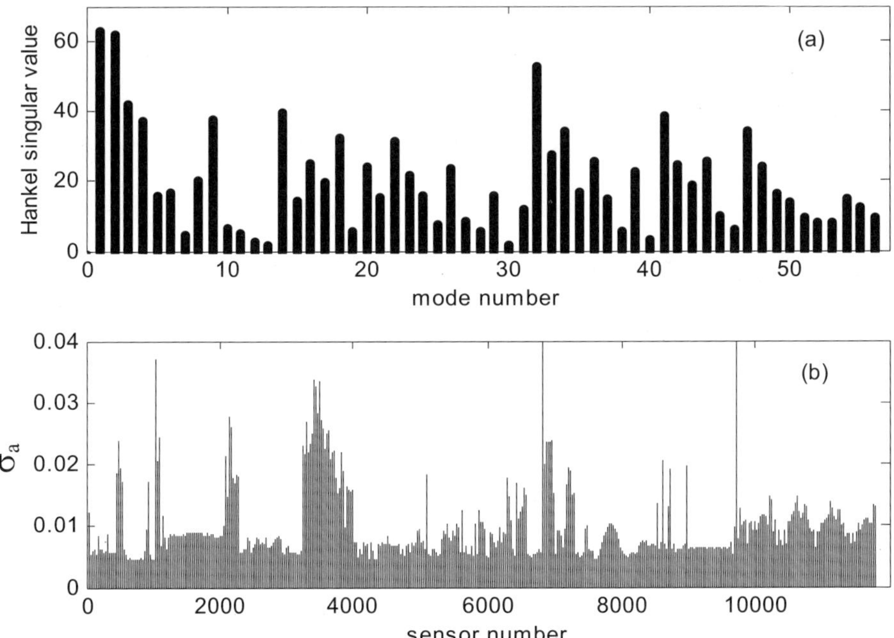

Figure 7.9. The International Space Station structure with four actuators: (a) The Hankel norms indicate the importance of each mode; and (b) the sensor indices for all modes indicate the importance of each sensor.

Figure 7.11 shows the membership index I, which has nonzero values for 341 locations. Figure 7.12(a),(b) indicates with the circles the selected sensor locations. It can be observed that many of the sensors are located in and around the control moment gyros (see Fig. 7.12) and the cable tray (see Fig. 1.9), since 13 out of the 56 modes involve extensive control moment gyro movement and nine are mostly cable tray modes. Many of the 56 modes are local modes that require concentrations of sensors at the particular locations seen in Fig. 7.12.

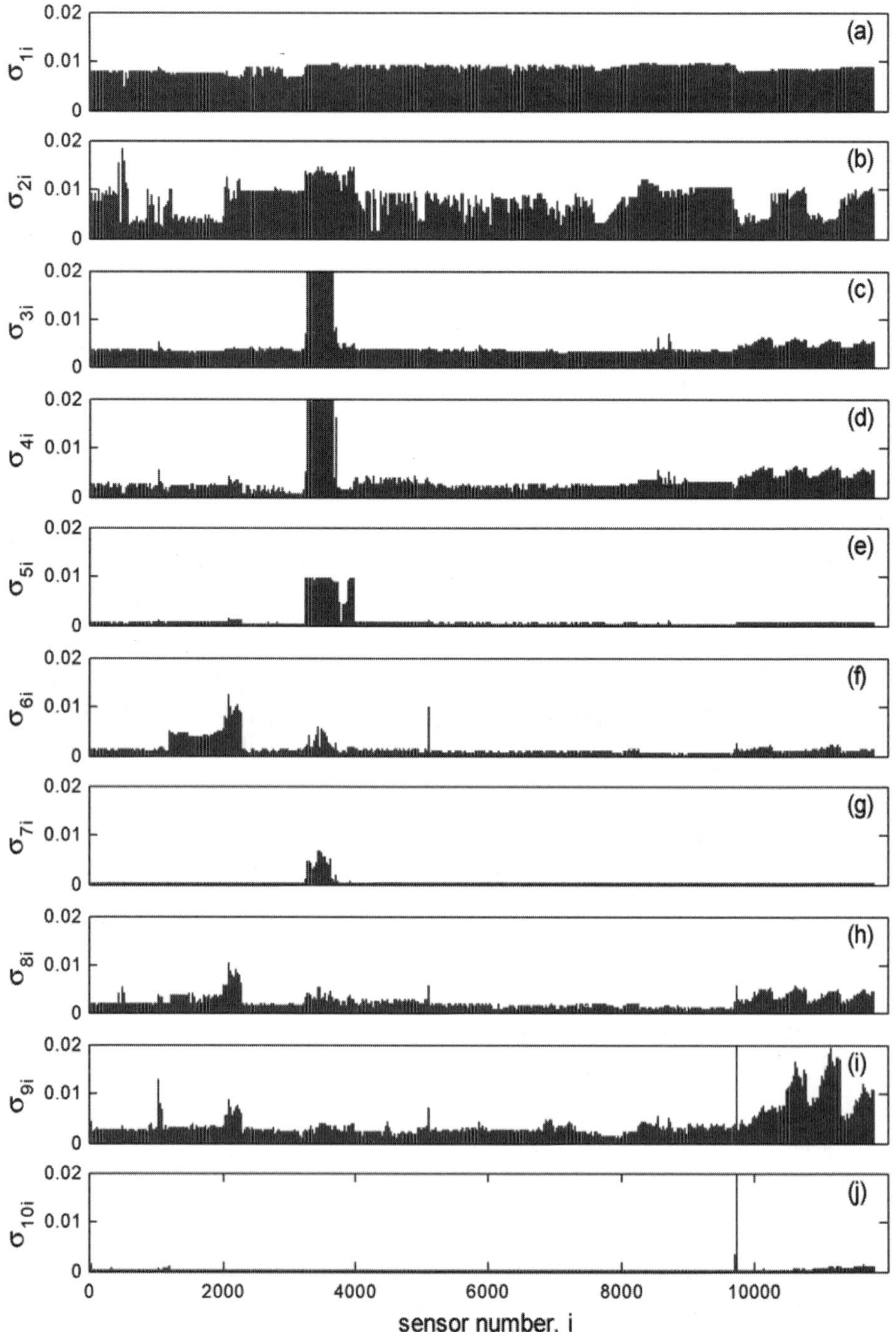

Figure 7.10. The placement indices for the first ten modes indicate the sensor importance for each mode: (a) Mode 1; (b) mode 2; (c) mode 3; (d) mode 4; (e) mode 5; (f) mode 6; (g) mode 7; (h) mode 8; (i) mode 9; and (j) mode 10.

In order to test the effectiveness of the procedure we compare the Hankel norms of each mode, for the structure with a full set of 11,804 sensors, and with the selected 341 sensors. The norms with the selected sensors should be proportional to the norms of the full set (they are always smaller than the norms of the full set, but proportionality indicates that each mode is excited and sensed comparatively at the same level). The norms are shown in Fig. 7.13, showing that the profile of the modal norms is approximately preserved.

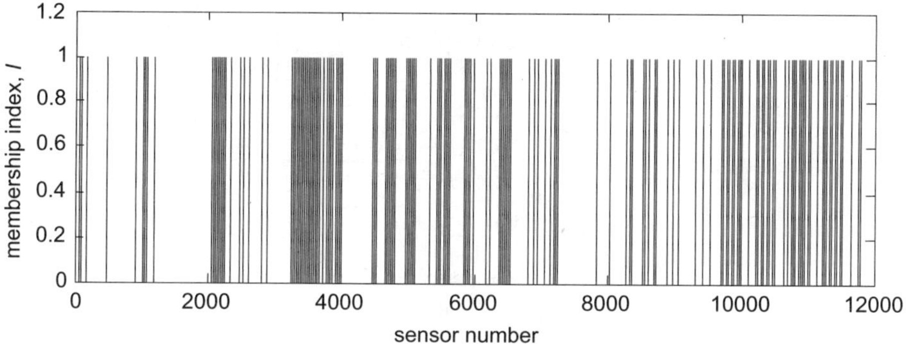

Figure 7.11. Membership index I for the International Space Station structure shows the selected locations of 341 sensors.

7.5 Placement for a Generalized Structure

The problem of actuator and sensor placement presented in this section refers to the more general problem which consists of the selection of actuators not collocated with disturbances, and sensors not collocated with the performance outputs. This problem has its origin in both structural testing and control.

7.5.1 Structural Testing and Control

The formulation of structural testing is based on a block diagram as in Fig. 3.10. In this diagram the structure input is composed of two inputs not necessarily collocated: the vector of disturbances (w) and the vector of actuator inputs (u). Similarly, the plant output is divided into two sets: the vector of the performance (z) and the vector of the sensor output (y). The actuator inputs include forces and torque applied during a test. The disturbance inputs include disturbances, noises, and commands, known and unknown, but not applied during the test. The sensor signals consist of structure outputs recorded during the test. The performance output includes signals that characterize the system performance, and is not generally measured during the test.

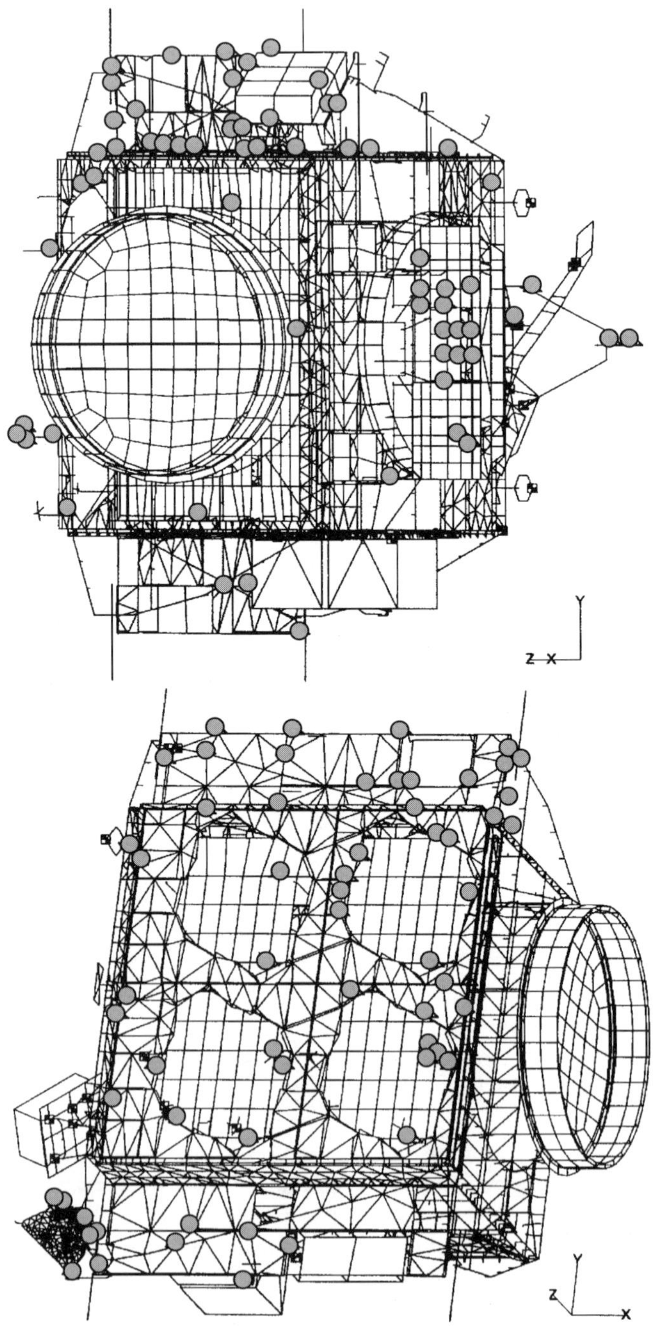

Figure 7.12. Two views of the sensor location (marked with ○) for the International Space Station structure.

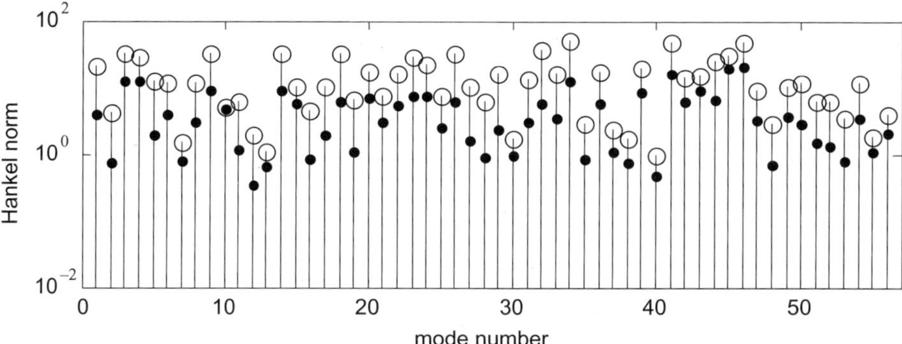

Figure 7.13. The modal Hankel norms of the full set of sensors (○), and the selected sensors (•) of the International Space Station truss: The norms of the selected sensors are proportional to the norms of the full set of sensors.

It is not possible in general to duplicate the dynamics of a real structure during testing. This happens not only due to physical restrictions or limited knowledge of disturbances, but also because the test actuators cannot be placed at the disturbance locations and sensors cannot be placed at the performance evaluation locations. Thus, to obtain the performance of the test item close to the performance of a structure in a real environment, we use the available (or candidate) locations of actuators and sensors and formulate the selection criteria to imitate the actual environment as close as possible.

The control design problem of a structure can be defined in a similar manner. The feedback loop is closed between the sensors and actuators of a structure. The actuators are placed within the allowable locations, and they are not necessarily collocated with the disturbance locations; sensors are placed at the sensor allowable locations, generally outside the locations of performance evaluation. In the control nomenclature, u is the control input, y is the plant output accessible to the controller, w is the vector of disturbances, and z is the vector of the performance output; for example, see [12].

7.5.2 Sensor and Actuator Properties

Consider a plant as in Fig. 3.10, with inputs w and u and outputs z and y. Let G_{wz} be the transfer matrix from w to z, let G_{wy} be the transfer matrix from w to y, let G_{uz} be the transfer matrix from u to z, and let G_{uy} be the transfer matrix from u to y. Let G_{wzi}, G_{uyi}, G_{wyi}, and G_{uzi} be the transfer functions of the ith mode. The following multiplicative property of modal norms holds, see (5.52),

$$\|G_{wzi}\|\|G_{uyi}\| \cong \|G_{wyi}\|\|G_{uzi}\|, \qquad \text{for} \qquad i = 1, \ldots, n, \tag{7.25}$$

where $\|.\|$ denotes either H_2, H_∞, or Hankel norms.

We show the impact of G_{uy} on the overall system performance using the following properties of the modal norms. Let G_i denote the transfer function of the ith mode, from the combined input $\{w, u\}$ to the combined output $\{z, y\}$. Its norm is as follows:

$$\|G_i\|^2 \cong \alpha \left(\left\|\begin{bmatrix} B_{mu} & B_{mw} \end{bmatrix}\right\|_2^2 \left\|\begin{bmatrix} C_{mz} \\ C_{my} \end{bmatrix}\right\|_2^2 \right) = \alpha\left(\|B_{mu}\|_2^2 + \|B_{mw}\|_2^2\right)\left(\|C_{mz}\|_2^2 + \|C_{my}\|_2^2\right)$$

$$= \alpha\left(\|B_{mu}\|_2^2 \|C_{mz}\|_2^2 + \|B_{mu}\|_2^2 \|C_{my}\|_2^2 + \|B_{mw}\|_2^2 \|C_{mz}\|_2^2 + \|B_{mw}\|_2^2 \|C_{my}\|_2^2\right)$$

$$= \|G_{uzi}\|^2 + \|G_{uyi}\|^2 + \|G_{wzi}\|^2 + \|G_{wyi}\|^2,$$

where $\alpha = \frac{1}{2\sqrt{\zeta_i \omega_i}}$ for the H_2 norm, $\alpha = \frac{1}{2\zeta_i \omega_i}$ for the H_∞ norm, and $\alpha = \frac{1}{4\zeta_i \omega_i}$ for the Hankel norm. From the above equation we obtain

$$\|G_i\|^2 \cong \|G_{uzi}\|^2 + \|G_{uyi}\|^2 + \|G_{wzi}\|^2 + \|G_{wyi}\|^2, \tag{7.26}$$

where $\|.\|$ denotes either H_2, H_∞, or Hankel norms. Consider S actuator locations, generating S inputs $\{u_1 \quad \ldots \quad u_S\}$. The actuators impact only the first two terms of the above equation, and the last two are constant. Denote these two terms as $\|G_{ui}\|^2$, i.e.,

$$\|G_{ui}\|^2 \cong \|G_{uzi}\|^2 + \|G_{uyi}\|^2. \tag{7.27}$$

From the definitions of norms (7.3) to (7.7) we obtain the following property:

Property 7.1(a). ***Additive Property of Actuators of a Generalized Structure.***

$$\|G_{ui}\|^2 \cong \alpha_{wi}^2 \sum_{k=1}^{S} \|G_{u_k yi}\|^2, \tag{7.28}$$

where $G_{u_k yi}$ is the transfer function of the ith mode from the kth actuator to the output y, and α_{wi} is the disturbance weight of the ith mode, defined as

$$\alpha_{wi} = \sqrt{1 + \frac{\left\|G_{wzi}\right\|^2}{\left\|G_{wyi}\right\|^2}}. \tag{7.29}$$

Proof. From (7.3) or (7.6), it follows that

$$\left\|G_{uzi}\right\|^2 \cong \sum_{k=1}^{S} \left\|G_{u_k zi}\right\|^2 \quad \text{and} \quad \left\|G_{uyi}\right\|^2 \cong \sum_{k=1}^{S} \left\|G_{u_k yi}\right\|^2,$$

where $G_{u_k zi}$ is the transfer function of the ith mode from the kth actuator to the performance z. Introducing the above equations to (7.27) we obtain

$$\left\|G_{ui}\right\|^2 \cong \sum_{k=1}^{S} \left(\left\|G_{u_k zi}\right\|^2 + \left\|G_{u_k yi}\right\|^2 \right).$$

Next, using (7.25) we obtain

$$\left\|G_{u_k zi}\right\| \cong \frac{\left\|G_{u_k yi}\right\| \left\|G_{wzi}\right\|}{\left\|G_{wyi}\right\|},$$

which, introduced to the previous equation, gives (7.28). ▣

Note that the disturbance weight α_{wi} does not depend on the actuator location. It characterizes structural dynamics caused by the disturbances w.

Similarly we obtain the additive property of the sensor locations of a general plant. Consider R sensor locations with R outputs $\{y_1 \ \ldots \ y_R\}$. The sensors impact only the second and fourth terms of (7.26) and the remaining terms are constant. Denote the second and fourth terms by $\left\|G_{yi}\right\|^2$, i.e.,

$$\left\|G_{yi}\right\|^2 \cong \left\|G_{wyi}\right\|^2 + \left\|G_{uyi}\right\|^2, \tag{7.30}$$

then the following property holds:

Property 7.1(b). ***Additive Property of Sensors of a Generalized Structure.***

$$\left\|G_{yi}\right\|^2 \cong \alpha_{zi}^2 \sum_{k=1}^{R} \left\|G_{uy_k i}\right\|^2, \tag{7.31}$$

where

$$\alpha_{zi} = \sqrt{1 + \frac{\|G_{wzi}\|^2}{\|G_{uzi}\|^2}} \tag{7.32}$$

is the performance weight of the ith mode.

Note that the performance weight α_{zi} characterizes part of the structural dynamics that is observed at the performance output. It does not depend on the sensor location.

7.5.3 Placement Indices and Matrices

Properties 7.1(a),(b) are the basis of the actuator and sensor search procedure of a general plant. The actuator index that evaluates the actuator usefulness in test is defined as follows:

$$\sigma_{ki} = \frac{\alpha_{ui}\|G_{u_k yi}\|}{\|G_u\|}, \tag{7.33}$$

where $\|G_u\|^2 = \|G_{uy}\|^2 + \|G_{uz}\|^2$, while the sensor index is

$$\sigma_{ki} = \frac{\alpha_{yi}\|G_{uy_k i}\|}{\|G_y\|}, \tag{7.34}$$

where $\|G_y\|^2 = \|G_{uy}\|^2 + \|G_{wy}\|^2$.

The indices are the building blocks of the actuator placement matrix Σ,

$$\Sigma = \begin{bmatrix} \sigma_{11} & \sigma_{12} & \dots & \sigma_{1k} & \dots & \sigma_{1S} \\ \sigma_{21} & \sigma_{22} & \dots & \sigma_{2k} & \dots & \sigma_{2S} \\ \dots & \dots & \dots & \dots & \dots & \dots \\ \sigma_{i1} & \sigma_{i2} & \dots & \sigma_{ik} & \dots & \sigma_{iS} \\ \dots & \dots & \dots & \dots & \dots & \dots \\ \sigma_{n1} & \sigma_{n2} & \dots & \sigma_{nk} & \dots & \sigma_{nS} \end{bmatrix} \begin{matrix} \\ \\ \\ \leftarrow i\text{th mode}, \\ \\ \\ \end{matrix} \tag{7.35}$$

$\uparrow$

kth actuator

or the sensor placement matrix

$$\Sigma = \begin{bmatrix} \sigma_{11} & \sigma_{12} & \cdots & \sigma_{1k} & \cdots & \sigma_{1R} \\ \sigma_{21} & \sigma_{22} & \cdots & \sigma_{2k} & \cdots & \sigma_{2R} \\ \cdots & \cdots & \cdots & \cdots & \cdots & \cdots \\ \sigma_{i1} & \sigma_{i2} & \cdots & \sigma_{ik} & \cdots & \sigma_{iR} \\ \cdots & \cdots & \cdots & \cdots & \cdots & \cdots \\ \sigma_{n1} & \sigma_{n2} & \cdots & \sigma_{nk} & \cdots & \sigma_{nR} \end{bmatrix} \begin{matrix} \\ \\ \\ \leftarrow i\text{th mode.} \\ \\ \\ \end{matrix} \tag{7.36}$$

$\uparrow$

kth sensor

The placement index of the kth actuator (sensor) is determined from the kth column of Σ. In the case of the H_2 norm it is the rms sum of the kth actuator indexes over all modes,

$$\sigma_k = \sqrt{\sum_{i=1}^{n} \sigma_{ik}^2}, \qquad k = 1,\ldots,S \text{ or } R \tag{7.37}$$

and in the case of the H_∞ and Hankel norms it is the largest index over all modes

$$\sigma_k = \max_i(\sigma_{ik}), \qquad i = 1,\ldots,n, \quad k = 1,\ldots,S \text{ or } R \tag{7.38}$$

This property shows that the index for the set of sensors/actuators is determined from the indexes of each individual sensor or actuator. This decomposition allows for the evaluation of an individual sensor/actuator through its participation in the performance of the whole set of sensors/actuators.

7.5.4 Placement of a Large Number of Sensors

For the placement of a large number of sensors the maximization of the performance index alone is not a satisfactory criterion. These locations can be selected using the correlation of each sensor modal norm. Define the kth sensor norm vector, which is composed of the squares of the modal norms

$$g_{uyk} = \begin{bmatrix} \left\| G_{uy_k 1} \right\|^2 \\ \left\| G_{uy_k 2} \right\|^2 \\ \vdots \\ \left\| G_{uy_k n} \right\|^2 \end{bmatrix}, \tag{7.39}$$

where $G_{uy_k i}$ denotes the transfer function of the ith mode at the kth sensor. The norm $\|.\|$ denotes the H$_2$, H$_\infty$, or Hankel norm. We select the sensor locations using the correlation coefficient r_{ik}, defined as follows:

$$r_{ik} = \frac{g_{uyi}^T g_{uyk}}{\|g_{uyi}\|_2 \|g_{uyk}\|_2}, \qquad i = 1,...,r, \quad k = i+1,...,r. \tag{7.40}$$

Denote a small positive number ε, say $\varepsilon = 0.01 - 0.20$. We define the membership index $I(k)$, $k = 1, \ldots, r$, as follows:

$$I(k) = \begin{cases} 0 & \text{if } r_{ik} > 1 - \varepsilon \text{ and } \sigma_k \le \sigma_j, \\ 1 & \text{elsewhere,} \end{cases} \tag{7.41}$$

for $k > j$ and r is the number of sensors. If $I(k) = 1$ the kth sensor is accepted and if $I(k) = 0$ the kth sensor is rejected (in this case the two locations j and k are either highly correlated or the jth location has higher performance σ_j).

From Property 7.1 the search procedure for the sensor placement follows:

1. The norms of the transfer functions G_{wzi}, $G_{uy_k i}$ are determined (for all modes and for each sensor) along with the norm of G_y (for all actuators and all sensors).
2. The performance σ_k of each sensor is determined from (7.34).
3. Check if the chosen location is highly correlated with the previously selected locations by determining the correlation coefficient r_{ik} from (7.40), and the membership index $I(k)$ from (7.41). Highly correlated sensors are rejected.

Example 7.5. Consider the 3D truss as in Fig. 1.3. The disturbance w is applied at node 7 in the horizontal direction. The performance z is measured as rates of all nodes. The input u is applied at node 26 in the vertical direction, and the candidate sensor locations are at nodes 5, 6, 7, 12, 13, 14, 19, 20, 21, 26, 27, and 28, in all three directions (a total of 36 locations). Using the first 50 modes, the task is to select a low number of sensors that would measure, as close as possible, the disturbance-to-performance dynamics.

First, we determine the H$_\infty$ norms of each mode of G_{wz}, G_{wy}, G_{uz}, and G_{uy}; they are presented in Fig. 7.14(a),(b). Next, we check (7.25). Indeed, it holds since the plots of $g_1(k) = \|G_{wzk}\|_\infty \|G_{uyk}\|_\infty$ and of $g_2(k) = \|G_{wyk}\|_\infty \|G_{uzk}\|_\infty$ overlap in Fig. 7.15.

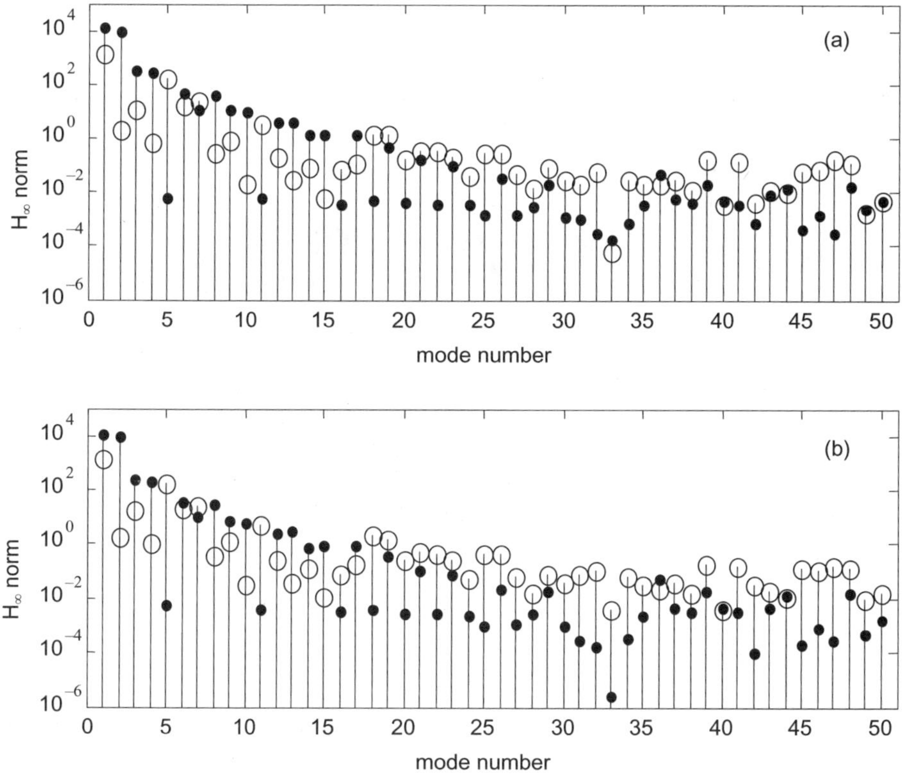

Figure 7.14. The H_∞ norms of the 3D truss modes: (a) G_{wz} (•) and G_{uy} (○); and (b) G_{wy} (•) and G_{uz} (○).

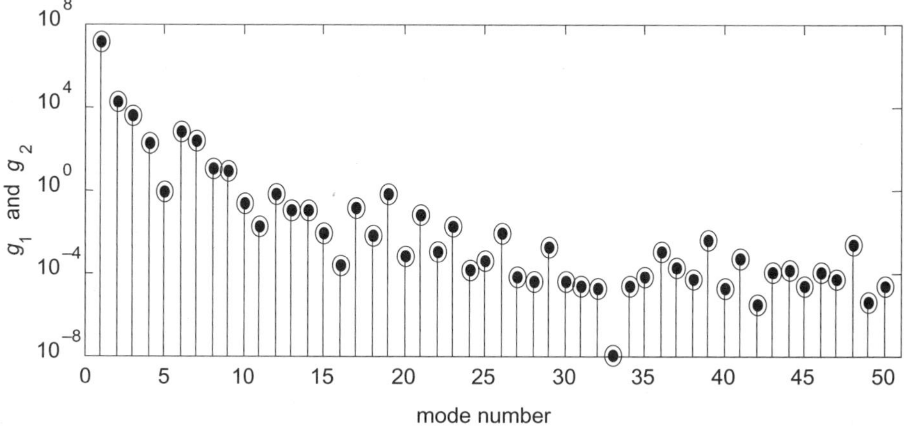

Figure 7.15. Overlapped plots of $g_1(k)$ (•) and $g_2(k)$ (○) show that (7.25) holds.

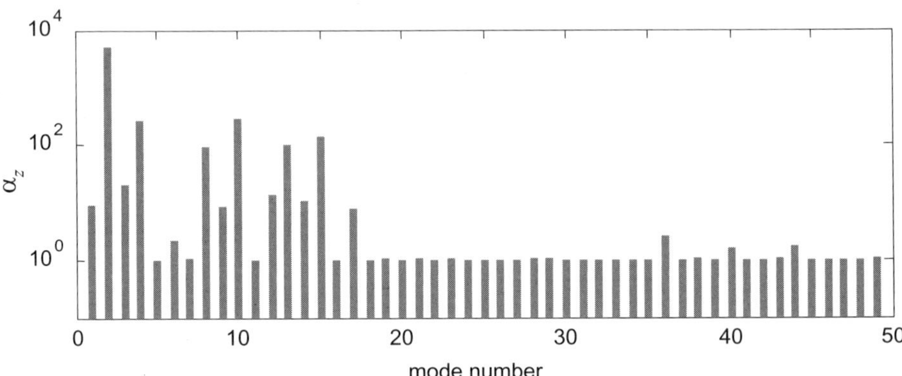

Figure 7.16. Modal weights for the 3D truss to accommodate disturbances in the generalized model.

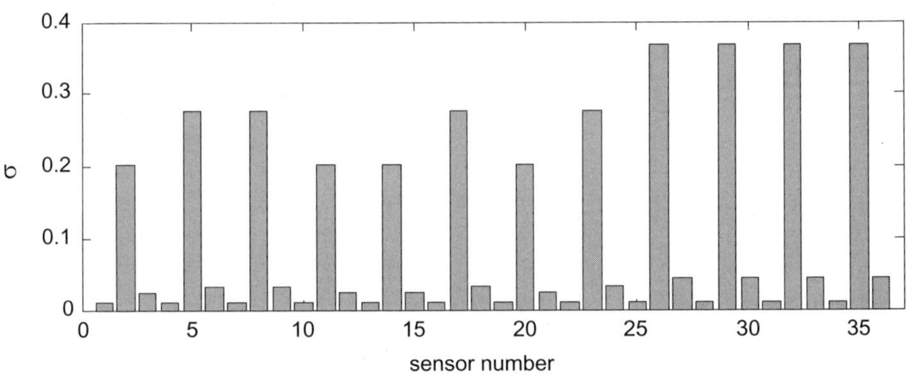

Figure 7.17. Sensor indices for the 3D truss show the importance of each sensor.

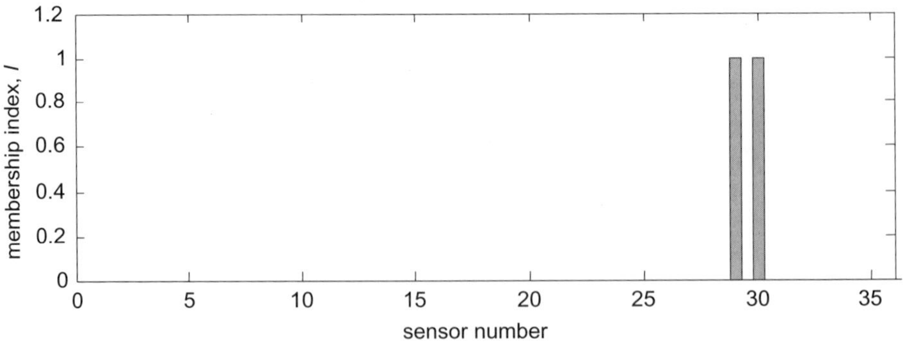

Figure 7.18. Membership index I for the 3D truss shows the selected sensor locations.

In the following we determine the sensor modal weights α_{yi} for each mode, which are shown in Fig. 7.16. We calculate the placement indices σ_k for each sensor from (7.38), and their plot is shown in Fig. 7.17. Note that there are sensors with a high value of σ_k, and they are highly correlated. Therefore we calculate the membership index $I(k)$ assuming $\varepsilon = 0.03$. The index is shown in Fig. 7.18. Its only nonzero values are for $k = 29$ and $k = 30$, that correspond to node 14 in the y- and z-directions. Thus the rate sensors at node 14 in the y- and z-directions are chosen for this particular task.

7.6 Simultaneous Placement of Actuators and Sensors

In this section we present a simultaneous selection of sensor and actuator locations; this is an extension of the actuator and sensor placement algorithm presented above. The latter algorithm describes either actuator placement for given sensor locations, or sensor placement for given actuator locations. The simultaneous placement is an issue of some importance, since fixing the locations of sensors while placing actuators (or vice versa) limits the improvement of system performance.

The algorithm consists of determination either H_2, H_∞, or Hankel norms for a single mode, single actuator, and single sensor. Based on these norms the sensor and actuator placement matrices are generated for each considered mode to evaluate sensor and actuator combinations, and to determine the simultaneous actuator and sensor locations that maximize each modal norm.

In this section the symbol $\| \,.\, \|$ will denote either the H_2, H_∞, or Hankel norms. For the set R of the candidate actuator locations, we select a subset r of actuators and, concurrently for the set S of the candidate sensor locations, we select a subset s of sensors. The criterion is the maximization of the system norm.

Recall that the norm $\|G_{ijk}\|$ characterizes the ith mode equipped with a jth actuator and kth sensor. Previously we defined the placement index for actuators and for sensors separately; see (7.8) and (7.10) or (7.12) and (7.14). Here we define the actuator and sensor placement index as follows:

$$\sigma_{ijk} = \frac{\|G_{ijk}\|}{\|G_{mi}\|} \tag{7.42}$$

for each mode, $i = 1, \ldots, n$.

The placement index σ_{ijk} is a measure of the participation of the jth actuator and kth sensor in the impulse response of the ith mode. Using this index the actuator and sensor placement matrix of the ith mode is generated,

$$\Sigma_i = \begin{bmatrix} \sigma_{i11} & \sigma_{i12} & \cdots & \sigma_{i1k} & \cdots & \sigma_{i1S} \\ \sigma_{i21} & \sigma_{i22} & \cdots & \sigma_{i2k} & \cdots & \sigma_{i2S} \\ \cdots & \cdots & \cdots & \cdots & \cdots & \cdots \\ \sigma_{ij1} & \sigma_{ij2} & \cdots & \sigma_{ijk} & \cdots & \sigma_{ijS} \\ \cdots & \cdots & \cdots & \cdots & \cdots & \cdots \\ \sigma_{iR1} & \sigma_{iR2} & \cdots & \sigma_{iRk} & \cdots & \sigma_{iRS} \end{bmatrix} \begin{matrix} \\ \\ \\ \leftarrow j\text{th actuator,} \\ \\ \\ \end{matrix} \qquad (7.43)$$

$$\uparrow$$
$$k\text{th sensor}$$

$i = 1,\ldots,n.$

For the ith mode the jth actuator index is the rms sum over all selected sensors,

$$\sigma_{aij} = \sqrt{\sum_{k=1}^{s} \sigma_{ijk}^2}. \qquad (7.44)$$

For the same mode the kth sensor index is the rms sum over all selected actuators

$$\sigma_{sik} = \sqrt{\sum_{j=1}^{r} \sigma_{ijk}^2}. \qquad (7.45)$$

These indices, however, cannot be readily evaluated, since in order to evaluate the actuator index one needs to know the sensor locations (which have not yet been selected) and vice versa. This difficulty can be overcome by using the property similar to (7.25). Namely, for the placement indices we obtain

$$\sigma_{ijk}\sigma_{ilm} \cong \sigma_{ijm}\sigma_{ilk}. \qquad (7.46)$$

This property can be proven by the substitution of the norms as in Chapter 5 into the definition of the index (7.42).

It follows from this property that, by choosing the two largest indices for the ith mode, say σ_{ijk} and σ_{ilm} (such that $\sigma_{ijk} > \sigma_{ilm}$), the corresponding indices σ_{ijm} and σ_{ilk} are also large. In order to show this, note that $\sigma_{ilm} \le \sigma_{ijm} \le \sigma_{ijk}$ holds, and $\sigma_{ilm} \le \sigma_{ilk} \le \sigma_{ijk}$ also holds, as a result of (7.46) and of the fact that $\sigma_{ijm} \le \sigma_{ijk}$ and $\sigma_{ilk} \le \sigma_{ijk}$. In consequence, by selecting individual actuator and sensor locations with the largest indices we automatically maximize the indices (7.44) and (7.45) of the sets of actuators and sensors.

We illustrate the determination of the locations of large indices with the following example: Let σ_{124}, σ_{158}, and σ_{163} be the largest indices selected for the first mode. They correspond to 2, 5, and 6 actuator locations, and 3, 4, and 8 sensor locations. They are marked black in Fig. 7.19. According to (7.46) the indices σ_{123}, σ_{128}, σ_{153}, σ_{154}, σ_{164}, and σ_{168} are also large. They are marked gray in Fig. 7.19. Now we see that the rms summation for actuators is over all selected sensors (3, 4, and 8), and the rms summation for sensors is for over all selected actuators (2, 5, and 6), and that both summations maximize the actuator and sensor indices.

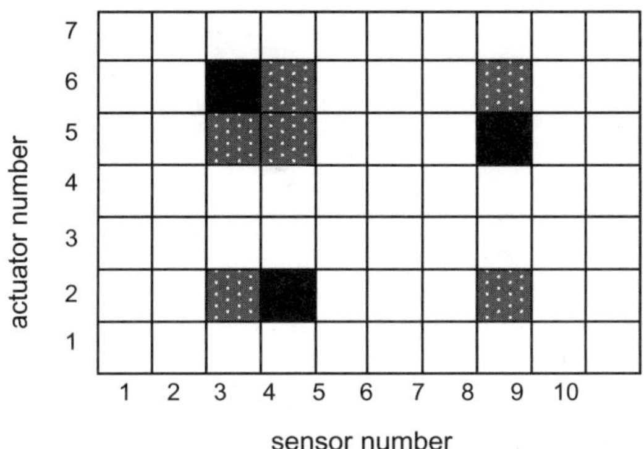

Figure 7.19. An example of the actuator and sensor placement matrix for the first mode: The largest indices are marked black, and the corresponding large indices are marked gray.

Example 7.6. Illustrate an actuator and sensor placement procedure with a clamped beam as in Figure 1.4. The beam is 150 cm long, cross-section of 1 cm^2, divided into 100 equal elements. The candidate actuator locations are the vertical forces at nodes 1 to 99, and the candidate sensor locations are the vertical rate sensors located at nodes 1 to 99. Using the H_2 norm, and considering the first four modes, we shall determine at most four actuator and four sensor locations (one for each mode).

Before we apply the placement procedure we check the accuracy of (7.46). For this purpose we choose the second mode, i.e., $i=2$, and select the following actuator and sensor locations: $j=k=3,\ l=m=q,$ and $q=1,\ldots,99$. For these parameters (7.46) is as follows:

$$\sigma_{233}\sigma_{2qq} \cong \sigma_{23q}\sigma_{2q3}, \qquad q=1,\ldots,99.$$

The plots of the left- and right-hand sides of the above equations are shown in Fig. 7.20, showing good coincidence.

In this example, $n=4$ and $R=S=99$. Using (7.42) and (7.43) we determine the actuator and sensor placement matrices for the first four modes and plot them in Fig. 7.21(a)–(d). The maximal values of the actuator and sensor index in the placement matrix determine the preferred location of the actuator and sensor for each mode. Note that for each mode four locations, two sensor locations and two actuator locations, have the same maximal value. Moreover, they are symmetrical with respect to the beam center; see Table 7.2. We selected no more than four collocated sensors and actuators for each mode. Namely, for mode 1—node 50; for mode 2—nodes 29 and 71; for mode 3—nodes 21, 50, and 79; and for mode 4—nodes 16, 40, 60, and 84.

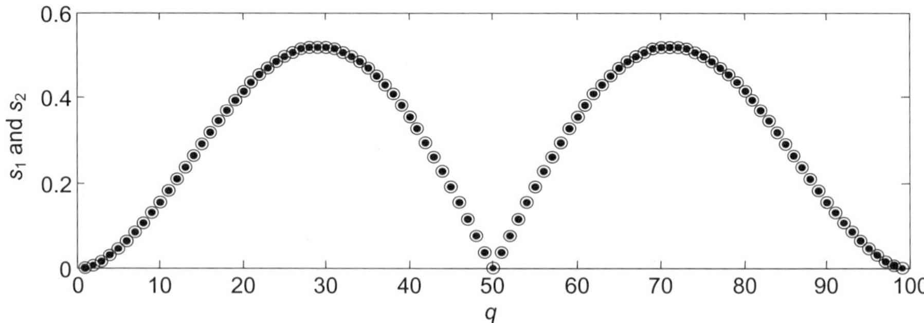

Figure 7.20. The verification of (7.49): ○ denotes $s_1 = \sigma_{233}\sigma_{2qq}$, and • denotes $s_2 = \sigma_{23q}\sigma_{2q3}$, and $s_1 = s_2$.

The above selection of actuators and sensors we performed for each mode individually. Let us investigate the actuator and sensor selection for the first mode, for the first two modes, for the first three modes, and for the first four modes. The H_2 indices for these combinations are shown in Fig. 7.22(a)–(d). The first index is, of course, identical with the index in Fig. 7.21(a). The second index shows the actuator and sensor location for the first two modes. Its maximum is at nodes 32 and 70; see Table 7.3. The third index shows the actuator and sensor location for the first three modes. Its maximum is at nodes 23, 50, and 77. The fourth index shows the actuator and sensor location for the first four modes. Its maximum is at nodes 19, 39, 61, and 81. Note that the locations of the above indices are shifted with respect to the locations for the individual modes.

Table 7.2. The best actuator and sensor locations for the individual modes.

	Locations: (Actuator, Sensor)
Mode 1	(50,50)
Mode 2	(29,29), (29,71), (71,29), (71,71)
Mode 3	(21,21), (21,50), (21,79), (50,21), (50,50), (50,79), (79,21), (79,50), (79,79)
Mode 4	(16,16), (16,40), (16,60), (16,84), (40,16), (40,40), (40,60), (40,84), (60,16), (60,40), (60,60), (60,84), (84,16), (84,40), (84,60), (84,84)

Table 7.3. The best actuator and sensor locations for the first four modes.

	Locations: (Actuator, Sensor)
Mode 1	(50,50)
Modes 1 and 2	(32,32), (32,70), (70,32), (70,70)
Modes 1, 2, and 3	(23,23), (23,50), (23,77), (50,23), (50,50), (50,77), (77,23), (77,50), (77,77)
Modes 1, 2, 3, and 4	(19,19), (19,39), (19,61), (19,81), (39,19), (39,39), (39,61), (39,81), (61,19), (61,39), (61,61), (61,81), (81,19), (81,39), (81,61), (81,81)

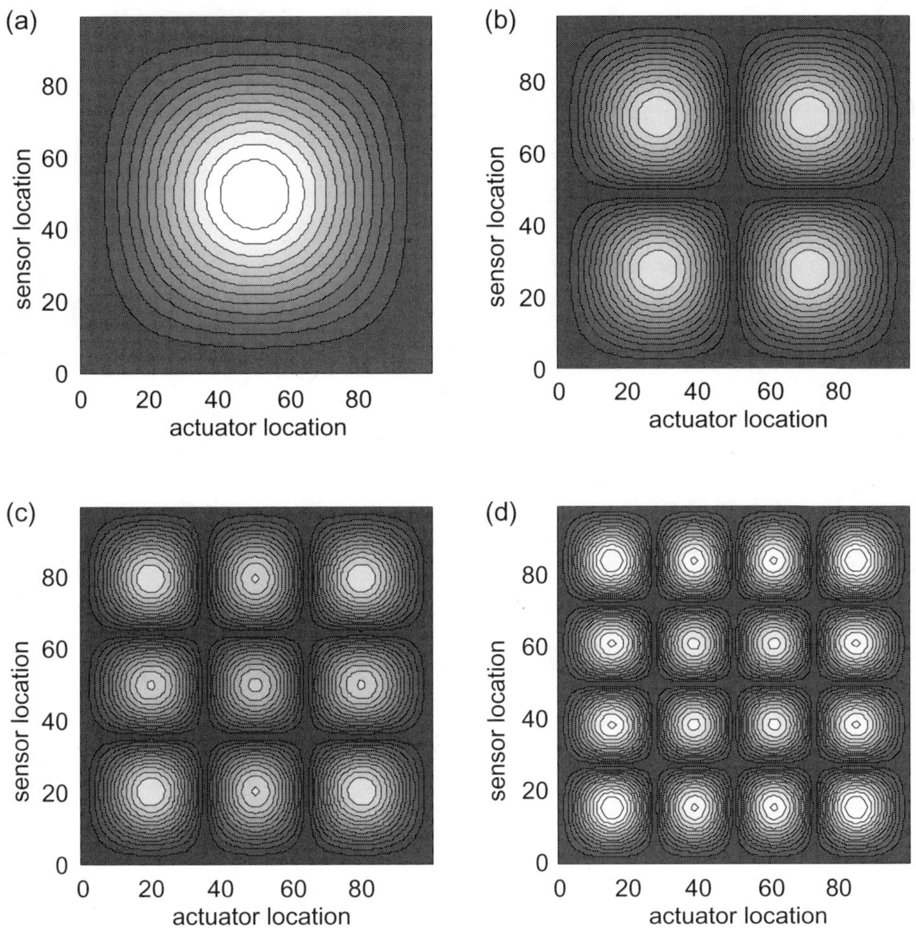

Figure 7.21. Actuator and sensor placement matrix using H_2 norm:

(a) For mode 1. The maximal placement indices, in light color, correspond to the following (actuator, sensor) locations: (50,50).

(b) For mode 2. The maximal placement indices, in light color, correspond to the following (actuator, sensor) locations: (29,29), (29,71), (71,71), (71,29).

(c) For mode 3. The maximal placement indices, in light color, correspond to the following (actuator, sensor) locations: (21,21), (21,50), (21,79), (50,21), (50,50), (50,79).

(d) For mode 4. The maximal placement indices, in light color, correspond to the following (actuator, sensor) locations: (16,16), (16,40), (16,60), (16,84), (40,16), (40,40), (40,60), (40,84), (60,16), (60,40), (60,60), (60,84), (84,16), (84,40), (84,60), (84,84).

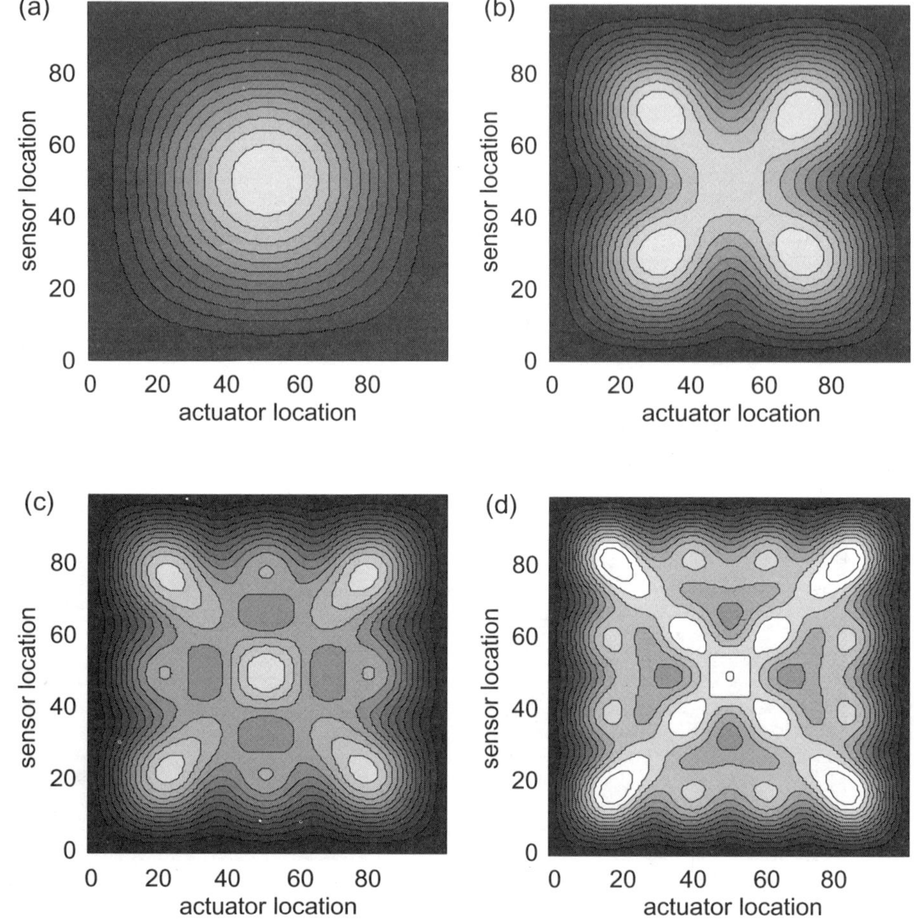

Figure 7.22. Actuator and sensor placement matrix using H_2 norm:

(a) For mode 1. The maximal placement indices, in light color, correspond to the following (actuator, sensor) locations: (50,50).

(b) For modes 1 and 2. The maximal placement indices, in light color, correspond to the following (actuator, sensor) locations: (32,32), (32,70), (70,70), (70,32).

(c) For modes 1, 2, and 3. The maximal placement indices, in light color, correspond to the following (actuator, sensor) locations: (23,23), (23,50), (23,77), (50,23), (50,50), (50,77).

(d) For modes 1, 2, 3, and 4. The maximal placement indices, in light color, correspond to the following (actuator, sensor) locations: (19,19), (19,39), (19,61), (19,81), (39,19), (39,39), (39,61), (39,81), (61,19), (61,39), (61,61), (61,81), (81,19), (81,39), (81,61), (81,81).

8
Modal Actuators and Sensors

how to excite and monitor selected modes

All coordinate systems are equal,
but some are more equal than others.
— *Edward Green*

In some structural tests it is desirable to isolate (i.e., excite and measure) a single mode. Such a technique considerably simplifies the determination of modal parameters, see [116]. This was first achieved by using the force appropriation method, also called the Asher method, see [107], or phase separation method, see [21]. In this method a spatial distribution and the amplitudes of a harmonic input force are chosen to excite a single structural mode. Modal actuators or sensors in a different formulation were presented in [38], [93], [75], and [114] with application to structural acoustic problems. In this chapter we present two techniques to determine gains and locations of actuators or sensors to excite and sense a target mode or a set of targeted modes.

In the first technique we determine actuator (or sensor) gains based on the relationship between the modal and nodal coordinates of the actuator or sensor locations; see [43]. This approach is distinct from the force appropriation method since it does not require harmonic input force. Rather, we determine the actuator locations and actuator gains, and the input force time history is irrelevant (modal actuator acts as a filter). The locations and gains, for example, can be implemented as a width-shaped piezoelectric film. Finally, in this approach we can excite and/or observe not only a single structural mode but also a set of selected modes.

The second technique—called an assignment technique—consists of the determination of the actuator (sensor) locations and gains to obtain a balanced system with the prescribed Hankel singular values. By setting the Hankel singular values equal to 1 for certain modes and to 0 for the remaining ones, the obtained sensors will "see" only modes associated with nonzero Hankel singular values. Just these sensors form a set of modal sensors. Similarly, by setting the Hankel singular values equal to 1 for certain modes and to 0 for the remaining, we obtain actuators that excite modes associated with nonzero Hankel singular values. Just these actuators form a set of modal actuators.

8.1 Modal Actuators and Sensors Through Modal Transformations

In this section we discuss the determination of actuator and sensor locations and gains such that they excite and sense selected structural modes. A structural model in this chapter is described by the second-order modal model, as in Subsection 2.2.2. In modal coordinates the equations of motion of each mode are decoupled; see (2.26). Thus, if the modal input gain is zero, the mode is not excited; if the modal output gain is zero, the mode is not observed. This simple physical principle is the base for the more specific description of the problem in the following sections.

8.1.1 Modal Actuators

The task in this section is to determine the locations and gains of the actuators such that n_m modes of the system are excited with approximately the same amplitude, where $1 \leq n_m \leq n$, and n is the total number of considered modes. We solve this task using the modal equations (2.19) or (2.26). Note that if the ith row, b_{mi}, of the modal input matrix, B_m, is zero, the ith mode is not excited. Thus, assigning entries of b_{mi} to either 1 or 0 we make the ith mode either excited or not. For example, if we want to excite the first mode only, B_m is a one-column matrix of a form $B_m = \begin{bmatrix} 1 & 0 & \cdots & 0 \end{bmatrix}^T$. On the other hand, if one wants to excite all modes independently and equally, one assigns a unit matrix, $B_m = I$.

Given the modal matrix B_m we derive the nodal matrix B_o from (2.23). We rewrite the latter equation as follows:

$$B_m = RB_o, \qquad \text{where} \qquad R = M_m^{-1}\Phi^T. \tag{8.1}$$

Matrix R is of dimensions $n \times n_d$. Recall that the number of chosen modes is $n_m \leq n$. If the selected modes are controllable, i.e., the rank of R is n_m, the least-squares solution of (8.1) is

$$B_o = R^+ B_m. \tag{8.2}$$

In the above equation R^+ is a pseudoinverse of R, $R^+ = V\Sigma^{-1}U^T$, where U, Σ, and V are obtained from the singular value decomposition of R, i.e., from $R = U\Sigma V^T$.

Note that a structure with a modal actuator excites n_m modes only (other modes are uncontrollable); therefore, the implementing modal actuator is equivalent to model reduction, where the structure has been reduced to n_m modes, or to $2n_m$ states.

The input matrix B_o in (8.2) that defines the modal actuator can be determined alternatively from the following equation:

$$B_o = M\Phi B_m, \tag{8.3}$$

which does not require a pseudoinverse. This is equivalent to (8.1). Indeed, let us left-multiply (8.3) by Φ^T to obtain $\Phi^T B_o = \Phi^T M \Phi B_m$ or $\Phi^T B_o = M_m B_m$. By left-multiplying the latter equation by M_m^{-1} we obtain (8.1).

Example 8.1. The Matlab code for this example is in Appendix B. Consider a clamped beam as in Fig. 1.4 divided into 60 elements (in order to enable the reader to use the beam data from Appendix C.2, the code in Appendix B deals with the beam divided into 15 elements). The vertical displacement sensors are located at nodes 1 to 59, and the single output is the sum of the sensor readings. Determine the actuator locations such that the second mode with 0.01 modal gain is excited, and the remaining modes are not excited. Consider the first nine modes.

In this case, the modal input matrix is, $B_m^T = \begin{bmatrix} 0 & 0.01 & 0 & 0 & 0 & 0 & 0 & 0 & 0 \end{bmatrix}^T$. Using it we determine a nodal input matrix B_o from (8.2). This contains gains of the vertical forces at nodes 1 to 59. The gain distribution of the actuators is shown in Fig. 8.1(a). Note that this distribution is proportional to the second mode shape. This distribution can be implemented as an actuator width proportional to the gain. Thus, an actuator that excites the second mode has the shape shown in Fig. 8.1(b).

Next, in Fig. 8.2 we present the magnitude of the transfer function for the input and the outputs defined as above. The plot shows clearly that only the second mode is excited. This is confirmed with the impulse response at node 24, Fig. 8.3(a), where only the second harmonic is excited. Figure 8.3(b) shows the simultaneous displacement of nodes 0 to 60 for the first nine time samples. They also confirm that only the second mode shape was excited.

If we want to excite the ith mode with certain amplitude, say, a_i, the H_∞ norm can be used as a measure of the amplitude of the ith mode. In the case of a single-input–single-output system the H_∞ norm of the ith mode is equal to the height of the ith resonance peak. In the case of multiple inputs (or outputs) the H_∞ norm of the ith mode is approximately equal to the root-mean-square (rms) sum of the *i*th resonance peaks corresponding to each input (or output). This is approximately determined as follows:

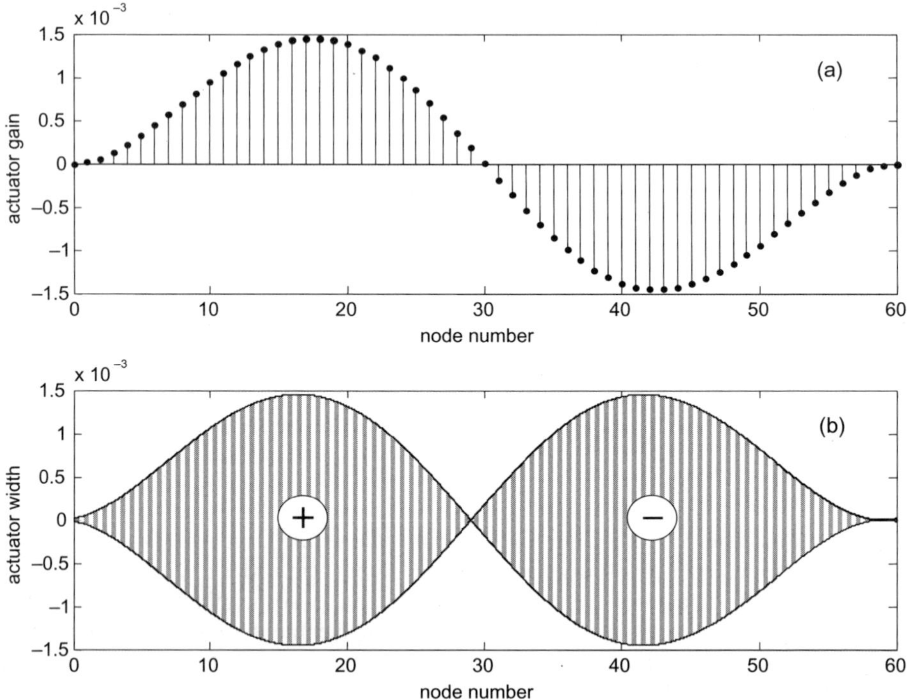

Figure 8.1. (a) Actuator gains and (b) the corresponding piezoelectric actuator width that excite the second mode.

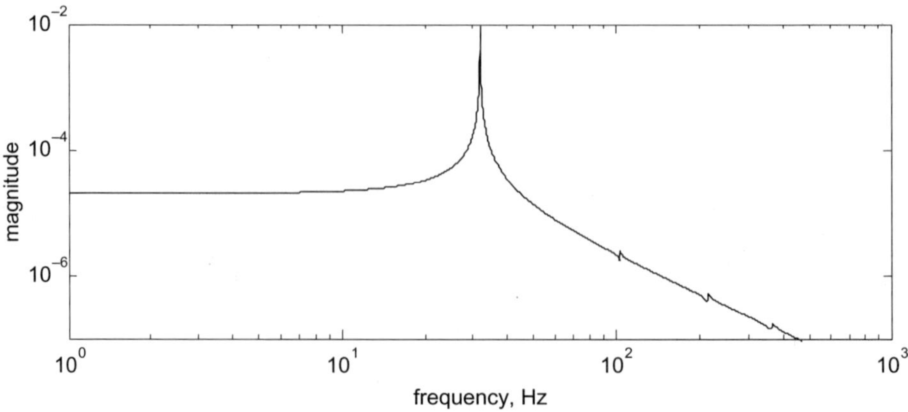

Figure 8.2. Magnitude of a transfer function with the second-mode modal actuator: Only the second mode is excited.

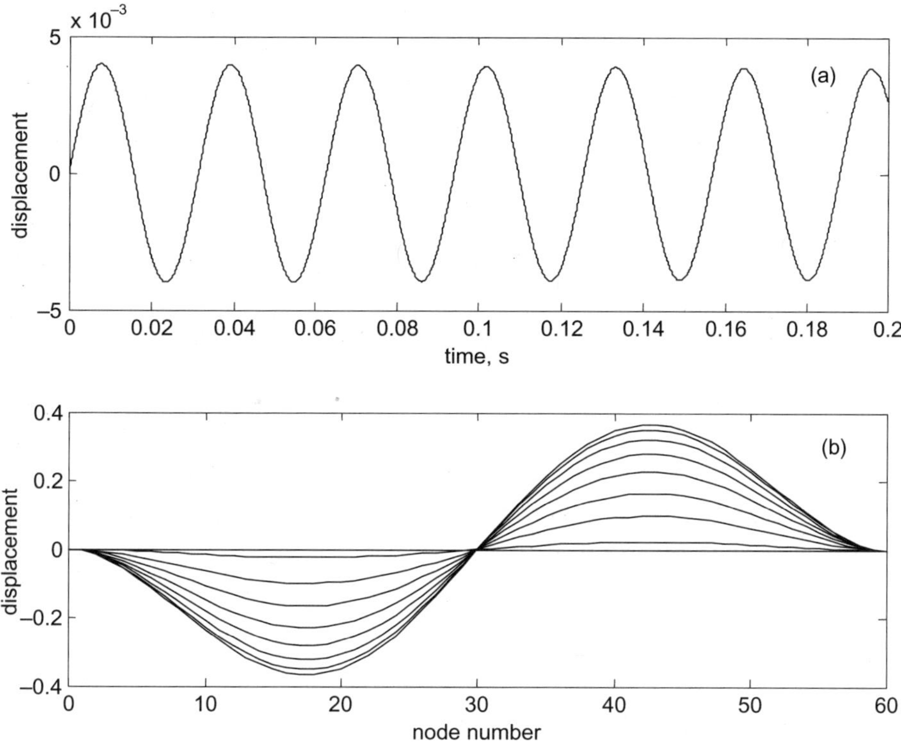

Figure 8.3. Beam with the second-mode modal actuator: (a) Impulse response at node 6 shows the harmonic motion with the second natural frequency; and (b) nodal displacements for the first nine time samples show the second mode shape.

$$\|G_i\|_\infty \cong \frac{\|b_{mi}\|_2 \|c_{mi}\|_2}{2\zeta_i \omega_i}; \tag{8.4}$$

see (5.21).

Assume a unity input gain for the current mode, i.e., $\|b_{mi}\|_2 = 1$, so that the current amplitude a_{oi} is

$$a_{oi} = \frac{\|c_{mi}\|_2}{2\zeta_i \omega_i}. \tag{8.5}$$

In order to obtain amplitude a_i we multiply a_{oi} by the weight w_i, such that

$$a_i = w_i a_{oi}. \tag{8.6}$$

Introducing (8.5) to the above equation we obtain

$$w_i = \frac{2a_i\zeta_i\omega_i}{\|c_{mi}\|_2}. \tag{8.7}$$

Define the weight matrix $W = \text{diag}(w_1, w_2, \ldots, w_n)$, then the matrix that sets the required output modal amplitudes is

$$B_{mw} = WB_m. \tag{8.8}$$

Example 8.2. Consider the same beam as in Example 8.1. Find a modal actuator that excites all nine modes with an amplitude of 0.01.

For this task the modal input matrix B_m is as follows: $B_m^T = 0.01 \times \begin{bmatrix}1 & 1 & 1 & 1 & 1 & 1 & 1 & 1 & 1\end{bmatrix}^T$. The weighting matrix we obtain from (8.7). The resulting gains of the nodal input matrix B_o are shown in Fig. 8.4(a); note that they do not follow any particular mode shape. The width of a piezoelectric actuator, that corresponds to the input matrix B_o and that excites all nine modes, is shown in Fig. 8.4(b).

In Fig. 8.5 we show the plot of the transfer function of the single-input system with the input matrix B_o. The plot shows that all nine modes are excited, with approximately the same amplitude of 0.01 cm. Figure 8.6(a) shows the impulse response at node 24. The time history consists of nine equally excited modes. Figure 8.6(b) shows the simultaneous displacement in the *y*-direction of all nodes. The rather chaotic pattern of displacement indicates the presence of all nine modes in the response.

8.1.2 Modal Sensors

The modal sensor determination is similar to the determination of modal actuators. The governing equation is derived from (2.24) and (2.25),

$$\begin{aligned} C_{mq} &= C_{oq}\Phi, \\ C_{mv} &= C_{ov}\Phi. \end{aligned} \tag{8.9}$$

If we want to observe a single mode only (say, the *i*th mode) we assume the modal output matrix in the form of $C_{mq} = \begin{bmatrix}0 & \ldots & 0 & 1 & 0 & \ldots & 0\end{bmatrix}$, where 1 stands at the *i*th position. If we want to observe n_m modes we assume the modal output matrix in the form of $C_{mq} = \begin{bmatrix}c_{q1}, c_{q2}, \ldots, c_{qn}\end{bmatrix}$, where $c_{qi} = 1$ for selected modes, otherwise, $c_{qi} = 0$. Next, we obtain the corresponding output matrix from (2.24):

$$C_{oq} = C_{mq}\Phi^{+}, \tag{8.10}$$

where Φ^{+} is the pseudoinverse of Φ. Similarly, we obtain the rate sensor matrix C_{ov} from the assigned modal rate sensor matrix C_{mv},

$$C_{ov} = C_{mv}\Phi^{+}. \tag{8.11}$$

Above we assumed that the assigned modes are observable, i.e., that the rank of Φ is n_m, where n_m is the number of the assigned modes.

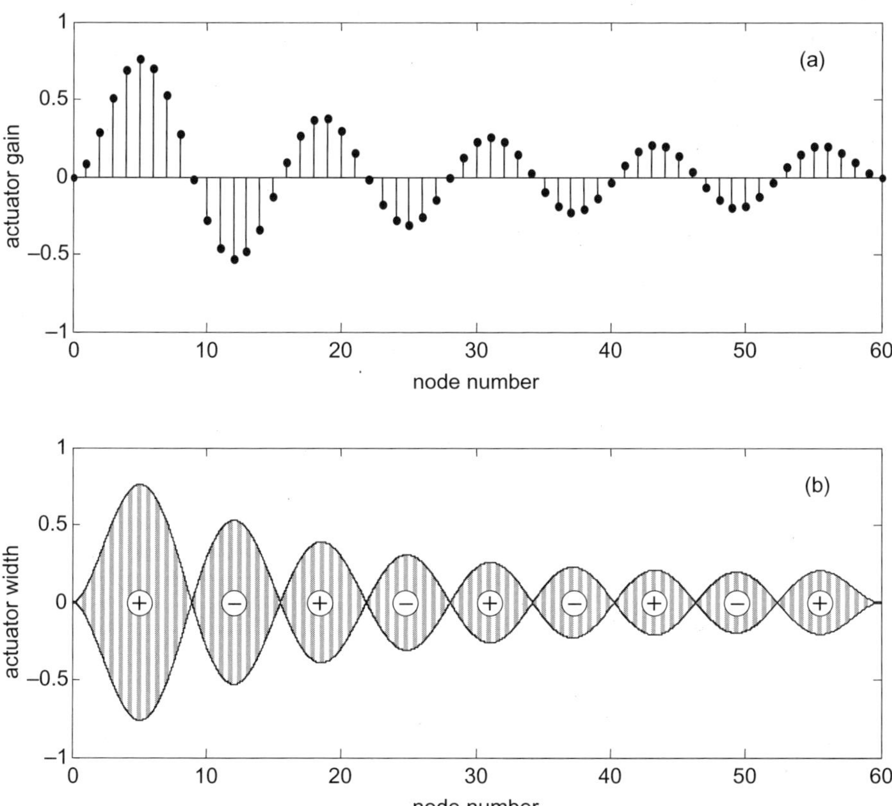

Figure 8.4. (a) Actuator gains; and (b) the corresponding piezoelectric actuator width that excites all nine modes.

Note that an output of a structure with a single modal sensor represents a single mode (other modes are not observable), therefore, the system has been reduced to a single mode, or to two states.

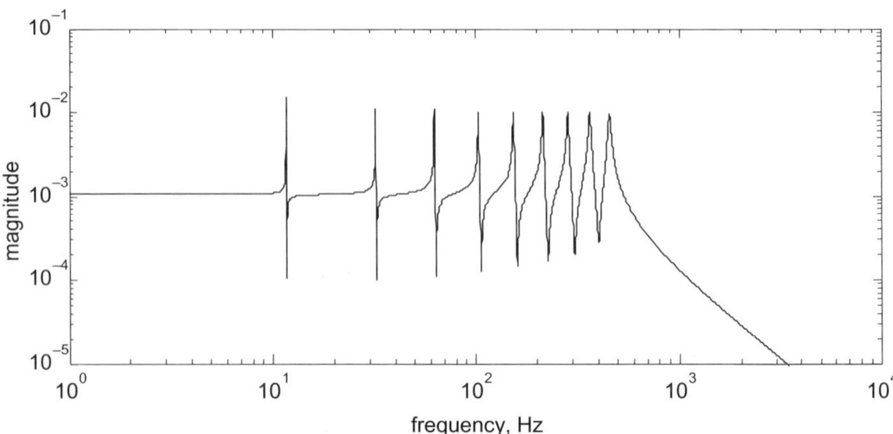

Figure 8.5. Magnitude of a transfer function for the nine-mode modal actuator shows the nine resonances of the excited modes.

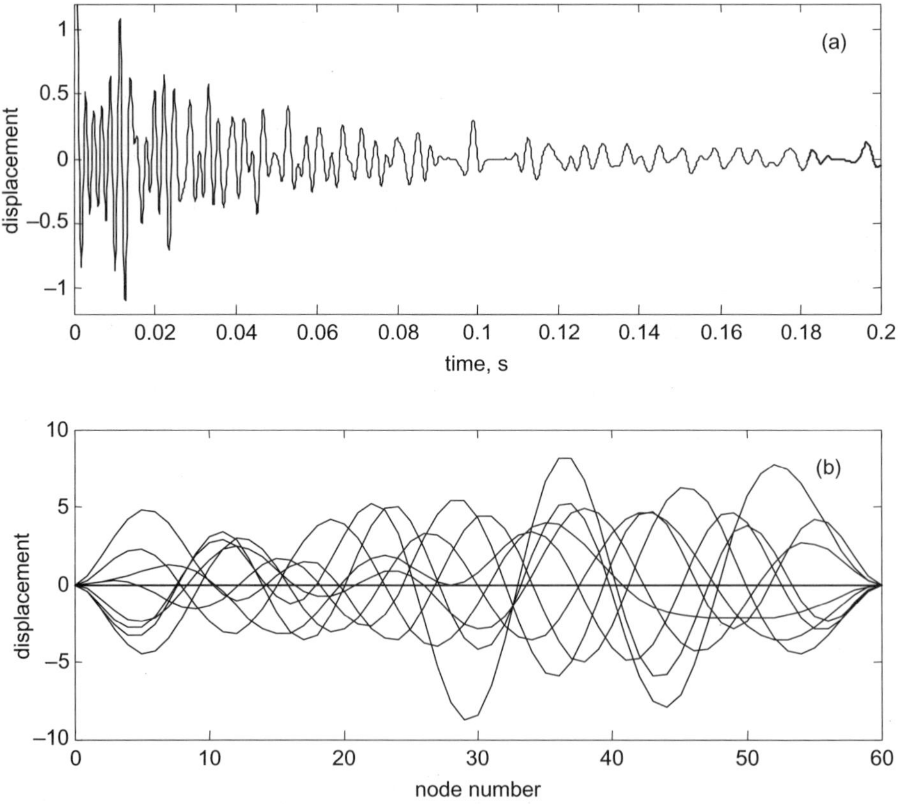

Figure 8.6. The beam with the nine-mode modal actuator: (a) Impulse response at node 6 includes nine modes motion; and (b) nodal displacements for the first nine time samples show no particular pattern.

The following are equations that determine modal sensors alternatively to (8.10) and (8.11):

$$C_{oq} = C_{mq} M_m^{-1} \Phi^T M,$$
$$C_{ov} = C_{mv} M_m^{-1} \Phi^T M. \qquad (8.12)$$

These equations are equivalent to (8.9). Indeed, let us right-multiply the first equation (8.12) by Φ obtaining $C_{oq}\Phi = C_{mq} M_m^{-1} \Phi^T M \Phi$, which gives $C_{oq}\Phi = C_{mq} M_m^{-1} M_m$ or $C_{oq}\Phi = C_{mq}$, i.e., the first equation (8.9). Similarly, we can show the equivalence of the second equation of (8.12) and (8.9).

We obtain multiple modes with assigned modal amplitudes a_i using the sensor weights, and the weighted sensors we obtain from (8.7). Namely, the ith weight is determined from the following equation:

$$\|c_{mi}\|_2 = \frac{2\zeta_i \omega_i a_i}{\|b_{mi}\|_2}, \qquad (8.13)$$

where a_i is the amplitude of the ith mode.

Example 8.3. Consider a beam from Fig. 1.4 with three vertical force actuators located at nodes 2, 7, and 12, and find the displacement output matrix C_{oq} such that the first nine modes have equal contribution to the measured output with amplitude 0.01.

The matrix C_{mq} that excites the first nine modes is the unit matrix of dimension 9, and of amplitude $a_i = 1$, i.e., $C_{mq} = 1 \times W \times I_9$. The gains that make the mode amplitudes approximately equal we determined from (8.13), and the output matrix C_{oq} we determined from (8.10). For this matrix the magnitudes of the transfer functions of the nine outputs in Fig. 8.7 show that all nine of them have a resonance peak of 1.0.

Example 8.4. Consider a beam from Fig. 1.4 with actuators as in Example 8.3, and find the nodal rate sensor matrix C_{ov} such that all nine modes, except mode 2, contribute equally to the measured output with an amplitude of 0.01.

The matrix C_{mv} that gives in the equal resonant amplitudes of 0.01 is as follows: $C_{mv} = 0.01 \times W \times [1 \quad 0 \quad 1 \quad 1 \quad 1 \quad 1 \quad 1 \quad 1 \quad 1]$, where the weight W is determined from (8.13), and the output matrix C_{ov} is obtained from (8.11). For this matrix the magnitude of the transfer function is shown in Fig. 8.8 (dashed line). This is compared with the magnitude of the transfer function for the output that contains all nine modes (solid line). It is easy to notice that the second resonance peak is missing in the plot.

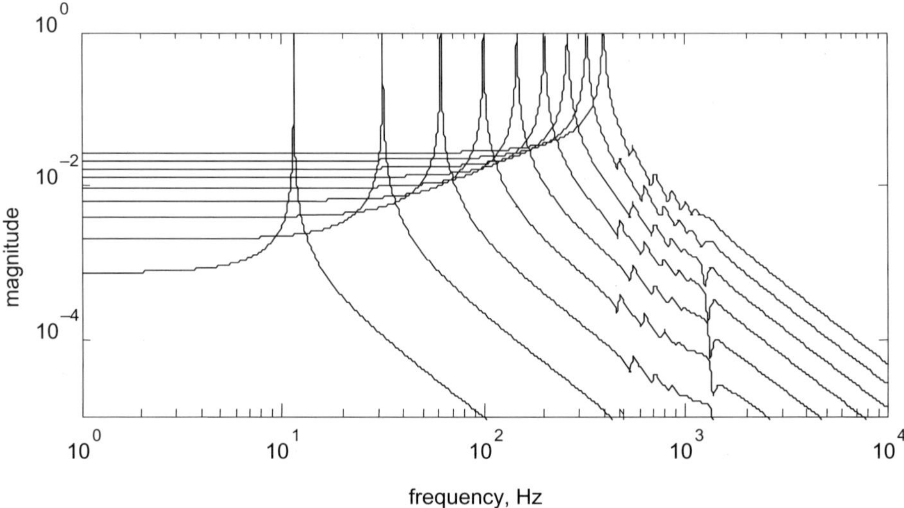

Figure 8.7. Magnitude of the transfer function with the nine single-mode sensors for the first nine modes shows a single resonance for each sensor.

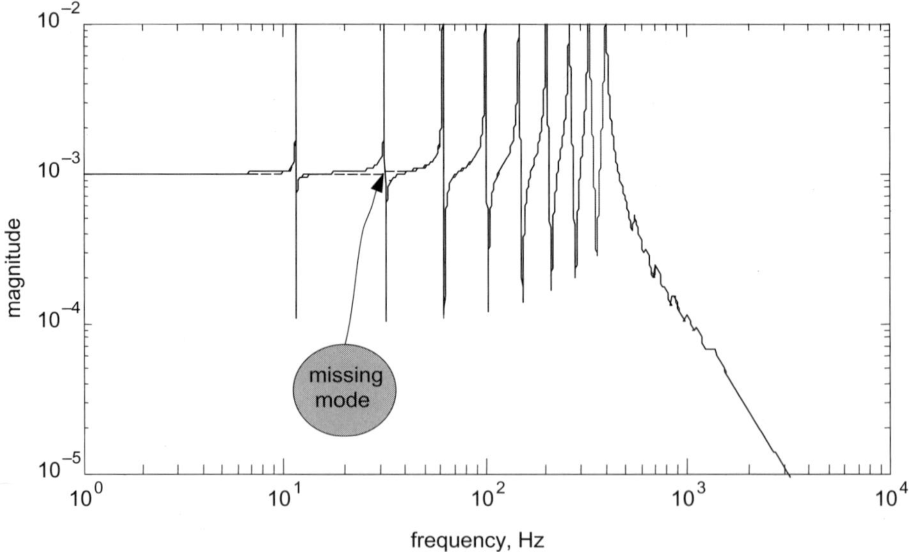

Figure 8.8. Magnitude of the transfer function with the nine-mode sensor (solid line), and for the eight-mode sensor (dashed line). The latter includes the first nine modes except the second one.

8.2 Modal Actuators and Sensors Through Grammian Adjustment

In the method presented above the modal actuator does not depend on the sensors' location. But, the sensors can be located such that the actuated mode can be unobservable. We can notice a similar situation in the modal sensor procedure: it is independent on the actuator location. But a sensed mode can be uncontrollable. The method proposed below allows us to avoid that undesirable situation.

It was shown in Chapter 4 that in modal coordinates the controllability and observability grammians are diagonally dominant; thus, their product is diagonally dominant as well, with approximate Hankel singular values on the diagonal. Each pair of Hankel singular values corresponds to the natural mode of a structure. If the pair of Hankel singular values is zero, the mode is not observable and not controllable. Thus, setting selected Hankel singular values to zero we can suppress the motion of certain modes. On the other hand, by appropriately scaling them we excite the required level of motion. Let us check how the scaling impacts the Hankel singular values (or Hankel norms) of a structure.

Consider a structure in modal representation (A_m, B_m, C_m), and a diagonal nonsingular matrix S in the following form:

$$S = \text{diag}(s_i I_2) = \text{diag}(s_1, s_1, s_2, s_2, ..., s_n, s_n). \tag{8.14}$$

Since A_m is block-diagonal, we have the following property:

$$S^{-1} A_m S = A_m. \tag{8.15}$$

Consider the controllability Lyapunov equation in modal coordinates

$$A_m W_c + W_c A_m^T + B_m B_m^T = 0, \tag{8.16}$$

and scale the input matrix to obtain B_{ms} such that

$$B_{ms} = S B_m. \tag{8.17}$$

We will show that the controllability grammian W_{cs} for the scaled input matrix is scaled as follows:

$$W_{cs} \cong S^2 W_c. \tag{8.18}$$

In order to show this let us consider a Lyapunov equation with the scaled input matrix, i.e.,

$$A_m W_{cs} + W_{cs} A_m^T + S B_m B_m^T S = 0.$$

We determine W_{cs} by multiplying the above equation from the left and right by S^{-1}, and inserting where necessary the identity matrix SS^{-1}, to obtain

$$S^{-1} A_m S S^{-1} W_{cs} S^{-1} + S^{-1} W_{cs} S^{-1} S A_m^T S^{-1} + B_m B_m^T = 0.$$

Using property (8.15) we simplify the above equation as follows:

$$A_m S^{-1} W_{cs} S^{-1} + S^{-1} W_{cs} S^{-1} A_m^T + B_m B_m^T = 0.$$

Comparing the above equation and (8.16) we find that $W_c = S^{-1} W_{cs} S^{-1}$ or, in other words,

$$W_{cs} = S W_c S. \tag{8.19}$$

Or, because W_c is diagonally dominant, $W_{cs} \cong S^2 W_c$.

Similar results we can obtain for the observability grammians in modal coordinates. Let W_o be an observability grammian obtained for the modal output matrix C_m, and let W_{os} be an observability grammian obtained for the scaled output matrix $C_{ms} = C_m S$ where the scaling matrix S is given above. Similarly to the controllability grammians we can show that the observability grammians are related as follows:

$$W_{os} = S W_o S \cong S^2 W_o. \tag{8.20}$$

The properties (8.18) and (8.20) allow us to scale the input and output matrices in order to obtain the required grammians. In particular, we will use them to obtain the required Hankel singular values. In modal coordinates the grammians are diagonally dominant; thus, the matrix of Hankel singular values (Γ) is obtained as

$$\Gamma \cong (W_c W_o)^{1/2}.$$

Let the input matrix be scaled with matrix S_c and the output matrix with S_o. The Hankel singular values for the scaled system are

$$\Gamma_s \cong (W_{cs} W_{os})^{1/2}.$$

Introducing the scaled grammians from (8.19) and (8.20) to the above equation we obtain

$$\Gamma_s \cong S_c S_o \Gamma, \tag{8.21}$$

since $\Gamma_s \cong (S_c W_c S_c S_o W_o S_o)^{1/2} \cong (S_c^2 S_o^2 W_c W_o)^{1/2} = S_c S_o (W_c W_o)^{1/2} = S_c S_o \Gamma$. In this derivation we used the properties $W_c S_c \cong S_c W_c$ and $W_o S_o \cong S_o W_o$, since S_c and S_o are diagonal and W_c and W_o are diagonally dominant.

Equation (8.21) is used to "shape" the Hankel singular values, or Hankel norms. Denote by Γ the Hankel singular values for the given (or initial) input and output matrices B and C. We require that the system has a new set of Hankel singular values denoted by the diagonal matrix Γ_s. From (8.21) we find that the scaling factors S_c and S_o will be as follows:

$$S_c S_o \cong \Gamma_s \Gamma^{-1}. \tag{8.22}$$

Note that if the system is controllable and observable the Hankel singular values matrix Γ is nonsingular. Now we have a freedom of scaling the input or output matrix, or both. If we decide to scale the input we assume $S_o = I$, if the output we assume $S_c = I$, if both we select S_c and S_o such that (8.22) is satisfied.

The algorithm is summarized as follows: Given a structure state-space representation (A,B,C). If the inputs and outputs locations are not known (matrices B and C) select them arbitrarily. The task is to find a new representation (A, B_n, C_n) such that its Hankel singular values are given by the positive semidefinite and diagonal matrix $\Gamma_s \geq 0$.

1. For a given initial state-space representation of a structure (A,B,C) find the corresponding modal representation (A_m, B_m, C_m). If the modal transformation is $x = Rx_m$, where x are current states and x_m are modal states, then

$$A_m = R^{-1}AR,$$

$$B_m = R^{-1}B,$$

$$C_m = CR.$$

2. Find the Hankel singular values Γ of the representation (A_m, B_m, C_m). This is done by determining the diagonally dominant controllability and observability grammians W_c and W_o from the Lyapunov equations (4.5), and the matrix Γ is obtained as

$$\Gamma \cong (W_c W_o)^{1/2}. \tag{8.23}$$

3. Determine the matrix $\Gamma_s \Gamma^{-1}$.

4. From (8.22) find either S_c or S_o. If we want to shape actuators assume $S_o = I$; thus, $S_c \cong \Gamma_s \Gamma^{-1}$. If we want shape sensors assume $S_c = I$; thus, $S_o \cong \Gamma_s \Gamma^{-1}$.
5. In the case of actuator shaping determine a new input matrix in modal coordinates

$$B_{nm} = S_c B_m = \Gamma_s \Gamma^{-1} B_m, \tag{8.24}$$

and in the case of sensor shaping determine a new output matrix in modal coordinates

$$C_{nm} = C_m S_o = C_m \Gamma_s \Gamma^{-1}. \tag{8.25}$$

6. Determine the new input and output matrices in the original (nodal) coordinates

$$\begin{aligned} B_n &= R B_{nm}, \\ C_n &= C_{nm} R^{-1}. \end{aligned} \tag{8.26}$$

Example 8.5. Consider a clamped beam as in Fig. 1.4 divided into $n = 60$ elements. Determine modal sensors for the first mode, and for the first five modes, that produce responses with amplitudes 1.

We obtained the state matrix A from the beam mass, stiffness, and damping matrices; see (2.35). The system has 177 degrees of freedom, or 354 states. The vertical input force is located at node 24. The preliminary sensor location is the vertical displacement at the same node. The matrix Γ_s is

$$\Gamma_s = \text{diag}([0.5, 0.5, 0, 0, \ldots, 0, 0])$$

in the first case (the amplitude is twice the Hankel singular values). For the second case, we have the following Hankel singular values:

$$\Gamma_s = \text{diag}([0.5, 0.5, 0.5, 0.5, 0.5, 0.5, 0.5, 0.5, 0.5, 0.5, 0, 0, \ldots, 0, 0]).$$

We follow the above algorithm. In the first step we determine the modal representation (A_m, B_m, C_m) and the modal transformation matrix R. For this representation we find the matrix of Hankel singular values by solving the Lyapunov equations (4.5), obtaining grammians W_c and W_o, and matrix Γ is obtained from (8.23). Next, the matrix $\Gamma_s \Gamma^{-1}$ is determined and is

$$\Gamma_s \Gamma^{-1} = \text{diag}(\, 5.62, 5.62, 0, 0, \ldots, 0, 0)$$

in the first case, and

$$\Gamma_s\Gamma^{-1} = \text{diag}(5.62, 5.62, 0.375, 0.375, 59.7, 59.7, 32.8, 32.8, 2882.7, 2830.7,\ 0,\ 0,\ \ldots,\ 0,\ 0)$$

in the second case.

Next we determine the modal sensor matrix C_{nm} from (8.25) and matrix C_n from (8.26). This matrix consists of two parts, displacement and velocity sensors, $C_n = [C_{nq} \quad C_{nv}]$. The velocity part was very small when compared with the displacement part, i.e., $C_{nv} \cong 0$, and the plot of C_{nq} is shown in Fig. 8.9 for the first case and in Fig. 8.10 for the second case. In both cases the horizontal displacements were virtually zero. The plots show vertical displacements. In-plane rotations were nonzero, but neglecting them has not changed the system response. The magnitudes of the transfer functions and impulse responses for Cases 1 and 2 are shown in Figs. 8.11 and 8.12, respectively. The transfer function for the first case shows a single mode excitation, and five equally excited modes in the second case. The impulse response shows a single harmonic excited in the first case, and multiple harmonics excited in the second case.

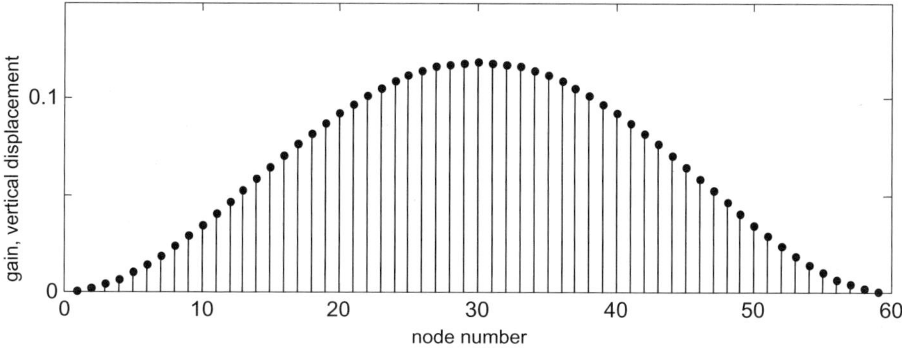

Figure 8.9. Sensor gains for single-mode modal filter.

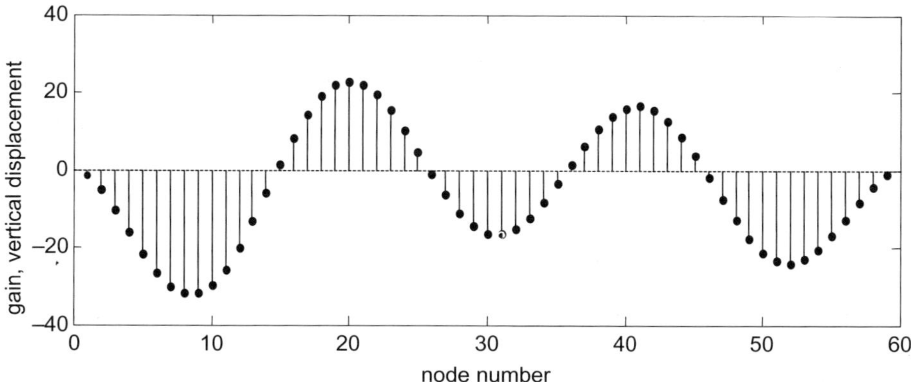

Figure 8.10. Sensor gains for five-mode modal filter.

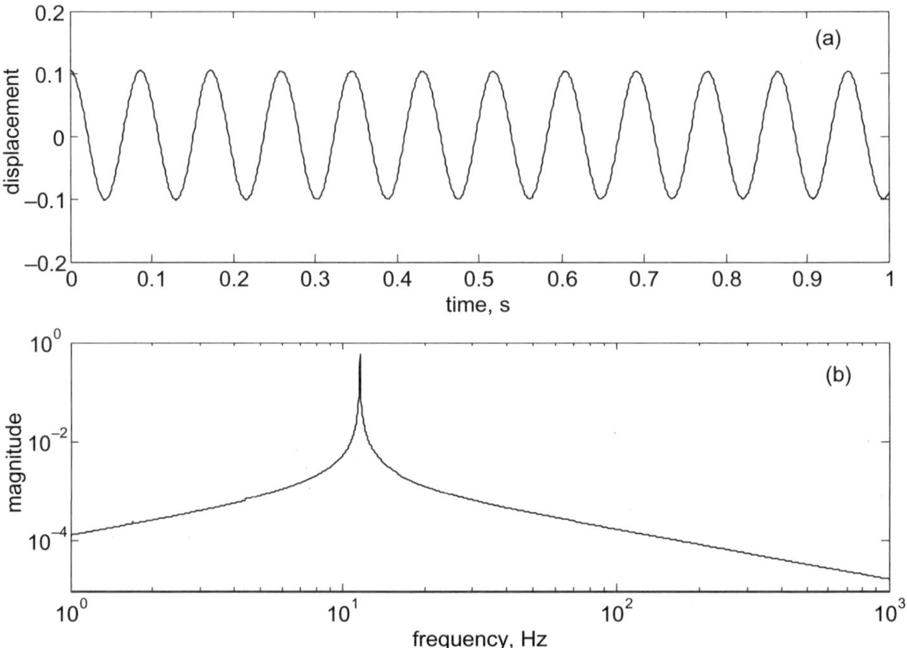

Figure 8.11. Single-mode modal filter: (a) Impulse response is a harmonic motion of natural frequency corresponding to the mode; and (b) magnitude of the transfer function shows a single resonant peak of the corresponding natural frequency.

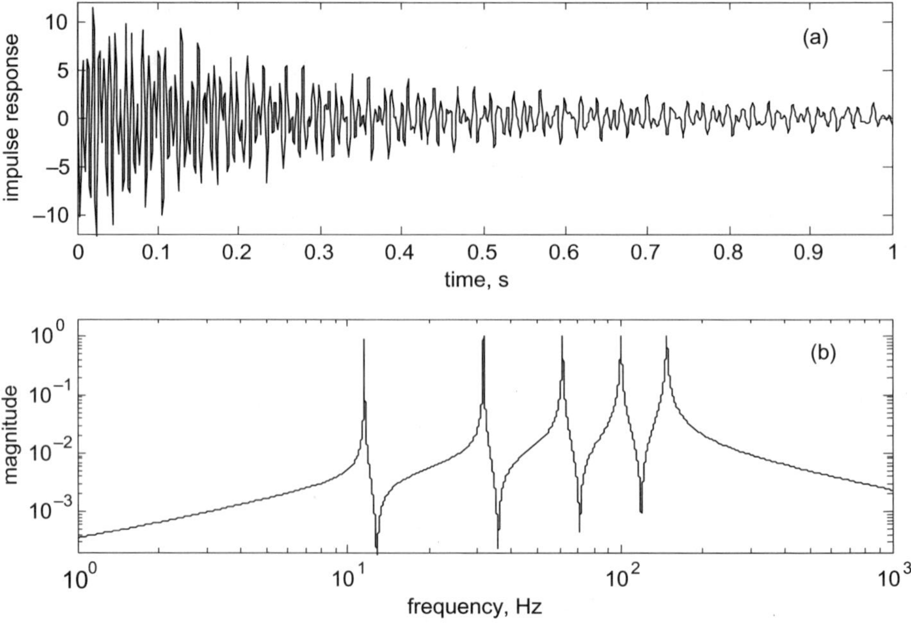

Figure 8.12. Five-mode modal filter: (a) Impulse response is composed of five harmonics; and (b) magnitude of the transfer function shows five resonance peaks.

9
System Identification

how to derive a model from field data

It is a capital mistake to theorize
before one has data.
—*Sherlock Holmes*

The LQG and H_∞ controllers, analyzed later in this book, are model-based ones, i.e., such that the plant model (used as an estimator) is a part of the controller. In this case the performance of the closed-loop system depends on the accuracy of the plant model. The accuracy is defined as a discrepancy between the dynamics of the actual plant and its model. For this reason, analytical models of a plant obtained, for example, from the finite-element model, are inaccurate and are acceptable in the simulation stages only. In implementation the test data are used to determine the accurate plant model—in a procedure known as system identification.

System identification is a fairly developed research field; the reader will find up-to-date identification methods in the comprehensive studies of Natke [111], Ljung [102], Juang [84], and Ewins [33], and get a good insight into the problem. Among the many identification procedures available, we describe here only one—the Eigensystem Realization Algorithm (ERA)—which gives the balanced (close to modal) state-space representation. The advantage of the ERA algorithm is that it does not require parametrization (the performance of various identification algorithms depends on the number of parameters to be identified which depends, in turn, on how the system model is represented). In addition, the modal/balanced representation gives an immediate answer to the question of the order of the identified system, as discussed in Chapter 6. The problem of system order in the identification procedure is an important one: for a structural model of too low order a significant part of the plant dynamics is missing; this may cause closed-loop instability due to spillover. A system of too high order, on the other hand, contributes to controller complexity, and may introduce unwanted dynamics and deteriorate the closed-loop system performance.

The ERA system identification is based on the realization method of Ho and Kalman [72]. This approach, developed by Juang into the ERA method, is widely used in flexible structure identification. The ERA method is described in [84]. The presentation below is based on derivations given in [84], [60], and [61].

9.1 Discrete-Time Systems

It is a common feature nowadays that for system identification purposes the input and output signals are recorded digitally—as discrete-time signals. For this reason, we depart in this chapter from the continuous-time models, and use the discrete-time state-space representation. The sampling time of the recorded signals is denoted by Δt, and the signal u at time $i\Delta t$ is denoted as u_i. With this notation the discrete-time state-space representation is given by the following difference equations; see (3.46):

$$\begin{aligned} x_{i+1} &= Ax_i + Bu_i, \\ y_i &= Cx_i + Du_i. \end{aligned} \tag{9.1}$$

For a system with s inputs and r outputs denote the controllability matrix of order p, $\mathcal{C}_p$ (also called the reachability matrix in the discrete-time case), and the observability matrix of order p, $\mathcal{O}_p$, see (4.9) and (4.13),

$$\mathcal{C}_p = \begin{bmatrix} B & AB & \cdots & A^{p-1}B \end{bmatrix}, \qquad \mathcal{O}_p = \begin{bmatrix} C \\ CA \\ \vdots \\ CA^{p-1} \end{bmatrix}, \quad \text{where } p \geq \max(s,r) \tag{9.2}$$

These matrices are of dimensions $n \times (s \times p)$ and $(r \times p) \times n$, respectively, where n is the assumed system order. It is also assumed that $s \times p > n$, and $r \times p > n$. In practice, the sizes $s \times p$ and $r \times p$ of these matrices are much larger than the system order n,

$$s \times p \gg n \qquad \text{and} \qquad r \times p \gg n, \tag{9.3}$$

in order to minimize the identification error caused by the measurement noise.

The controllability grammian, $W_c(p)$, over the time interval $T = [0 \; p\Delta t]$, and the observability grammian, $W_o(p)$, over the same interval, are defined as follows:

$$W_c(p) = \mathcal{C}_p \mathcal{C}_p^T \qquad \text{and} \qquad W_o(p) = \mathcal{O}_p \mathcal{O}_p^T. \tag{9.4}$$

9.2 Markov Parameters

Consider the impulse response of a system described by (9.1) and assume the initial condition zero, $x_0 = 0$. We describe the impulse input by the following sequence: $u_0 = 1$ and $u_i = 0$ for $i = 1,2,3,4,\ldots$. For this input the solution of (9.1) is

$$\begin{aligned} x_0 &= 0, \\ y_0 &= Du_0 = D, \\ x_1 &= Ax_0 + Bu_0 = B, \\ y_1 &= Cx_1 + Du_1 = CB, \\ x_2 &= Ax_1 + Bu_1 = AB, \\ y_2 &= Cx_2 + Du_2 = CAB, \\ x_3 &= Ax_2 + Bu_2 = A^2B, \\ y_3 &= Cx_3 + Du_3 = CA^2B, \\ &\vdots \\ x_k &= Ax_{k-1} + Bu_k = A^{k-1}B, \\ y_k &= Cx_k + Du_k = CA^{k-1}B. \end{aligned} \tag{9.5}$$

We see that at the moment $t = k\Delta t$ and the impulse response is $y_k = CA^{k-1}B$. The matrices $h_k = CA^{k-1}B$, $k = 1,2,3,\ldots$, are known as the Markov parameters of a system. They will be used for the identification of a structural model, since the system matrices *A, B, C,* and *D* are implanted into the Markov parameter sequence.

9.3 Identification Algorithm

The presented algorithm is based on the measured impulse responses of a system, and derived from the Markov and Hankel matrices. The matrices h_k, of dimension $r \times s$, $k = 0,1,2,\ldots$, such that

$$h_k = CA^kB, \tag{9.6}$$

are called Markov matrices or Markov parameters. For the discrete-time systems they have the following physical interpretation: the *i*th column of the *k*th Markov matrix h_k represents the impulse response at the time $k\Delta t$ with a unit impulse at the *i*th input. Thus, in many cases, the Markov matrices can be directly measured or obtained from the input–output time records (see the next section) and, therefore, are often used in system identification.

The base for the identification algorithm is the Hankel matrix, H_1, and the shifted Hankel matrix, H_2, which are defined as follows:

$$H_1 = \begin{bmatrix} h_1 & h_2 & \cdots & h_p \\ h_2 & h_3 & \cdots & h_{p+1} \\ \cdots & \cdots & \cdots & \cdots \\ \cdots & \cdots & \cdots & \cdots \\ h_p & h_{p+1} & \cdots & h_{2p-1} \end{bmatrix} \quad \text{and} \quad H_2 = \begin{bmatrix} h_2 & h_3 & \cdots & h_{p+1} \\ h_3 & h_4 & \cdots & h_{p+2} \\ \cdots & \cdots & \cdots & \cdots \\ \cdots & \cdots & \cdots & \cdots \\ h_{p+1} & h_{p+2} & \cdots & h_{2p} \end{bmatrix}. \tag{9.7}$$

Their dimensions are $(r \times p) \times (s \times p)$. It is easy to note that the Hankel matrix and the shifted Hankel matrix are obtained from the controllability and observability matrices, namely,

$$\begin{aligned} H_1 &= \mathcal{O}_p \mathcal{C}_p, \\ H_2 &= \mathcal{O}_p A \mathcal{C}_p. \end{aligned} \tag{9.8}$$

These matrices do not depend on system coordinates. Indeed, let the new representation, x_n, be a linear combination of the representation x, i.e., $x_n = Rx$, then $A_n = RAR^{-1}$, $\mathcal{C}_{np} = R\mathcal{C}_p$, and $\mathcal{O}_{np} = \mathcal{O}_p R^{-1}$; therefore, the Hankel matrices in the new coordinates are the same as in the original ones, since

$$\begin{aligned} H_{n1} &= \mathcal{O}_{np}\mathcal{C}_{np} = \mathcal{O}_p R^{-1} R \mathcal{C}_p = \mathcal{O}_p \mathcal{C}_p = H_1, \\ H_{n2} &= \mathcal{O}_{np} A_n \mathcal{C}_{np} = \mathcal{O}_p R^{-1} R A R^{-1} R \mathcal{C}_p = \mathcal{O}_p A \mathcal{C}_p = H_2. \end{aligned}$$

In the identification algorithm we do not know, of course, the controllability or observability matrices, $\mathcal{C}_p$ and $\mathcal{O}_p$. However, we have access to the measured impulse responses; consequently, the Hankel matrices, H_1 and H_2, are known. The basic idea in the identification procedure given below is to decompose H_1 similarly to the first equation (9.8), as in [84], [60], and [61],

$$H_1 = PQ. \tag{9.9}$$

The obtained matrices Q and P serve as the new controllability and observability matrices of the system. Therefore, replacing $\mathcal{C}_p$ and $\mathcal{O}_p$ in the second equation (9.8) with Q and P, respectively, one obtains

$$H_2 = PAQ. \tag{9.10}$$

But, if matrices P and Q are full rank, we obtain the system matrix A from the last equation as

$$A = P^+ H_2 Q^+, \tag{9.11}$$

where P^+ and Q^+ are the pseudoinverses of P and Q, respectively,

$$\begin{aligned} P^+ &= (P^T P)^{-1} P^T, \\ Q^+ &= Q^T (QQ^T)^{-1}, \end{aligned} \tag{9.12}$$

such that

$$P^+ P = I \qquad \text{and} \qquad QQ^+ = I. \tag{9.13}$$

Having determined A, the matrix B of the state-space representation (A,B,C,D) is easily found as the first s columns of Q (this follows from the definition of the controllability matrix, (9.2)); therefore,

$$B = QE_s, \qquad \text{where} \qquad E_s = \begin{bmatrix} I_s & 0 & \cdots & 0 \end{bmatrix}^T. \tag{9.14}$$

Similarly, the first r rows of P give the output matrix C,

$$C = E_r^T P, \qquad \text{where} \qquad E_r = \begin{bmatrix} I_r & 0 & \cdots & 0 \end{bmatrix}^T. \tag{9.15}$$

The determination of the feed-through matrix D we will explain later.

The decomposition (9.9) of the Hankel matrix H_1 is not unique. It could be, for example, the Cholesky, LU, or QR decompositions. However, using the singular value decomposition we obtain the identified state-space model (A,B,C,D) in the balanced representation. Indeed, denote the singular value decomposition of the Hankel matrix H_1 as follows:

$$H_1 = V \Gamma^2 U^T, \tag{9.16}$$

where

$$\begin{aligned} &\Gamma = \operatorname{diag}(\gamma_1, \gamma_2, \ldots, \gamma_m), \\ &UU^T = I \qquad \text{and} \qquad V^T V = I, \end{aligned} \tag{9.17}$$

and $m \le \min(s \times p, r \times p)$. Comparing (9.9) and (9.16), we obtain the controllability and observability matrices in the form

$$Q = \Gamma U^T \qquad \text{and} \qquad P = V\Gamma;$$

hence, their pseudoinverses, in this case, are as follows:

$$Q^{+} = U\Gamma^{-1} \qquad \text{and} \qquad P^{+} = \Gamma^{-1}V^{T}.$$

But, from (9.4), it follows that

$$\begin{aligned} W_c(p) &= QQ^T = \Gamma^2, \\ W_o(p) &= P^T P = \Gamma^2, \end{aligned} \tag{9.18}$$

i.e., that the controllability and observability grammians over the time interval $T = [0 \;\; p\Delta t]$ are equal and diagonal, hence the system is balanced over the interval T. This fact has a practical meaning: the states of the identified system are equally controlled and observed. But weakly observed and weakly controlled states can be ignored, since they do not contribute significantly to the system dynamics. They are usually below the level of measurement noise. Thus, using the singular value decomposition of the Hankel matrix H_1 one can readily determine the order of the identified state-space representation (A,B,C,D): the states with small singular values can be truncated. Of course, the measurement and system noises have an impact on the Hankel singular values, and this problem is analyzed in [84].

Beside the noise impact on the identification accuracy, one has to carefully determine the sampling time Δt: it should be small enough to include the system bandwidth, but not too small, in order to ease the computational burden. The size of the record p should satisfy the conditions of (9.3). The procedure identifies the model such that the model response fits the plant response for the time $T = [0 \;\; p\Delta t]$. Thus, too-short records can produce a model which response fits the plant response within the time segment T, and departs outside T, which makes the model unstable. This we illustrate in the following examples.

9.4 Determining Markov Parameters

From measurements one obtains the input and output time histories, rather than the Markov parameters themselves (the exceptions are impulse response measurements). However, the above presented algorithm identifies the state-space representation from the Hankel matrices, which are composed of Markov parameters. Therefore, in this section we describe how to obtain the Markov parameters from the input and output measurements.

In order to do this, denote the Markov matrix H that contains $p+1$ Markov parameters

$$H = \begin{bmatrix} D & CB & CAB & \cdots & CA^{p-1}B \end{bmatrix} = \begin{bmatrix} h_o & h_1 & h_2 & \cdots & h_p \end{bmatrix}. \tag{9.19}$$

Denote also the output measurement matrix Y,

$$Y = \begin{bmatrix} y_o & y_1 & y_2 & \cdots & y_q \end{bmatrix}, \tag{9.20}$$

and the input measurement matrix in the following form:

$$U = \begin{bmatrix} u_o & u_1 & u_2 & \cdots & u_p & \ldots & u_q \\ 0 & u_o & u_1 & \cdots & u_{p-1} & \cdots & u_{q-1} \\ 0 & 0 & u_o & \cdots & u_{p-2} & \cdots & u_{q-2} \\ \cdots & \cdots & \cdots & \cdots & \cdots & \cdots & \cdots \\ 0 & 0 & 0 & \cdots & u_o & \cdots & u_{q-p} \end{bmatrix}. \tag{9.21}$$

We show below (using the state equations (9.1)) that the relationship between the H, Y, and U matrices is as follows:

$$Y = HU. \tag{9.22}$$

Namely, for $t = 0,\ \Delta t,\ 2\Delta t,\ \ldots,$ and zero initial conditions $(x_o = 0)$ we obtain

$$\begin{aligned} y_o &= Du_o, \\ x_1 &= Bu_o, \\ y_1 &= Cx_1 + Du_1 = CBu_o + Du_1, \\ x_2 &= Ax_1 + Bu_1 = ABu_o + Bu_1, \\ y_2 &= Cx_2 + Du_2 = CABu_o + CBu_1 + Du_2, \\ x_3 &= Ax_2 + Bu_2 = A^2Bu_o + ABu_1 + Bu_2, \\ y_3 &= Cx_3 + Du_3 = CA^2Bu_o + CABu_1 + CBu_2 + Du_3. \end{aligned}$$

Continuing for $i = 3,4,\ldots,q,\ q \geq p$, we combine equations for y_i into the system of equations (9.22). It was also assumed that for sufficiently large enough p one obtained $A^p \cong 0$.

If the matrix U is of full rank (enough data samples are collected so that there are more independent equations than unknowns) the solution of (9.22) is as follows:

$$H = YU^+, \qquad \text{where} \qquad U^+ = U^T(UU^T)^{-1}. \tag{9.23}$$

Matrix M contains all the Markov parameters necessary for the system identification procedure, and the first component is the feed-through matrix D.

For noisy input and output data we determine the Markov parameters using the averaging, or correlation, matrices as follows. By right-multiplying (9.22) by U^T and averaging, we obtain

$$\mathrm{E}(YU^T) = H\,\mathrm{E}(UU^T), \tag{9.24}$$

where E(.) is an averaging operator. However, $R_{yu} = \mathrm{E}(YU^T)$ is the correlation matrix between the input and output, and $R_{uu} = \mathrm{E}(UU^T)$ is the autocorrelation matrix of the input. Thus, the above equation now reads as follows:

$$R_{yu} = HR_{uu}. \tag{9.25}$$

If R_{uu} is nonsingular, then

$$H = R_{yu}R_{uu}^{-1}; \tag{9.26}$$

otherwise,

$$H = R_{yu}R_{uu}^{+}. \tag{9.27}$$

Let the data records Y and U be divided into N segments, Y_i and U_i, $i=1,2,\ldots,N$, then we obtain the correlation matrices as follows:

$$R_{yu} = \frac{1}{N}\sum_{i=1}^{N} Y_i U_i^T \qquad \text{and} \qquad R_{uu} = \frac{1}{N}\sum_{i=1}^{N} U_i U_i^T. \tag{9.28}$$

The difference between (9.22) and (9.25) lies in the fact that the first equation uses raw data while the second uses averaged (smoothed) data, and that the size of the matrix to be inverted is much larger in (9.22) than in (9.25). Indeed, for s inputs, p Markov parameters, and q data samples (note that $q \gg s$, and $q \gg p$) the size of a matrix to be inverted in the first case (U) is $sp \times q$, while in the second case (R_{uu}) it is $sp \times sp$.

9.5 Examples

In this section we perform the identification of the models of a simple structure (in order to illustrate the method in a straightforward manner); the 2D truss—a more complicated structure, and the Deep Space Network antenna where the model is identified from the available field data.

9.5.1 A Simple Structure

The Matlab code for this example is in Appendix B. Analyze a simple system with $k_1 = 10$, $k_2 = 50$, $k_3 = 50$, $k_4 = 10$, $m_1 = m_2 = m_3 = 1$, and with proportional damping matrix, $D = 0.005K + 0.1M$. The input is applied to the third mass and the output is the velocity of this mass. Identify the system state-space representation using the step response. The sampling time is $\Delta t = 0.1$ s.

We apply the unit step force at mass 3 at $t = 0.1$ s. We measure the velocity of mass 3 and its plot is shown in Fig. 9.1(a). First, we determine the Markov parameters from (9.23) using matrices U and Y. Matrix U is defined in (9.21). Since the input is the unit step, its entries are as follows: $u_0 = 0$, $u_1 = 1$, $u_2 = 1$, $u_3 = 1$, etc. We measured 300 samples; thus, $q = 300$ in (9.21). The matrix Y is composed of the output measurements, as defined in (9.20). We would like to determine 30 Markov parameters; thus, $p = 30$ in (9.20). The solution of (9.23) gives the Markov parameters, as plotted in Fig. 9.1(b).

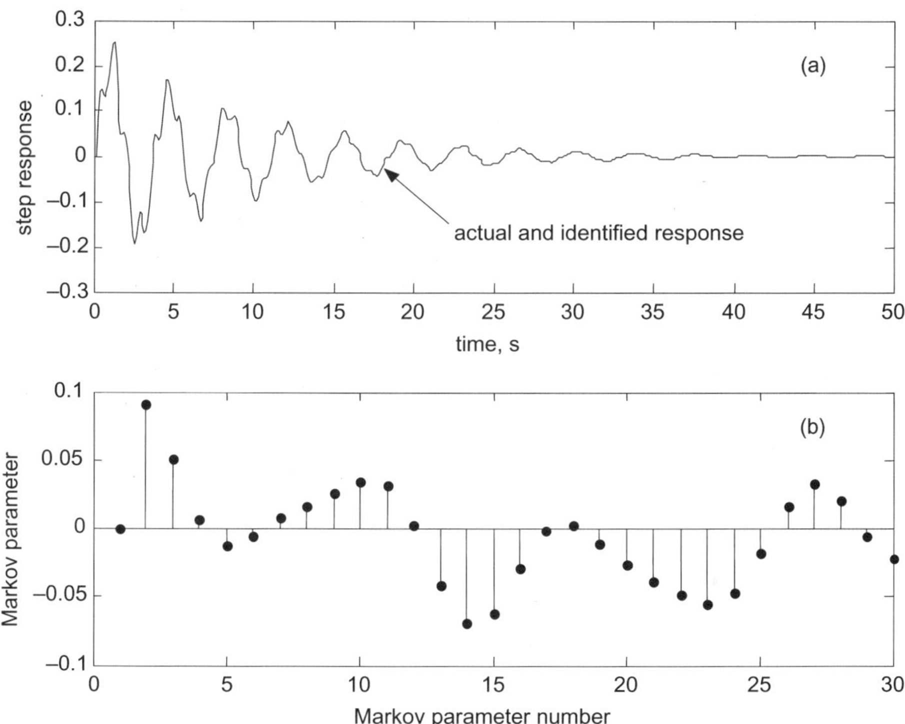

Figure 9.1. Simple system test: (a) Velocity of mass 3 due to step force at the same mass is used to identify the system state-space representation; and (b) Markov parameters of the system that correspond to the step response.

Having calculated the Markov parameters we determine the state-space representation of the system. Note first that the feed-through matrix D is zero, since the first Markov parameter is zero. Second, we form the Hankel matrices H_1 and H_2 from the already determined Markov parameters, as in (9.7). We decompose the matrix H_1 into P and Q matrices, as in (9.9). Next, we obtain the state matrix A from (9.11), the input matrix B from (9.14) (or as a first column of Q), and the output matrix C from (9.15) (or as a first row of P).

The remaining problem is the order of the matrix A. It can be assumed comparatively large, but by doing this the noise dynamics can be included in the model. In order to determine the minimal order of the state-space representation we use the singular value decomposition of the Hankel matrix H_1; see (9.16). The singular values of H_1 are the Hankel norms of the system that denote the importance of each state. The plots of the Hankel singular values of H_1 are shown in Fig. 9.2. It can be seen from the plot that the Hankel singular values for the first six states are nonzero, and the remaining Hankel singular values are zero. Thus, the minimal order of the identified state-space representation is 6.

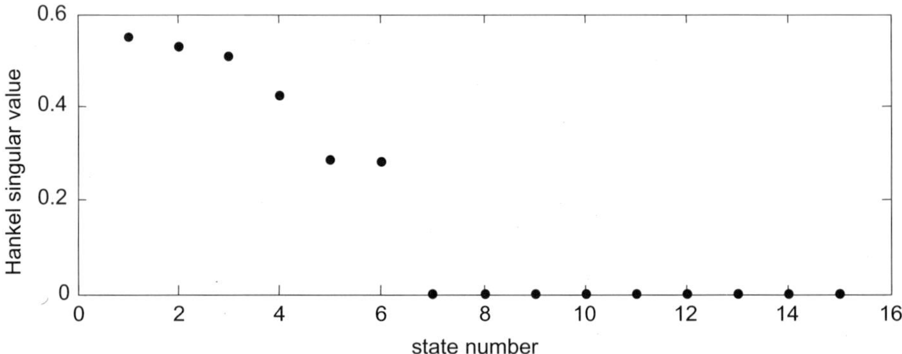

Figure 9.2. Hankel singular values of the identified model of the simple system.

The obtained model is a form of a discrete-time state-space representation. The final step is to convert it to the continuous-time representation, which is as follows:

$$A = \left[\begin{array}{cc:cc:cc} -1.4592 & 2.9565 & -6.1662 & 0.4949 & 0.7802 & 0.8767 \\ -2.9565 & 0.0371 & 0.1709 & 0.2999 & -0.3370 & -0.0568 \\ \hdashline 6.1662 & 0.1709 & 0.6390 & 3.5649 & -0.5755 & -0.1282 \\ 0.4949 & -0.2999 & -3.5649 & 0.0840 & -0.6860 & -1.0312 \\ \hdashline 0.7802 & 0.3370 & 0.5755 & -0.6860 & -0.0116 & -12.2759 \\ -0.8767 & -0.0568 & -0.1282 & 1.0312 & 12.2759 & -0.0991 \end{array}\right],$$

$$B = \left[\begin{array}{c} -2.9880 \\ -0.2581 \\ \hdashline 0.6855 \\ 0.9733 \\ \hdashline 0.8774 \\ -0.6380 \end{array}\right],$$

$$C = \left[\begin{array}{cc:cc:cc} -0.2779 & -0.0168 & 0.0065 & 0.0850 & 0.0831 & -0.0083 \end{array}\right].$$

We simulated the step response of the obtained system. The "measured" (used for system identification) and simulated responses are compared in Fig. 9.1(a). They virtually overlap.

The above was an example of ideal—or nonnoisy—measurements. Since all measurements contain noise, let us consider a situation where the measured step response is corrupted with an additive white noise of standard deviation of 0.003. This measurement is shown in Fig. 9.3(a). Again, we obtain the Markov parameters, see Fig.9.3b, which are slightly different from those in the nonnoisy case. From the Markov parameters we obtain the Hankel matrices H_1 and H_2, and determine the state-space representation. The Hankel singular values of H_1 are shown in Fig. 9.4. Now we see that the zero-valued Hankel singular values from the previous case become nonzero. Still they are small enough to determine that the system order is 6. However, larger measurement noise will cause even larger singular values corresponding to the noise, and makes it impossible to distinguish between "system" states and "noise" states. The step response of the identified model is plotted in Fig. 9.3(a), showing good coincidence with the measurements.

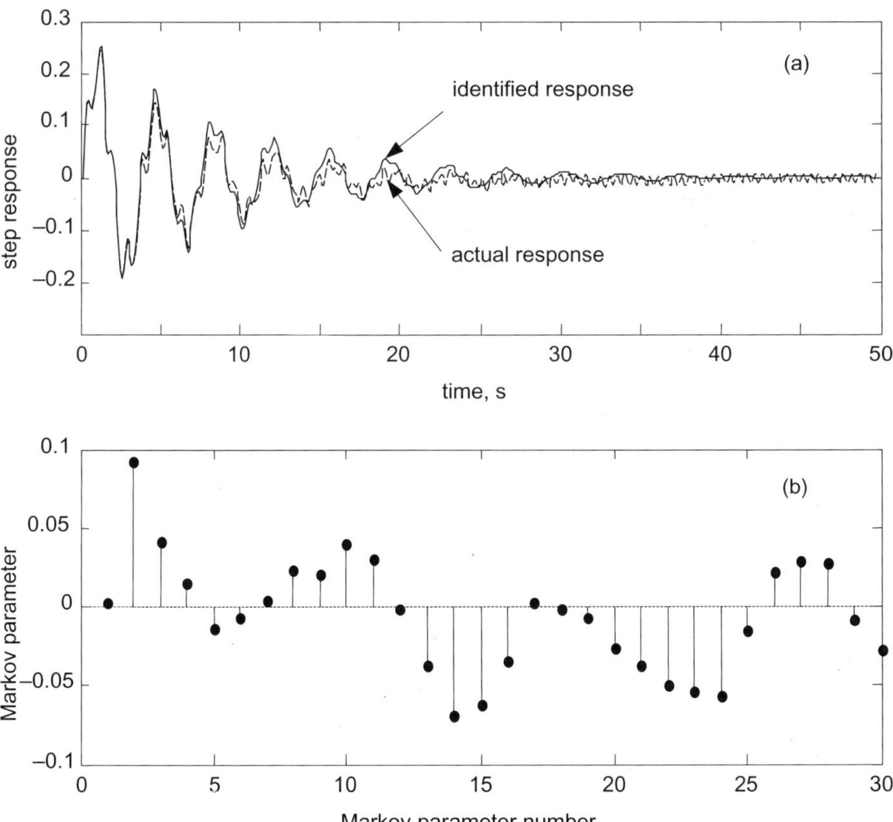

Figure 9.3. (a) Noisy velocity measurements of mass 3 and the response of the identified system; and (b) Markov parameters obtained from the noisy response.

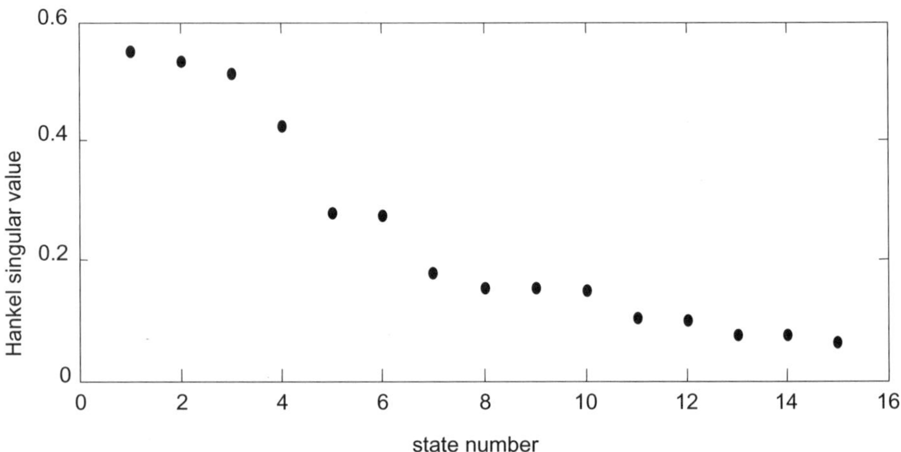

Figure 9.4. Hankel singular values of the identified model of the noisy system.

9.5.2 The 2D Truss

Analyze the 2D truss from Fig. 1.2, with proportional damping matrix, $D=10^{-5}K+3M$. The input force is applied at node 10, in the vertical direction. The output is the velocity of this node in the vertical direction. Identify the system state-space representation using the step response. The sampling time is $\Delta t = 0.001$ s.

We measure the velocity of node 10 and its plot is shown in Fig. 9.5(a). Next, we determine the Markov parameters from (9.23) using matrices U and Y. Matrix U is defined in (9.21). Since the input is the unit step, its entries are as follows: $u_0 = 0$, $u_1 = 1$, $u_2 = 1$, $u_3 = 1$, ... We measured the input and output for 4 s; thus, 4000 samples were gathered and $q = 4000$ in (9.21). The matrix Y is composed of the output measurements, as defined in (9.20). We would like to determine 160 Markov parameters; thus, $p = 160$ in (9.20). The solution of (9.23) gives the Markov parameters, as plotted in Fig. 9.5(b).

Having calculated the Markov parameters we determine the state-space representation of the system from (9.11), (9.14), and (9.15). After determining A, B, and C the minimal order of the matrix A shall be determined using the singular value decomposition of the Hankel matrix H_1. The plots of the Hankel singular values are shown in Fig. 9.6. It can be seen from the plot that the Hankel singular values for the first six states are larger than 0.05, and the remaining Hankel singular values are small (smaller than 0.05). Thus, the minimal order of the identified state-space representation is 6. The plots of the overlapped actual step response, and the identified reduced-order model are shown in Fig. 9.5(a). The magnitudes of the transfer function of the actual and identified reduced-order models are shown in Fig. 9.7, showing close coincidence.

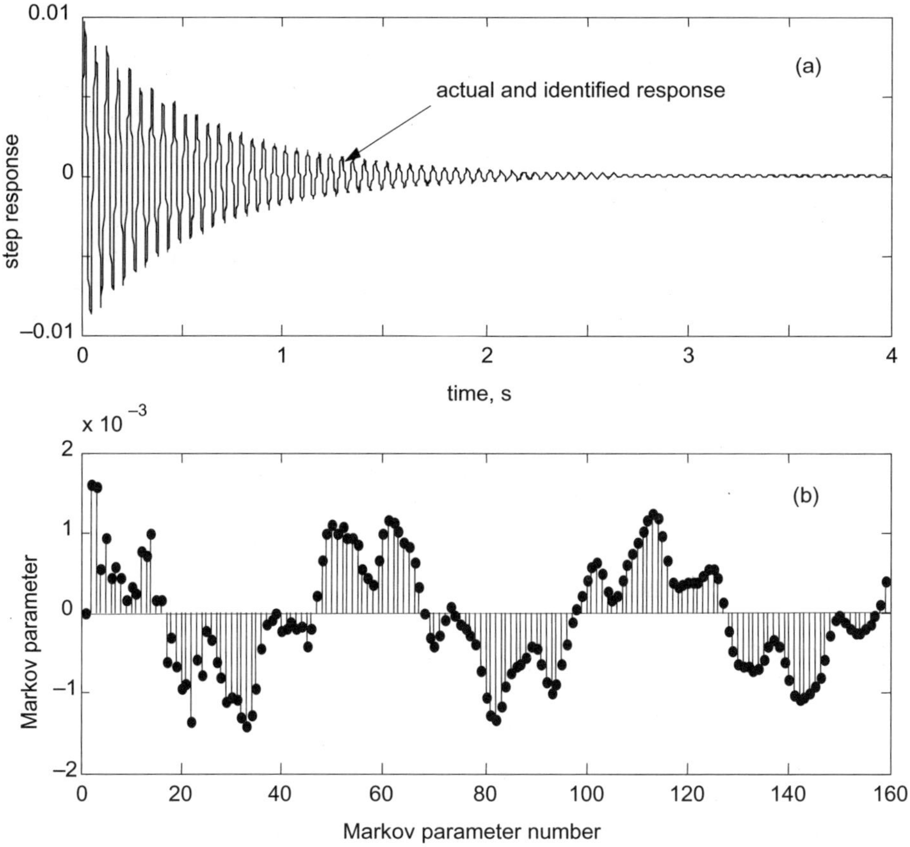

Figure 9.5. Truss identification: (a) Velocity of node 10 used in system identification; and (b) the Markov parameters obtained from the response.

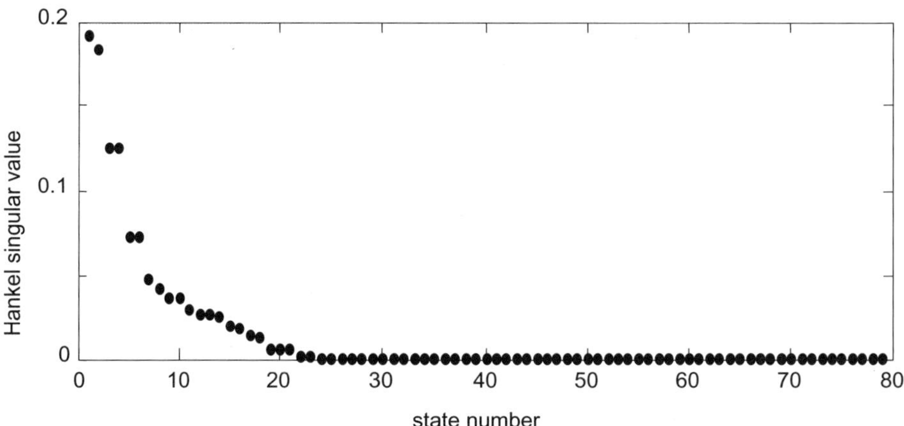

Figure 9.6. Hankel singular values of the identified truss model.

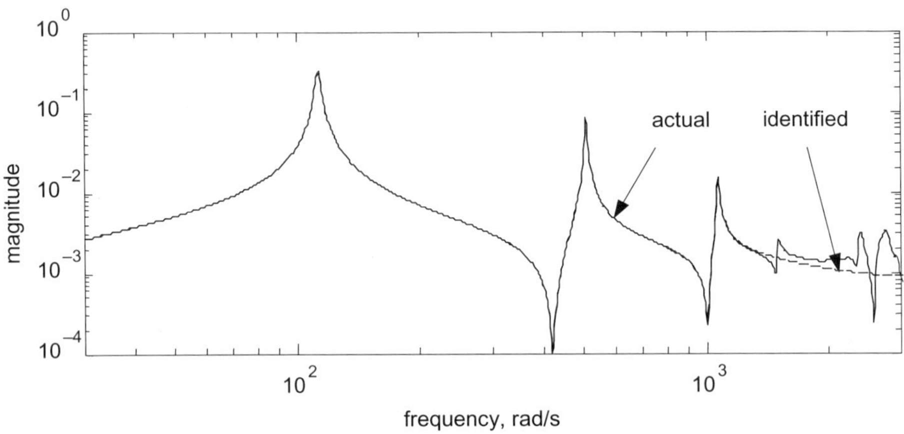

Figure 9.7. The magnitudes of the truss transfer function: Actual and identified reduced-order models.

9.5.3 The Deep Space Network Antenna

We describe the determination of the antenna open-loop model from the field test data. The block diagram of the antenna open-loop system is given in Fig. 1.6. It consists of two inputs (azimuth and elevation rate commands), and four outputs (azimuth and elevation encoders, and elevation and cross-elevation pointing errors). We describe the determination of the model between the elevation rate command and elevation encoder. The input signal is white noise (a voltage proportional to the rate command) sampled at the rate 30.6 Hz, collecting 20,000 samples, which means that the actual test took about 11 min. The noise is shown in Fig. 9.8(a), and its first 10 s in Fig. 9.8(b). The antenna response at the elevation encoder is shown in Fig. 9.9(a), and its first 10 s in Fig. 9.9(b). Note that the output is much slower than the input.

From the measured input (u) and output (y) signals we determine the transfer function using the Matlab function $p = \text{spectrum}(u,y,nn)$. The integer $nn = 2048$ is the length of a section that the input and output signals are divided by, in order to smooth the transfer function. The plot of the magnitude of the transfer function is shown in Fig. 9.10 (dotted line).

Next, from the measured input and output signals we identify the state-space model of the antenna using computer program OKID [85], [84], that uses the ERA algorithm. Since the order of the model is not known off-hand, we select a significantly high-order model, of order 41. For this model we determine the Hankel singular values, which are shown in Fig. 9.11. We evaluated that the Hankel singular values for states 16 and larger are small enough to neglect these states. The reduced model thus consists of 15 states. The transfer function of the reduced model is shown in Fig. 9.10 (solid line). The figure shows close coincidence between the measured and identified transfer functions.

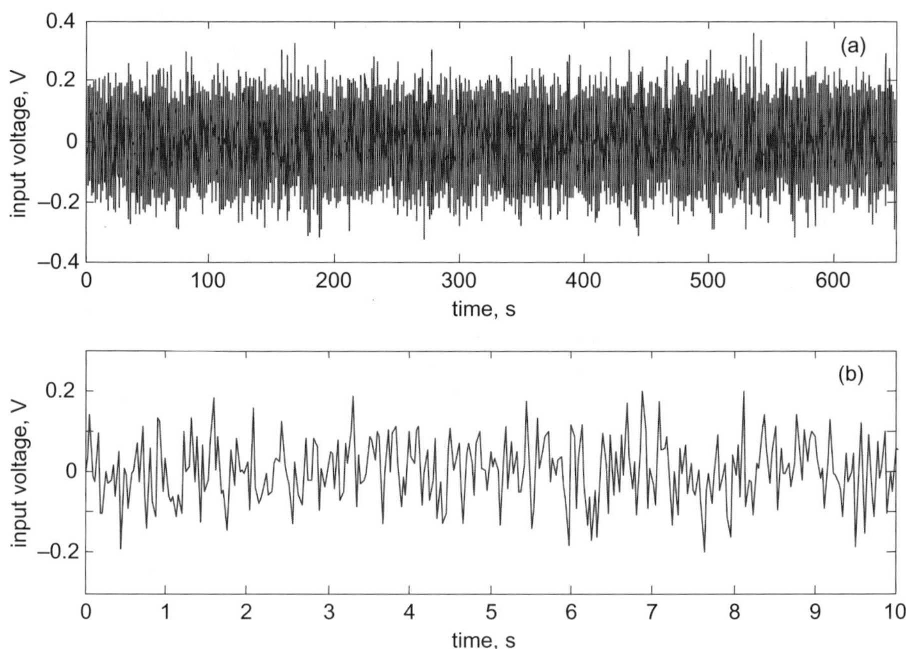

Figure 9.8. The white noise input to the antenna: (a) Full record; and (b) 10 s sample.

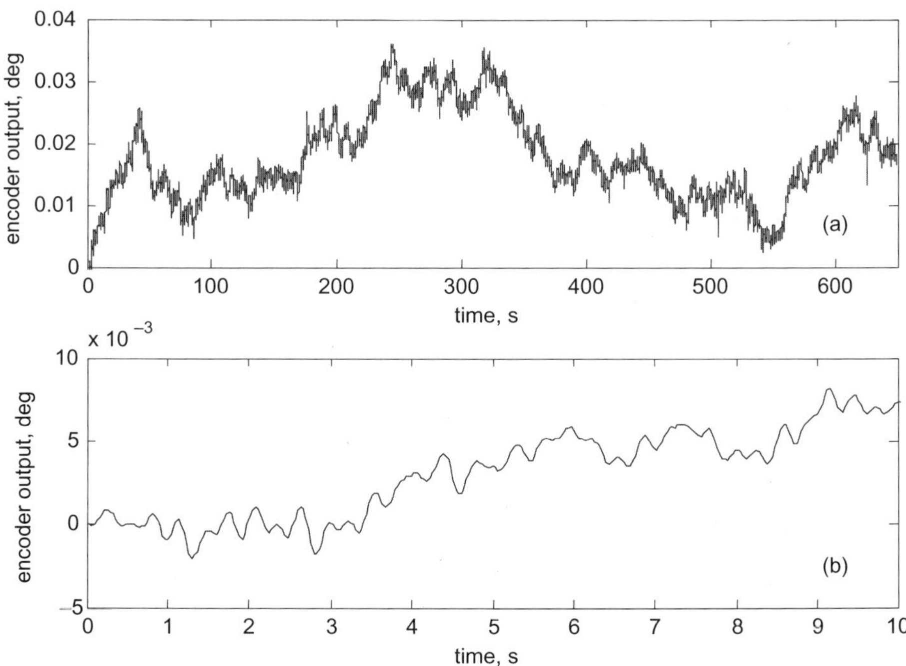

Figure 9.9. The antenna encoder response to the white noise input: (a) Full record; and (b) 10 s sample showing low-frequency vibrations.

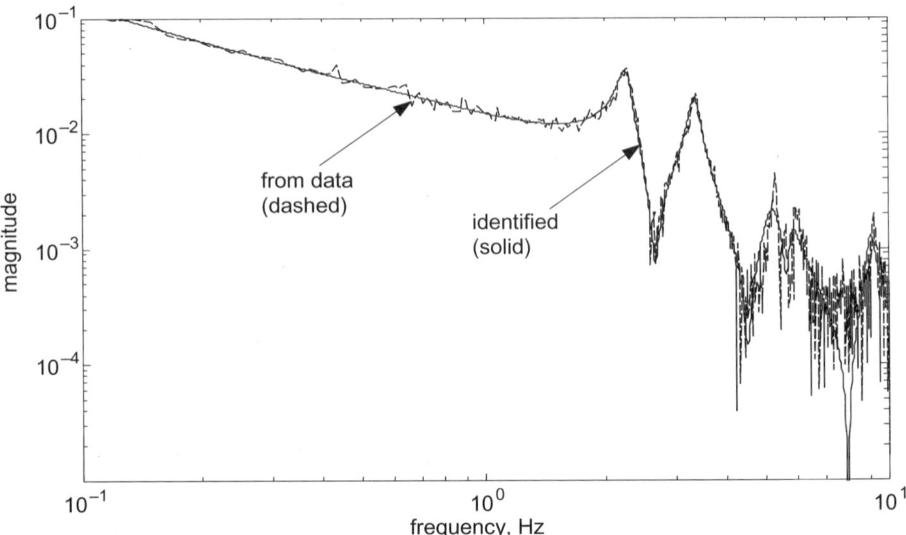

Figure 9.10. Magnitudes of the antenna transfer function: Obtained from field data (dashed line), and from the identified model (solid line).

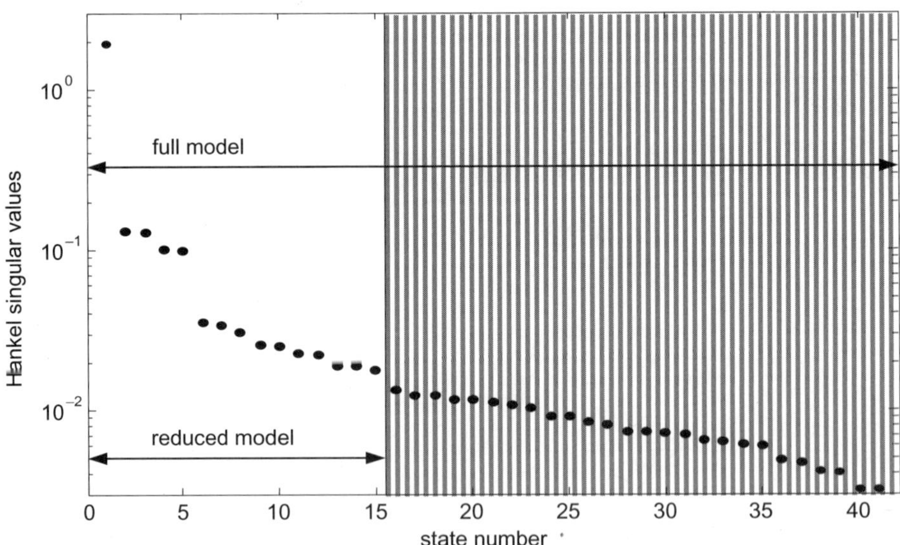

Figure 9.11. Hankel singular values of the full and reduced antenna models.

10
Collocated Controllers

how to take the first step in structural control

Who cares how it works,
just as long as it gives the right answer?
—*Jeff Scholnik*

Collocated controllers have their sensors collocated with actuators. They are a special case of the dissipative controllers, which are designed based on the passivity principle. In this book we select the collocated controllers as a first step to controller design, since they are simple, always stable, and some of their properties are similar to the more advanced controllers described later in this book. A good introduction to the collocated control of structures—but from a different point of view—can be found in the book by Preumont [120].

The most direct approach to controller design is to implement a proportional gain between the input and output. This approach, although simple, seldom gives a superior performance, since the performance enhancement is tied to the reduction of the stability margin. However, if some conditions are satisfied, a special type of proportional controller is obtained—a dissipative one. As stated by Joshi [83, p. 45] "the stability of dissipative controllers is guaranteed regardless of the number of modes controlled (or even modeled), and regardless of parameter errors." Therefore, for safety reasons, they are the most convenient candidates for implementation. However, the simplicity of the control law does not simplify the design. For example, in order to obtain the required performance a multi-input–multi-output controller with a large number of inputs and outputs has to be designed. Determining the gains for this controller is not an obvious task. In this chapter we investigate the properties of the collocated controllers, and show how to design collocated controllers for flexible structures in order to meet certain objectives.

10.1 A Low-Authority Controller

In the following we distinguish between the low- and high-authority controllers. This distinction allows us to design controllers that significantly suppress the flexible vibrations of structures (which is done by the low-authority controller), and to follow a command precisely (which is done by the high-authority controller).

The control forces that act on a structure can be divided into tracking forces and damping forces. The tracking forces move the structure to follow a target and the damping forces act on the structure to suppress vibrations. Typically, the tracking forces are significantly larger than the damping forces. For this reason a structural controller can be divided into low- and high-authority controllers. The low-authority controller is the one that uses a limited input (torque, force) to control the vibrations of a system. In the case of flexible structures the limited input introduces additional damping to the system. This action does not considerably influence the global motion of the entire structure, which requires powerful actuators of the high-authority controllers. Accordingly, the control system action on a flexible structure can be separated into two stages: stage one, when damping is added to a structure and vibrations are suppressed showing faster decay; and stage two, of "total" system motion where the damping is little affected.

In the frequency domain, the first stage is characterized by the suppression of the resonance peaks, while the off-resonance transfer function is little affected; see Fig. 10.1 for a simple structure example. Further increase of gains increases significantly the control input; see Fig. 10.2. The input however is limited due to physical constraints, and this feature may explain the usefulness of the low-authority controller for structures: using limited, or small input power, it can efficiently control the vibrations.

Another look at the low-authority controller is to observe the root-locus. The feedback gains move the structural poles. A typical root-locus pattern for a structure is shown in Fig. 10.3. The poles for low controller gains move in a horizontal pattern, i.e., the control gains mostly impact the real part of the poles. Comparing the root location in Fig. 2.1 we see that the structural damping increases, while the natural frequencies are not impacted. For higher gains, however, the root-locus drifts from the horizontal pattern, and the natural frequencies change significantly. The first phase of horizontal movement is caused by the gains of a low-authority controller.

The impact of low-authority controller configuration on the control system is analyzed further in this and the following chapters.

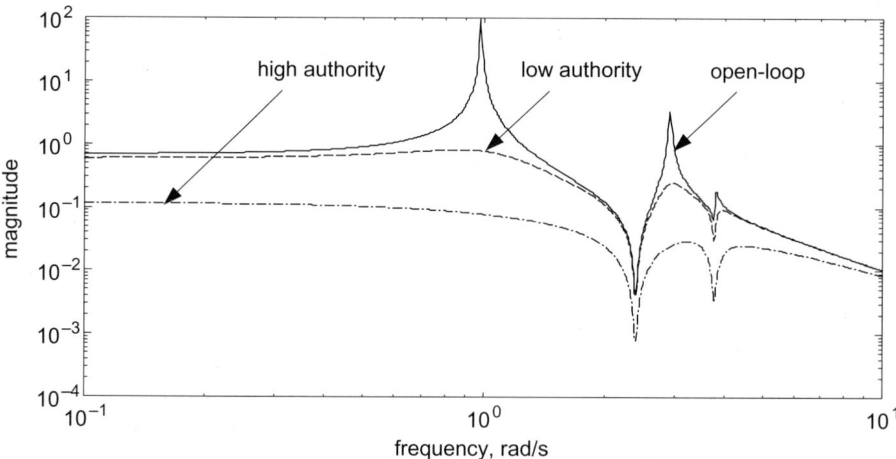

Figure 10.1. Magnitude of the transfer function of the simple system: Open-loop, closed-loop with low-authority controller, and closed-loop with high-authority controller. Low-authority controller suppresses resonance peaks, while the action of the high authority includes a wide frequency spectrum.

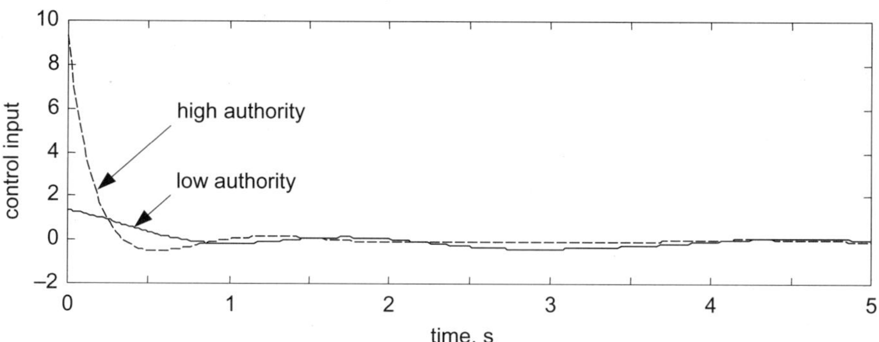

Figure 10.2. Control input of the simple system, for low- and high-authority controllers: The high-authority controller requires a strong input.

10.2 Dissipative Controller

Dissipative controllers and their properties are based on Popov's theory of hyperstability [117], subsequently developed as a positive real property of the control systems [2], [10], and as the dissipative (passive) property of the systems [135], [136], [27]. The terms dissipative, passive, positive real, and hyperstable systems are synonyms, and their inter-relations are discussed by Wen [133]. In this chapter we call the above systems dissipative systems.

Consider a square stable plant (A, B, C), i.e., a linear system with the number of inputs equal to the number of outputs. An open-loop square system with simple poles is dissipative, see [2], if there exists a symmetric positive definite matrix P and a matrix Q that satisfy the following equations:

$$\begin{aligned} A^T P + PA &= -Q^T Q, \\ B^T P &= C. \end{aligned} \tag{10.1}$$

The system is strictly dissipative if $Q^T Q$ is positive definite.

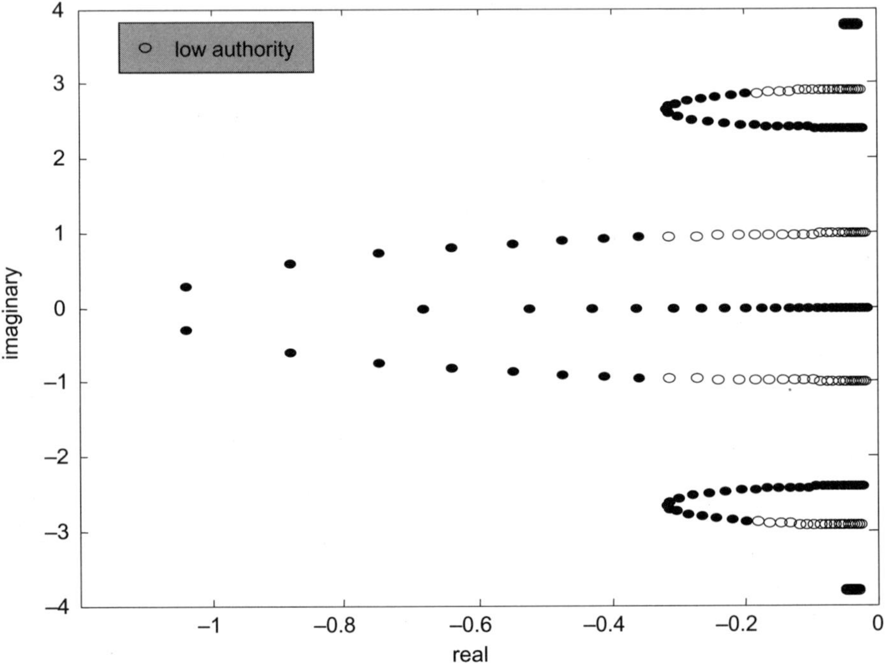

Figure 10.3. Root-locus of a simple structure for a low-authority controller: Poles of the low-authority controller vary significantly only their imaginary parts (damping) while the real parts (natural frequencies) remain almost unchanged.

The above definition allows for the simple determination of a dissipative system (at least in theory). Given A and B, we select the matrix Q. Next, we solve the first of (10.1) for P, and find the output matrix C from the second equation (10.1).

We are going to discuss three particular cases of the dissipative systems. In the first case, when $Q = B^T$ is chosen we obtain

$$\begin{aligned} P &= W_c, \\ C &= B^T W_c, \end{aligned} \tag{10.2}$$

and W_c is the controllability grammian. In this case, the actuators are weighted proportionally to the system controllability grammian.

In the second case we choose the matrix $Q = CW_o^{-1}$, where W_o is the observability grammian. Thus,

$$P = W_o^{-1}, \quad B = W_o C^T. \tag{10.3}$$

In this case, the sensors are weighted proportionally to the system observability grammian.

In the third case, we select $Q = (-A - A^T)^{1/2}$; therefore, one obtains

$$P = I, \quad B = C^T. \tag{10.4}$$

In this particular case the actuators and sensors are collocated. This case is most frequently used, since it requires simple actuator and sensor collocation to guarantee the closed-loop system stability.

The guaranteed stability of the closed-loop system is the most useful property of the dissipative system. It was shown by Desoer and Vidyasagar [27] and by Benhabib, Iwens, and Jackson [10] that, for the square and strictly dissipative plant and the square and dissipative controller (or vice versa: the square and dissipative plant and the square and strictly dissipative controller), the closed-loop system is asymptotically stable. In particular, if the feedback gain matrix is positive definite, the closed-loop system is asymptotically stable.

10.3 Properties of Collocated Controllers

As a corollary, consider a system with the state-space representation (A, B, C), which has collocated sensors and actuators, that is, $C = B^T$. In this case, a closed-loop system with the proportional feedback gain

$$u = -Ky \tag{10.5}$$

is stable, for $K = \text{diag}(k_i)$, i = 1, ..., r and $k_i > 0$. This particularly useful configuration can be used only if there is the freedom to choose the collocated sensors and actuators, and if the number of available sensors and actuators is large enough to satisfy the performance requirements. It should be stressed that the stability property allows one to design simple and stable controllers, regardless of the plant parameter variations. However, one has to be aware that stability does not imply performance, and sometimes the performance of these controllers can be poor, as reported by Hyland [77].

It should also be noted that structures imply some limitation on collocation. Consider a structure with the state-space representation (A, B, C) as in (2.38). In this representation the upper half of matrix B is equal to zero regardless of the configuration of the applied forces. Thus, in order to satisfy the collocation requirement, the left half of C must be equal to zero. But the displacement measurements are located in this part (while the right part locates the rate measurements). Consequently, a flexible structure is dissipative if the force inputs and the rate outputs are collocated. Thus, for structures, to each actuator corresponds a collocated sensor, but the opposite is not necessarily true. Therefore, when designing collocated controllers it is beneficial to choose the actuators first, and subsequently determine the sensor locations by introducing $C = B^T$. In this way a physically realizable dissipative system is obtained. In the case when the outputs are determined first and the inputs are chosen afterward as $B = C^T$, one still deals with a dissipative system, but not necessarily a physically realizable one. Note also that the collocation of force actuators and rate sensors is a sufficient, but by no means necessary, condition of dissipativeness. For example, if the weighted collocation is used, the system with displacement sensors is dissipative.

For flexible structures we will consider the low-authority controllers. Let the plant have r inputs and outputs. Denote by $K = \mathrm{diag}(k_i)$, $k_i > 0$, $i = 1, \ldots, r$, the gain matrix of a collocated controller, then its closed-loop matrix is $A_c = A - BKB^T$. Let A be in the modal form 2 and let b_i be the ith column of B. The collocated controller is of low authority if for the closed-loop matrix A_c one obtains $\mathrm{eig}(A_c) \cong \mathrm{eig}(A - \sum_{i=1}^{r} k_i \mathrm{diag}(b_i b_i^T))$. In other words, for the low-authority controller one can replace BKB^T with its diagonal terms. For the flexible structures the collocated controller has the following property:

Property 10.1. *Relationship Between A, B, and C for the Low-Authority Collocated Controller.* *For* $\max(k_i) \le k_o$ *and a controllable and observable flexible system there exists* $k_o > 0$ *such that the collocated controller is of low authority. Furthermore, if A is in the almost-balanced modal form 2, the following holds:*

$$BB^T = C^T C \cong -\Gamma(A + A^T) = \mathrm{diag}(\gamma_1\alpha_1, \gamma_1\alpha_1, \gamma_2\alpha_2, \gamma_2\alpha_2, \ldots, \gamma_n\alpha_n, \gamma_n\alpha_n) \quad (10.6)$$

or, for the ith block, it can be written as

$$B_i B_i^T = C_i^T C_i \cong -\gamma_i (A_i + A_i^T) = \gamma_i \alpha_i I_2, \quad (10.7)$$

where $\alpha_i = 2\zeta_i\omega_i$, B_i *is the ith two-row block of B, and* C_i *is the two-column block of C.*

Proof. Note that $b_i K b_j^T$ is the *ij*th term of BKB^T. Since $(b_i K b_j^T)^2 \leq (b_i K b_i^T)(b_j K b_j^T)$ therefore, for A in the modal form 2 and for small gain K such that $\max(k_i) \leq k_o$, the off-diagonal terms of BKB^T do not influence the eigenvalues of A_c and they can be ignored. Equations (10.6) and (10.7) follow from the Lyapunov equations (4.5). ▣

10.4 Root-Locus of Collocated Controllers

Here we present the relationship between the controller gains and the closed-loop pole locations. In order to determine the properties of the collocated controllers in modal coordinates, consider further the dissipativity conditions (10.1) for a structure in the modal coordinates 2. Consider also a feedback as in (10.5). In this case the closed-loop equations are as follows:

$$\begin{aligned} \dot{x} &= (A - BKB^T)x + Bu_o, \\ y &= Cx, \end{aligned} \tag{10.8}$$

where u_o is a control command ($u_o \equiv 0$ in the case of vibration suppression). Since the matrix A is in the modal form 2 and K is diagonal, $K = \operatorname{diag}(k_1, ..., k_r)$, then in the modal coordinates with collocated sensors and actuators we obtain the closed-loop matrix $A_c = A - BKB^T$ in the form

$$A_c = A - \sum_{j=1}^{r} k_j b_j b_j^T, \tag{10.9}$$

where b_j is the *j*th column of B. In the modal coordinates matrix A_c is block-diagonal, that is, $A_c = \operatorname{diag}(A_{c1}, ..., A_{cn})$, where A_{ci} is the *i*th 2×2 block. For this block (10.9) is as follows:

$$A_{ci} \cong A_i - \sum_{j=1}^{r} k_j b_{ji} b_{ji}^T, \tag{10.10}$$

where b_{ji} is the *i*th block of the *j*th column of B. In this equation the cross terms $b_{jk} b_{ji}^T$ (for $k \neq i$) are omitted as negligible for the low-authority controllers in the modal coordinates; see Property 10.1. Also, from Property 10.1, the following holds: $b_{ji}b_{ji}^T \cong -\gamma_{ji}(A_i + A_i^T)$, where γ_{ji} is the ith Hankel singular value obtained for the jth column of B, i.e., for the triplet (A, b_j, b_j^T). Thus, (10.10) is now

$$A_{ci} \cong A_i + 2k_j\gamma_{ji}(A_i + A_i^T). \tag{10.11}$$

For A_i as in (2.53), we obtain $A_i + A_i^T = -2\zeta_i\omega_i I_2$, and rewrite (10.11) as follows:

$$A_{ci} \cong \begin{bmatrix} -\beta_i\zeta_i\omega_i & -\omega_i \\ -\omega_i & -\beta_i\zeta_i\omega_i \end{bmatrix}, \tag{10.12}$$

with the parameter β_i given as

$$\beta_i = 1 + 2\sum_{j=1}^{r} k_j\gamma_{ji}. \tag{10.13}$$

Comparing the closed-loop matrix as in (10.12) and the open-loop matrix as in (2.53), we see that β_i is a measure of the shift of the ith pair of poles. Denote the closed-loop pair of poles $(\lambda_{cri} \pm j\lambda_{cii})$ and the open-loop pair $(\lambda_{ori} \pm j\lambda_{oii})$, then it follows from (10.12) that they are related

$$(\lambda_{cri} \pm j\lambda_{cii}) \cong (\beta_i\lambda_{ori} \pm j\lambda_{oii}), \qquad i = 1, \ldots, n, \tag{10.14}$$

or the real part of the poles (modal damping) changes by factor β_i,

$$\lambda_{cri} \cong \beta_i\lambda_{ori}, \qquad i = 1, \ldots, n, \tag{10.15}$$

while the imaginary part (natural frequency) remains almost unchanged

$$\lambda_{cii} \cong \lambda_{oii}, \qquad i = 1, \ldots, n. \tag{10.16}$$

The above equation shows that the real part of the ith pair of poles is shifted, while the imaginary part is stationary. The shift is proportional to the gain of each input, and to the ith Hankel singular values associated with each input.

Equations (10.12) and (10.13) set the basic limitation for the dissipative controller design. To be precise, the number of inputs (and outputs) limits the number of controlled modes (or controlled pairs of poles). In order to illustrate this, we assume a single-input–single-output system. In this case, $\beta_{1i} = 1 + 2k_1\gamma_{1i}$ and the scalar gain k_1 is the only free parameter available for the design. Thus, only one pole can be shifted to the required position. If more than one pair should be shifted, their placement would be a least-squares compromise, which typically would be nonsatisfactory. Thus, in order to avoid this rough approximation, it is often required for the dissipative controllers to have a large number of sensors and actuators to meet the required performance criteria.

The pole-shift factor β_i is also interpreted as a ratio of the variances of the open-loop (σ_{oi}^2) and closed-loop (σ_{ci}^2) states excited by the white noise input, i.e.,

$$\beta_i \cong \frac{\sigma_{oi}^2}{\sigma_{ci}^2}. \tag{10.17}$$

Since $\beta_i \geq 1$, it is therefore a relative measure of the noise suppression of the closed-loop system with respect to the open-loop system. This interpretation follows from the closed-loop Lyapunov equation

$$(A - BKB^T)W_{cc} + W_{cc}(A - BKB^T)^T + BB^T = 0, \tag{10.18}$$

where W_{cc} is the closed-loop controllability grammian. For the ith pair of variables the above equation is as follows:

$$\left(A_i - \sum_{j=1}^{r} k_j B_{ji} B_{ji}^T \right) w_{cci} + w_{cci} \left(A_i - \sum_{j=1}^{r} k_j B_{ji} B_{ji}^T \right)^T + B_i B_i^T \cong 0. \tag{10.19}$$

Introducing (10.7), after some algebra, we obtain

$$w_{cci} + 2w_{cci} \left(\sum_{j=1}^{r} k_j \gamma_{ji} \right) - w_{oci} \cong 0, \tag{10.20}$$

where $w_{oci} \cong \gamma_i$ is the diagonal entry of the open-loop controllability grammian. Finally, we obtain

$$\frac{w_{oci}}{w_{cci}} = \frac{\sigma_{oi}^2}{\sigma_{ci}^2} \cong 1 + 2\sum_{j=1}^{r} k_j \gamma_{ji} = \beta_i. \tag{10.21}$$

Based on (10.13), (10.14), and (10.17) we develop a tool for the pole placement of the dissipative controllers. The task is to determine gains k_j, $j = 1,\ldots,r$, such that the selected poles are placed at the required location (or as close as possible in the least-squares sense). Equivalently, the task is to determine gains k_j, j = 1, ..., r, such that the input noise of the selected modes is suppressed at ratio β_i. The approach follows from (10.13), since one can determine the gains such that q poles are shifted by β_i, $i=1,\ldots,q$, i.e., $\lambda_{cri} \cong \beta_i \lambda_{ori}$, or the noise can be suppressed by β_i, i.e., $\sigma_{oi}^2 \cong \beta_i \sigma_{ci}^2$. Define the gain vector k,

$$k = \begin{bmatrix} k_1 & k_2 & \ldots & k_r \end{bmatrix}, \tag{10.22}$$

so that we rewrite (10.13) as

$$d\beta \cong Gk, \tag{10.23}$$

where $d\beta$ is the vector of the pole shifts

$$d\beta = \begin{bmatrix} \beta_1 - 1 \\ \beta_2 - 1 \\ \vdots \\ \beta_q - 1 \end{bmatrix}, \tag{10.24}$$

and G is the matrix of the system Hankel singular values for each actuator and sensor location

$$G = 2\begin{bmatrix} \gamma_1 & \gamma_2 & \ldots & \gamma_r \end{bmatrix} = 2\begin{bmatrix} \gamma_{11} & \gamma_{21} & \cdots & \gamma_{r1} \\ \gamma_{12} & \gamma_{22} & \cdots & \gamma_{r2} \\ \cdots & \cdots & \cdots & \cdots \\ \gamma_{1q} & \gamma_{2q} & \cdots & \gamma_{rq} \end{bmatrix}, \tag{10.25}$$

where $\gamma_i = \begin{bmatrix} \gamma_{i1} & \gamma_{i2} & \ldots & \gamma_{iq} \end{bmatrix}^T$ is the set of Hankel singular values for the ith actuator/sensor location, and γ_{ij} is the jth Hankel singular value for the ith actuator/sensor location.

The least-squares solution of (10.23) is as follows:

$$k \cong G^+ d\beta, \tag{10.26}$$

where G^+ is the pseudoinverse of G. The set of equations (10.23) is either overdetermined ($q > r$, or rank(G) = r), or square ($q = r =$ rank(G)), or underdetermined ($q < r$, or rank(G) = q), see [68]. The form of the pseudoinverse depends on the number of inputs and outputs r, and the number of poles shifted, q, i.e., on the rank of the matrix G.

10.5 Collocated Controller Design Examples

Two examples of modal collocated controller design are presented: the controller design for the simple flexible system, and for the 2D truss structure.

10.5.1 A Simple Structure

The Matlab code for this example is in Appendix B. The system is shown in Fig. 1.1, with masses $m_1 = m_2 = m_3 = 1$, stiffness $k_1 = 10$, $k_2 = k_4 = 3$, $k_3 = 4$, and the damping matrix D as a linear combination of the mass and stiffness matrices, $D = 0.004K + 0.001M$. The input force is applied to mass m_3 and the output is the rate of the same mass. The poles of the open-loop system are

$$\lambda_{o1,o2} = -0.0024 \pm j0.9851,$$
$$\lambda_{o3,o4} = -0.0175 \pm j2.9197,$$
$$\lambda_{o5,o6} = -0.0295 \pm j3.8084.$$

The system Hankel singular values are as follows:

$$\gamma_1 = \left[63.6418,\ 63.6413,\ 4.9892,\ 4.9891,\ 0.2395,\ 0.2391\right]^T.$$

There are two tasks:

- Shift the first pole by increasing its real part twofold, and leave the other poles stationary; and
- increase the real parts of the first and second pole twofold, and leave the third pole stationary.

In the first part we need to increase the first pole damping twofold and leave the other poles stationary. For this increase we require the following factors: $\beta_1 = 2$ and $\beta_2 = \beta_3 = 1$; therefore, $d\beta = \begin{bmatrix} 1 & 1 & 0 & 0 & 0 & 0 \end{bmatrix}^T$. For this case, $G = 2\gamma_1$; thus, we obtain the gain k = 0.0078 from (10.26). For this gain we compute the closed-loop eigenvalues

$$\lambda_{c1,c2} = -0.0049 \pm j0.9851,$$
$$\lambda_{c3,c4} = -0.0189 \pm j2.9197,$$
$$\lambda_{c5,c6} = -0.0296 \pm j3.8084,$$

and from this result we see that the actual pole shifts are $\beta_1 = 1.9939$, $\beta_2 = 1.0779$, and $\beta_3 = 1.0037$, which are close to the required ones.

Next, we consider the second part, a design that increases the first and second pole damping twofold and leaves the third stationary. In this case, $\beta_1 = \beta_2 = 2$ and $\beta_3 = 1$ as required; therefore, $d\beta = [1 \quad 1 \quad 1 \quad 1 \quad 0 \quad 0]^T$. We obtain the gain $k = 0.0084$ from (10.26) and, consequently, we compute the closed-loop eigenvalues for this gain:

$$\lambda_{c1,c2} = -0.0051 \pm j0.9851,$$
$$\lambda_{c3,c4} = -0.0190 \pm j2.9197,$$
$$\lambda_{c5,c6} = -0.0296 \pm j3.8084.$$

Comparing the open- and closed-loop poles, we see that the actual shifts, $\beta_1 = 2.0718$, $\beta_2 = 1.0840$, and $\beta_3 = 1.0040$, are almost the same as in the first case. Thus, we hardly meet the requirements. This case shows that for the underdetermined problem (the number of inputs is smaller than the number of poles to be shifted), the obtained least-squares result is the best, but not satisfactory, result.

10.5.2 The 2D Truss

The 2D truss is presented in Fig. 1.2, with the damping matrix proportional to the mass and stiffness matrix, $D = 0.3M + 0.00002K$. Control forces are applied at node 4, directed horizontally, and at node 10, directed vertically. The rate output is collocated with the force. The system has 16 modes. The task is to suppress the two most controllable and observable modes by increasing their damping 60 times.

We obtain the required feedback gain from (10.26). In order to use this equation note that in this case $\beta_1 = \beta_2 = 60$ and the remaining β's are equal to 1. Let γ_1 and γ_2 be vectors of the Hankel singular values for the first input and output, and for the second input and output, respectively. Then $G = 2\ [\gamma_1 \quad \gamma_2]$. For this case $d\beta = [59 \quad 59 \quad 59 \quad 59 \quad 0 \quad 0 \quad \ldots \quad 0]^T$, and for these data we obtain from (10.26) the gain matrix $k = \text{diag}(4.3768,\ 385.0546)$.

For this gain we determine the closed-loop poles, and the pole shift was obtained as a ratio of real parts of the closed- and open-loop poles, as in definition (10.14), i.e., $\beta_i = \lambda_{cri} / \lambda_{ori}$. The plot of β_i in Fig. 10.4 shows that $\beta_1 = 58.94$ and $\beta_2 = 57.46$ are close to the assigned value of 60. The damping of the two poles increased 60 times, while the other poles changed insignificantly.

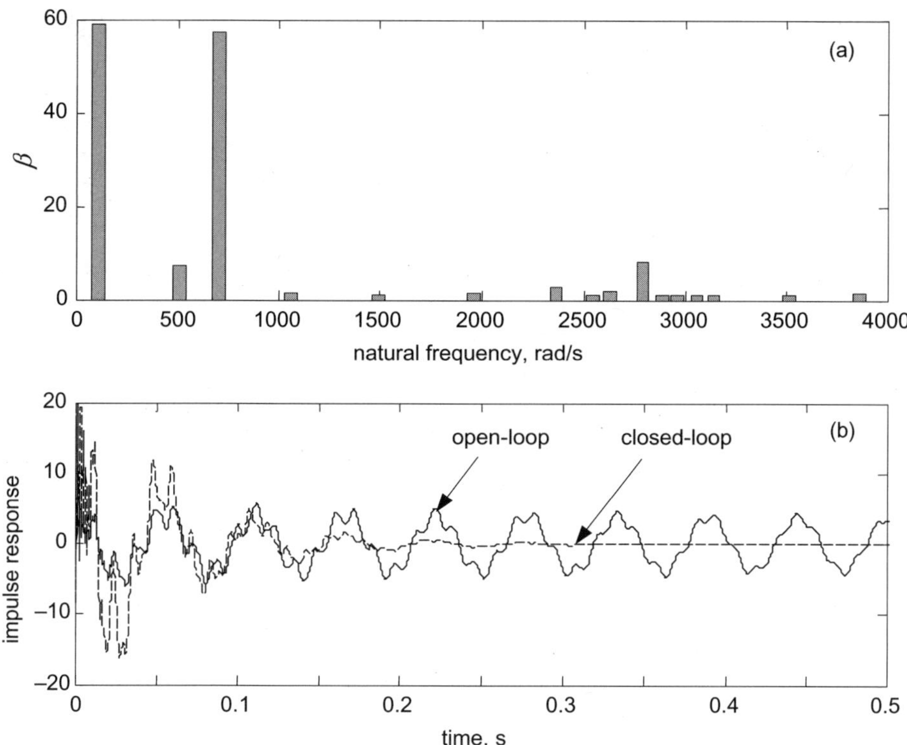

Figure 10.4. A dissipative controller for the 2D truss: (a) Factor β_i shows the two most controllable and observable modes; and (b) open-loop (solid line) and closed-loop (dashed line) impulse responses show the increased damping of the closed-loop system.

11
LQG Controllers

how to design an advanced feedback loop

In theory, there is no difference between theory and practice.
But, in practice, there is.
—Jan L.A. van de Snepscheut

The control issues, as applied to structures, include precise positioning or tracking. It is expected that the positioning and tracking requirements should be satisfied for structures with natural frequencies within the controller bandwidth and within the disturbance spectra. LQG (**L**inear system, **Q**uadratic cost, **G**aussian noise) controllers can typically meet these conditions and they are often used for tracking and disturbance rejection purposes. A good insight into the problems of analysis and design of LQG controllers can be obtained from the books by Kwakernaak and Sivan [91], Maciejowski [104], Anderson and Moore [3], Furuta, Sano, and Atherton [41], Lin [100], Skogestad and Postlethwaite [129], Dorato, Abdallah, and Cerone [28], Burl [13], and Fairman [34].

Two issues in LQG controller design are of special importance: the determination of the weights of the performance index—to satisfy the performance requirements and controller order reduction—to reduce the control implementation complexity. The first issue—weight determination—ultimately impacts the closed-loop system performance, in terms of the tracking accuracy and the disturbance rejection properties. If the weights of the LQG performance index are inappropriately chosen, the LQG controller performance will not satisfy the requirements. The selection of weights is most often not an easy task. As stated by Lin [100, p. 93] "It takes a great deal of experience to transform design requirements and objectives to the performance index that will produce the desired performance." Our task is to replace experience with analytical tools.

The second issue—controller order reduction—impacts the implementation in terms of complexity and accuracy of the controller software. These problems are especially important for structures, since the structural models are typically of high order, making order reduction a necessity. The order of the controller is equal to the order of the plant, which is in most cases unacceptably high.

In this chapter both problems, weight determination and controller order reduction, are solved in the modal coordinates, using unique structural properties in modal coordinates.

11.1 Definition and Gains

A block diagram of an LQG control system is shown in Fig. 11.1. It consists of a stable plant or structure (G) and controller (K). The plant output y is measured and supplied to the controller. Using the output y the controller determines the control signal u that drives the plant. The inside structure of the plant and controller is shown in Fig. 11.2. The plant is described by the following state-space equations:

$$\begin{aligned} \dot{x} &= Ax + Bu + v, \\ y &= Cx + w, \end{aligned} \tag{11.1}$$

as shown in Fig. 11.2. In the above description the plant state vector is denoted x. The plant is perturbed by random disturbances, denoted v, and its output is corrupted by noise w. The noise v, called process noise, has covariance $V = E(vv^T)$, the noise w is called measurement noise, and its covariance is $W = E(ww^T)$. Both noises are uncorrelated, i.e., $E(vw^T) = 0$, where $E(.)$ is the expectation operator. Without loss of generality, it is assumed that the covariance of the measurement noise is unity, i.e., $W = I$.

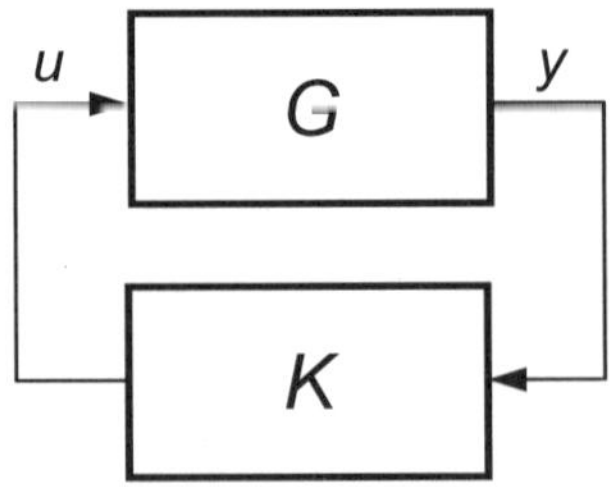

Figure 11.1. The LQG closed-loop system: G—plant (structure), K—controller, u—actuator input, and y—the sensed output.

The controller is driven by the plant output y. The controller produces the control signal u that drives the plant. This signal is proportional to the plant estimated state denoted $\hat{x}$, and the gain between the state and the controlled signal u is the controller gain (K_c). We use the estimated state $\hat{x}$ rather than the actual state x, since typically the latter is not available from measurements. The estimated state is obtained from the estimator, which is part of the controller, as shown in Fig. 11.2. The estimator equations follow from the block-diagram in Fig. 11.2:

$$\dot{\hat{x}} = A\hat{x} + Bu + K_e(y - C\hat{x}). \tag{11.2}$$

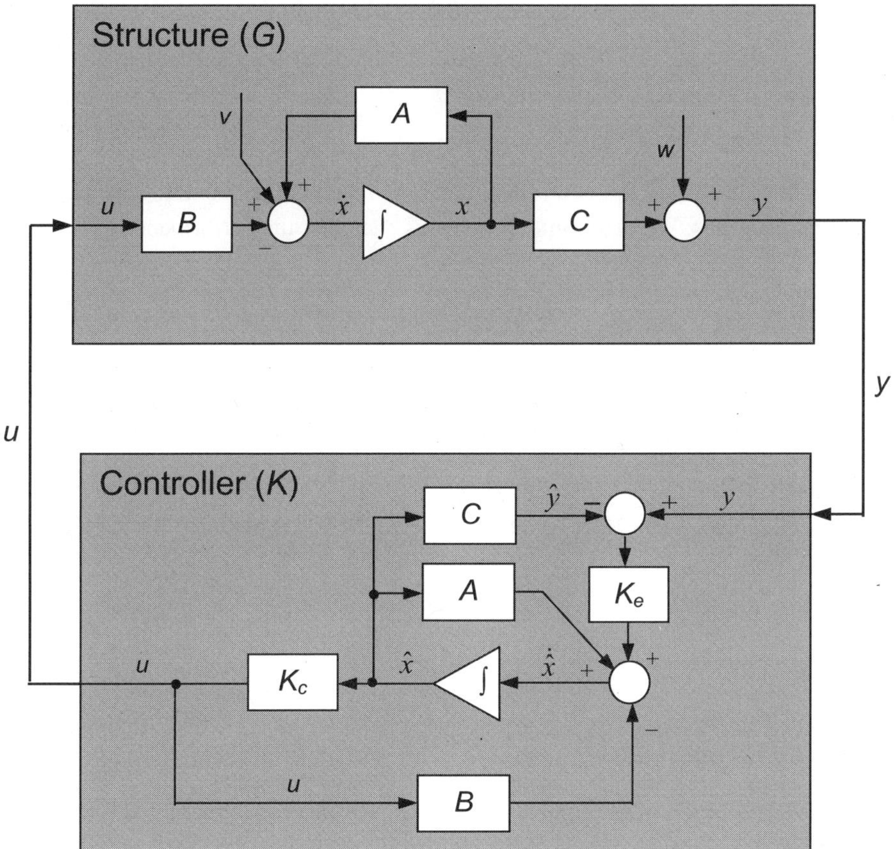

Figure 11.2. The inner structure of the LQG closed-loop system.

Assuming that the plant model is known exactly, we see that the estimated state is an exact copy of the actual state, except for the initial (transient) dynamics. From the above equation we see that in order to determine the estimator we have to determine the estimator gain, K_e.

Using (11.2) and the block-diagram in Fig. 11.2 we derive the controller state-space equations from input y to output u:

$$\dot{\hat{x}} = (A - BK_c - K_eC)\hat{x} + K_e y,$$
$$u = -K_c\hat{x}. \tag{11.3}$$

From these equations we obtain the controller state-space representation $(A_{lqg}, B_{lqg}, C_{lqg})$,

$$A_{lqg} = A - BK_c - K_eC,$$
$$B_{lqg} = K_e, \tag{11.4}$$
$$C_{lqg} = -K_c.$$

In the above equations the controller gain (K_c) and the estimator gain (K_e) are unknown quantities. We determine these gains such that the performance index J,

$$J^2 = E\left(\int_0^\infty (x^T Qx + u^T Ru)\, dt \right) \tag{11.5}$$

is minimized. In the above equation R is a positive definite input weight matrix and Q is a positive semidefinite state weight matrix. We assumed further that $R = I$ without loss of generality.

It is well known (see [91], [3]) that the minimum of J is obtained for the feedback

$$u = -K_c\hat{x} \tag{11.6}$$

with the gain matrix,

$$K_c = B^T S_c, \tag{11.7}$$

and S_c is the solution of the controller algebraic Riccati equation (called CARE)

$$A^T S_c + S_c A - S_c BB^T S_c + Q = 0. \tag{11.8}$$

The optimal estimator gain is given by

$$K_e = S_e C^T, \tag{11.9}$$

where S_e is the solution of the filter (or estimator) algebraic Riccati equation (called FARE)

$$AS_e + S_e A^T - S_e C^T CS_e + V = 0. \tag{11.10}$$

The above is a formal procedure to design the LQG controller (i.e., to determine the gains K_c and K_e). By saying formal we mean that the index J is known in advance, and that the weighting matrix Q is also known. But the performance of the closed-loop system is rather seldom specified through J or Q. It is rather defined through the closed-loop parameters, such as bandwidth, the root-mean-square of the system response to disturbances, or settling time and overshoot. These quantities are reflected in the weighting matrix Q, but not in an explicit way—and the dependence is not an obvious one. However, approximate relationships between weights and closed-loop performance can be derived for structures, giving guidelines as to how to determine the weights that shape the closed-loop system performance that meets the requirements. We do this by relating the weights with the closed-loop pole locations and the reduction of the disturbance noise. The following sections will lead to the rational design of the LQG controller for structures.

11.2 The Closed-Loop System

The state-space equations of the open-loop system are given by (11.1), and the state-space equations of the LQG follow from (11.4),

$$\begin{aligned} \dot{\hat{x}} &= (A - K_e C - BK_c)\hat{x} + K_e y, \\ u &= -K_c \hat{x}. \end{aligned} \tag{11.11}$$

Defining a new state variable

$$x_o = \begin{Bmatrix} x \\ \varepsilon \end{Bmatrix}, \tag{11.12}$$

where $\varepsilon = x - \hat{x}$, we obtain the closed-loop state-space equations in the form:

$$\begin{aligned} \dot{x}_o &= A_o x_o + B_o v, \\ z &= C_o x_o, \end{aligned} \tag{11.13}$$

where

$$\begin{aligned} A_o &= \begin{bmatrix} A - BK_c & BK_c \\ 0 & A - K_e C \end{bmatrix}, \\ B_o &= \begin{bmatrix} I \\ I \end{bmatrix}, \\ C_o &= \begin{bmatrix} C & 0 \end{bmatrix}, \end{aligned} \tag{11.14}$$

is the closed-loop triple.

11.3 The Balanced LQG Controller

The solutions of the CARE and FARE depend on the states we choose. Among the multiple choices there exists a state-space representation such that the CARE and FARE solutions are equal and diagonal, see [82], [113], and [49], assuming that the system is controllable and observable. In this case we obtain

$$S_c = S_e = \mathrm{M} = \mathrm{diag}(\mu_1, \mu_2, ..., \mu_N), \tag{11.15}$$

where $\mu_1 \geq \mu_2 \geq \cdots \geq \mu_N > 0$ and M is a diagonal positive definite $\mathrm{M} = \mathrm{diag}(\mu_i)$, $i = 1,\ldots,N$, $\mu_i > 0$. A state-space representation with condition (11.15) satisfied is called an LQG balanced representation and μ_i, $i = 1,\ldots, N$, are its LQG singular (or characteristic) values.

Let R be the transformation of the state x such that $x = R\overline{x}$. Then the solutions of CARE and FARE in the new coordinates are as follows:

$$\begin{aligned} \overline{S}_c &= R^T S_c R, \\ \overline{S}_e &= R^{-1} S_e R^{-T}, \end{aligned} \tag{11.16}$$

and the weighting matrices are

$$\begin{aligned} \overline{Q}_c &= R^T Q_c R, \\ \overline{Q}_e &= R^{-1} Q_e R^{-T}. \end{aligned} \tag{11.17}$$

The transformation R to the LQG-balanced representation is obtained as follows:

- For a given state-space representation (A,B,C), find the solutions S_c and S_e of CARE and FARE. Decompose S_c and S_e as follows:

$$\begin{aligned} S_c &= P_c^T P_c, \\ S_e &= P_e P_e^T. \end{aligned} \tag{11.18}$$

- Form a matrix H, such that

$$H = P_c P_e. \tag{11.19}$$

- Find the singular value decomposition of H,

$$H = V \mathrm{M} U^T. \tag{11.20}$$

- Obtain the transformation matrix either as

$$R = P_e U \mathrm{M}^{-1/2} \tag{11.21}$$

or

$$R = P_c^{-1} V \mathrm{M}^{1/2}. \tag{11.22}$$

Proof. By inspection. We introduce R to (11.16) to show that (11.15) is satisfied. ▣

We give in Appendix A.12 the Matlab function *bal_LQG*, which transforms a representation (A, B, C) to the LQG balanced representation (A_b, B_b, C_b).

11.4 The Low-Authority LQG Controller

For LQG controllers we modify the definition of the low-authority controller of a structure as known from Chapter 10. Let (A, B, C) be the open-loop modal representation of a flexible structure (in the modal form 1 or 2), and let $A_{c1} = A - BB^T S_c$, $A_{c2} = A - S_e C^T C$ be the closed-loop matrices where S_c and S_e are the solutions of the CARE and FARE equations, respectively. The LQG controller is of low authority if its closed-loop matrices have the following property:

$$\mathrm{eig}(A_{c1}) = \mathrm{eig}(A - BB^T S_c) \cong \mathrm{eig}(A - \mathrm{diag}(BB^T) S_c) \tag{11.23}$$

and

$$\mathrm{eig}(A_{c2}) = \mathrm{eig}(A - S_e C^T C) \cong \mathrm{eig}(A - S_e\, \mathrm{diag}(C^T C)). \tag{11.24}$$

In other words, for the low-authority controller, BB^T and $C^T C$ can be replaced with their diagonal terms.

The low-authority LQG controller has the following property:

Property 11.1. *Relationship Between A, B, and C for the Low-Authority LQG Controller.* *Let $\|S_c\|_2 \leq s_o$ and $\|S_e\|_2 \leq s_o$. For a controllable and observable flexible system there exists $s_o > 0$ such that the controller is of low authority. Furthermore, if A is in the modal form 1, one can use the following replacement for BB^T (or $C^T C$):*

$$\begin{aligned} BB^T &\cong -W_c(A + A^T) = \mathrm{diag}(0, 2w_{c1}\alpha_1, 0, 2w_{c2}\alpha_2, ..., 0, 2w_{cn}\alpha_n), \\ C^T C &\cong -W_o(A + A^T) = \mathrm{diag}(0, 2w_{o1}\alpha_1, 0, 2w_{o2}\alpha_2, ..., 0, 2w_{on}\alpha_n), \end{aligned} \tag{11.25}$$

or, for the ith block,

$$B_iB_i^T \cong -w_{ci}(A_i + A_i^T) = w_{ci}\begin{bmatrix} 0 & 0 \\ 0 & 2\alpha_i \end{bmatrix},$$
$$C_i^TC_i \cong -w_{oi}(A_i + A_i^T) = w_{oi}\begin{bmatrix} 0 & 0 \\ 0 & 2\alpha_i \end{bmatrix}, \tag{11.26}$$

where $\alpha_i = 2\zeta_i\omega_i$. *If A is in the modal form 2 the following replacement is used:*

$$BB^T \cong -W_c(A + A^T) = \operatorname{diag}(w_{c1}\alpha_1, w_{c1}\alpha_1, w_{c2}\alpha_2, w_{c2}\alpha_2, ..., w_{cn}\alpha_n, w_{cn}\alpha_n),$$
$$C^TC \cong -W_o(A + A^T) = \operatorname{diag}(w_{o1}\alpha_1, w_{o1}\alpha_1, w_{o2}\alpha_2, w_{o2}\alpha_2, ..., w_{on}\alpha_n, w_{on}\alpha_n), \tag{11.27}$$

or, for the ith block,

$$B_iB_i^T \cong -w_{ci}(A_i + A_i^T) = w_{ci}\alpha_i I_2,$$
$$C_i^TC_i \cong -w_{oi}(A_i + A_i^T) = w_{oi}\alpha_i I_2, \tag{11.28}$$

where B_i *is the ith two-row block of B and* C_i *is the two-column block of C.*

Proof. Denote by b_i the ith row of B. Note that for the positive-semidefinite matrix BB^T one obtains $(b_ib_j^T)^2 \le (b_ib_i^T)(b_jb_j^T)$, i.e., that the off-diagonal terms do not exceed the geometric mean value of the corresponding diagonal terms. Therefore, if A is in modal form 1 or 2, and for small S_c such that $\|S_c\|_2 \le s_o$, the off-diagonal terms of BB^T do not influence the eigenvalues of A_{c1}, i.e., $\operatorname{eig}(A_{c1})$ $= \operatorname{eig}(A - BB^TS_c)$. If the matrix BB^T is obtained from the Lyapunov equations (4.5) and replaced by its diagonal terms, one obtains (11.25)–(11.28). Similar applies to the eigenvalues of $A - S_eCC^T$. ▣

We illustrate the low-authority LQG controller for a simple structure as in Fig. 1.1. In Fig. 11.3 we mark "○" the root-locus for the first and the second mode and for the increasing values of the matrix S_c, and with "•" we mark the approximate root-locus, using the diagonal part of BB^T. The figure shows good agreement between the exact and approximate roots for small S_c.

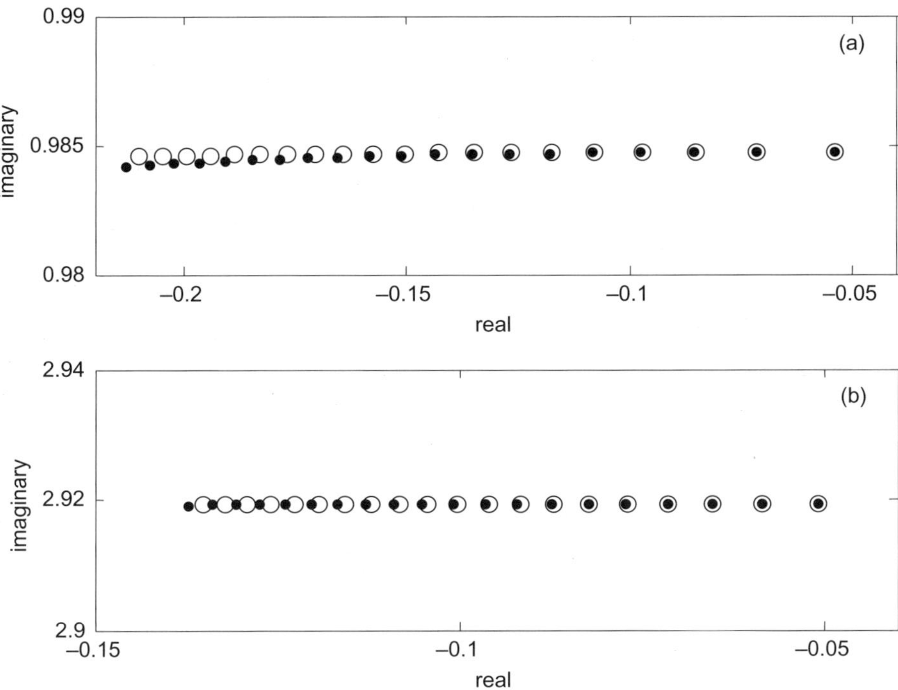

Figure 11.3. The exact (○) and approximate (•) root-locus of a simple system show horizontal movement and good coincidence for low gain: (a) First pole; and (b) second pole.

11.5 Approximate Solutions of CARE and FARE

The design of the LQG controller seems to be a straightforward task since it goes as follows: for given weights Q and V we obtain the gains K_c and K_e from (11.7) and (11.9), and the controller representation from (11.4). However, from the implementation point of view, this approach is not appropriately defined, since the design process typically starts from the definition of the required closed-loop system performance, such as the norm of the tracking error, or the location of the closed-loop poles. Thus, we have to find appropriate weights Q and V that meet the performance requirements. This task does not have an analytical solution in general, and is frequently solved using a trial-and-error approach. In the following sections we solve this problem using the properties of flexible structures and the low-authority controllers.

For the LQG design in modal coordinates we use diagonal weight matrices Q and V. This significantly simplifies the design process, and can be justified as follows: Consider the term $E\int_0^{\infty} x^T Q x\,dt$, which represents the participation of weight Q in the performance index J. It is evaluated as follows:

$$E\int_0^\infty x^T Qx\, dt = E\int_0^\infty \sum_{i,j=1}^N q_{ij}x_i x_j\, dt, \tag{11.29}$$

where x_i is the ith component of x. For a positive-definite matrix Q one obtains

$$q_{ij}^2 \le q_i q_j$$

and the states in modal coordinates are almost orthogonal, that is,

$$\left(E\int_0^\infty x_i x_j\, dt\right)^2 \ll \left(E\int_0^\infty x_i^2\, dt\right)\left(E\int_0^\infty x_j^2\, dt\right).$$

Introducing two previous equations to (11.29), one obtains

$$E\int_0^\infty x^T Qx\, dt \cong E\int_0^\infty \sum_{i=1}^N q_{ii}x_i x_j\, dt = E\int_0^\infty x^T Q_d x\, dt, \tag{11.30}$$

where Q_d is a diagonal matrix that consists of the diagonal entries of Q.

Due to the duality of Q and V, the same applies to the matrix V. That is, in modal coordinates the full matrix V can be replaced with its diagonal part V_d, and the system performance remains almost unchanged.

Next, based on Property 11.1, we will show that the low-authority controllers in the modal representation produce diagonally dominant solutions of the CARE and FARE equations.

Property 11.2(a). *Approximate Solution of CARE.* *Assume a diagonal weight matrix $Q = \operatorname{diag}(q_i I_2)$, $i = 1,\ldots,n$, then there exist $q_i \le q_{oi}$ where $q_{oi} > 0$, $i = 1,\ldots,n$, such that*

(a)
$$S_c \cong \operatorname{diag}(s_{ci} I_2) \tag{11.31}$$

is the solution of (11.8), *and*

(b)
$$s_{ci} \cong \frac{\beta_{ci} - 1}{2w_{ci}}, \qquad \beta_{ci} = \sqrt{1 + \frac{2q_i w_{ci}}{\zeta_i \omega_i}}. \tag{11.32}$$

Proof. (a) Note that for $Q = 0$ we obtain $S_c = 0$. For small Q the CARE transforms into the Lyapunov equation $A^T S_c + S_c A + Q = 0$. For a modal matrix A and diagonal Q the solution of this equation is diagonally dominant, see Property 4.1. Thus, there exist $q_i \le q_{oi}$ where $q_{oi} > 0$, $i = 1,\dots,n$, such that (11.31) holds.

(b) For diagonally dominant S_c, (11.8) turns into a set of the following equations:

$$s_{ci}(A_i + A_i^T) - s_{ci}^2 B_i B_i^T + q_i I_2 \cong 0, \qquad i = 1,\dots,n.$$

For a low-authority controller in modal coordinates we obtain $B_i B_i^T \cong -w_{ci}(A_i + A_i^T)$ and $A_i + A_i^T = -2\zeta_i \omega_i I_2$, see (11.28). Therefore the above equation is now in the following form:

$$s_{ci}^2 + \frac{s_{ci}}{w_{ci}} - \frac{q_i}{2\zeta_i \omega_i w_{ci}} \cong 0, \qquad i = 1,\dots,n$$

There are two solutions of the above equation, but for a stable system and for $q_i = 0$ it is required that $s_{ci} = 0$, therefore (11.32) is the unique solution of the above equation. ▣

A similar result is obtained for the FARE equation.

Property 11.2(b). *Approximate Solution of FARE.* *For a diagonal V, $V = \operatorname{diag}(v_i I_2)$, $i = 1,\dots,n$, there exist $v_i \le v_{oi}$ where $v_{oi} > 0$, $i = 1,\dots,n$, such that*

(a)
$$S_e \cong \operatorname{diag}(s_{ei} I_2) \tag{11.33}$$

is the solution of (11.8), *where*

(b)
$$s_{ei} \cong \frac{\beta_{ei} - 1}{2 w_{oi}}, \quad \text{where} \quad \beta_{ei} = \sqrt{1 + \frac{2 v_i w_{oi}}{\zeta_i \omega_i}}. \tag{11.34}$$

From (11.31)–(11.34) we determine the LQG singular values as a geometric mean of s_{ci} and s_{ei}, $\mu_i = \sqrt{s_{ci} s_{ei}}$, i.e.,

$$\mu_i \cong \frac{\sqrt{(\beta_{ci} - 1)(\beta_{ei} - 1)}}{2\gamma_i}, \qquad i = 1,\dots,n. \tag{11.35}$$

11.6 Root-Locus

Using the diagonally dominant solutions of CARE and FARE we determine the relationship between the weights and the pole location and noise suppression, which is a useful tool in controller design.

Property 11.3(a). ***LQG Root-Locus.*** *Let the weight* Q *be*

$$Q = \operatorname{diag}(0, 0, ..., q_i I_2, ..., 0, 0); \tag{11.36}$$

then for the low-authority controller ($q_i \leq q_{oi}$) *the closed-loop pair of flexible poles* ($\lambda_{cri} \pm j\lambda_{cii}$) *relates to the open-loop poles* ($\lambda_{ori} \pm j\lambda_{oii}$) *as follows:*

$$(\lambda_{cri} \pm j\lambda_{cii}) \cong (\beta_{ci}\lambda_{ori} \pm j\lambda_{oii}), \qquad i = 1, ..., n, \tag{11.37}$$

or, the real part of the poles is changed by factor β_{ci},

$$\lambda_{cri} \cong \beta_{ci}\lambda_{ori}, \qquad i = 1, ..., n, \tag{11.38}$$

while the imaginary part of the closed-loop poles remains almost unchanged

$$\lambda_{cii} \cong \lambda_{oii}, \qquad i = 1, ..., n, \tag{11.39}$$

where β_{ci} *is defined in* (11.32).

Proof. For small weight q_i the matrix A of the closed-loop system is diagonally dominant, i.e., $A_c \cong \operatorname{diag}(A_{ci})$, $i = 1, ..., n$, and $A_{ci} = A_i - B_i B_i^T s_{ci}$. Introducing the first of (11.26), we obtain

$$A_{ci} \cong A_i + 2s_{ci}\gamma_i(A_i + A_i^T)$$

and introducing A_i as in (3.2) to the above equation we have

$$A_{ci} \cong \begin{bmatrix} -\beta_{ci}\zeta_i\omega_i & -\omega_i \\ \omega_i & -\beta_{ci}\zeta_i\omega_i \end{bmatrix}$$

with β_{ci} as in (11.32). ▣

This result implies that the weight Q as in (11.36) shifts the ith pair of complex poles of the flexible structure, and leaves the remaining pairs of poles almost unchanged. Only the real part of the pair of poles is changed (just moving the pole apart from the imaginary axis and stabilizing the system), and the imaginary part of the poles remains unchanged.

The above proposition has additional interpretations. Note that the real part of the ith open-loop pole is $\lambda_{oi} = -\zeta_i \omega_i$, and that the real part of the ith closed-loop pole is $\lambda_{ci} = -\zeta_{ci}\omega_i$, see Fig. 2.1; note also that the height of the open-loop resonant peak is $\alpha_{oi} = \kappa / 2\zeta_i \omega_i$, where κ is a constant, and the closed-loop resonant peak is $\alpha_{ci} = \kappa / 2\zeta_{ci}\omega_i$. From (11.37) we obtain $\beta_{ci} = \lambda_{cri} / \lambda_{ori}$; hence,

$$\beta_{ci} = \frac{\zeta_{ci}}{\zeta_i} = \frac{\alpha_{oi}}{\alpha_{ci}} \tag{11.40}$$

is a ratio of the closed- and open-loop damping factors, or it is a ratio of the open- and closed-loop resonant peaks. Therefore, if a suppression of the ith resonant peak by the factor β_{ci} is required, the appropriate weight q_i is determined from (11.32), obtaining

$$q_i \cong \frac{(\beta_{ci}^2 - 1)\zeta_i \omega_i}{2 w_{ci}}. \tag{11.41}$$

The variable β_{ci} is also interpreted as a ratio of the variances of the open-loop (σ_{oi}^2) and closed-loop (σ_{ci}^2) states excited by the white noise input

$$\beta_{ci} \cong \frac{w_{oci}}{w_{cci}} = \frac{\sigma_{oi}^2}{\sigma_{ci}^2}. \tag{11.42}$$

This interpretation follows from the closed-loop Lyapunov equations

$$(A - BB^T S_c)W_c + W_c(A - BB^T S_c)^T + BB^T = 0,$$

which for the ith pair of variables is as follows:

$$(A_i - B_i B_i^T s_{ci})w_{cci} + w_{cci}(A_i - B_i B_i^T s_{ci})^T + B_i B_i^T \cong 0.$$

Introducing (11.28) gives

$$w_{cci} + 2 w_{cci} w_{oci} s_{ci} - w_{oci} \cong 0$$

or

$$\frac{w_{oci}}{w_{cci}} \cong 1 + s_{ci} w_{oci} = \beta_{ci}. \tag{11.43}$$

The plots of β_{ci} with respect to the weight q_i and for the controllability factor of the ith mode, $w_{ci} = 1$ are shown in Fig. 11.4. We obtain the same plot with respect to w_{ci} for $q_i = 1$.

The estimator poles are shifted in a similar manner.

Property 11.3(b). ***Root-Locus of the Estimator.*** *Denote*

$$V = \text{diag}(0,0,...,v_i I_2,...,0,0), \tag{11.44}$$

then for moderate weights $(v_i \leq v_{oi})$, *and the estimator pair of poles* $(\lambda_{eri} \pm j\lambda_{eii})$ *relates to the open-loop poles* $(\lambda_{ori} \pm j\lambda_{oii})$ *as follows:*

$$(\lambda_{eri}, \ \pm j\lambda_{eii}) \cong (\beta_{ei}\lambda_{ori}, \ \pm j\lambda_{oii}), \qquad i = 1,...,n, \tag{11.45}$$

or, the real part of the poles is changed by factor β_{ei},

$$\lambda_{eri} \cong \beta_{ei}\lambda_{ori}, \qquad i = 1,...,n, \tag{11.46}$$

while the imaginary part of the closed-loop poles remains almost unchanged

$$\lambda_{eii} \cong \lambda_{oii}, \qquad i = 1,...,n, \tag{11.47}$$

where β_{ei} *is defined in* (11.34).

The above applies for low-authority controllers, i.e., controllers that modify only moderately the system natural frequencies, as defined by Aubrun and Marguiles, see [5] and [6]. The controller authority is limited by the values q_{oi} and v_{oi} such that one has $q_i \leq q_{oi}$ and $v_i \leq v_{oi}$. The limiting values q_{oi} and v_{oi} are not difficult to determine. There are several indicators that the weight q_i approaches q_{oi} (or that v_i approaches v_{oi}). Namely, q_{oi} is the weight at which the ith pair of complex poles of the plant departs significantly from the horizontal trajectory in the root-locus plane and approaches the real axis, see Fig. 10.3. Alternatively, it is a weight at which the ith resonant peak of the plant transfer function disappears (the peak is flattened). A similar result applies to the estimator weights v_{oi}.

11.7 Almost LQG-Balanced Modal Representation

We will show that for the diagonally dominant solutions of CARE and FARE in modal coordinates ($S_c \cong \text{diag}(s_{ci}I_2)$ and $S_e \cong \text{diag}(s_{ei}I_2)$, i = 1, ..., n), we obtain an approximately balanced solution (M) of CARE and FARE in a straightforward manner, by taking a geometric mean of CARE and FARE solutions, i.e.,

$$\text{M} \cong (S_c S_e)^{1/2} = \text{diag}(\mu_i I_2), \tag{11.48}$$

where

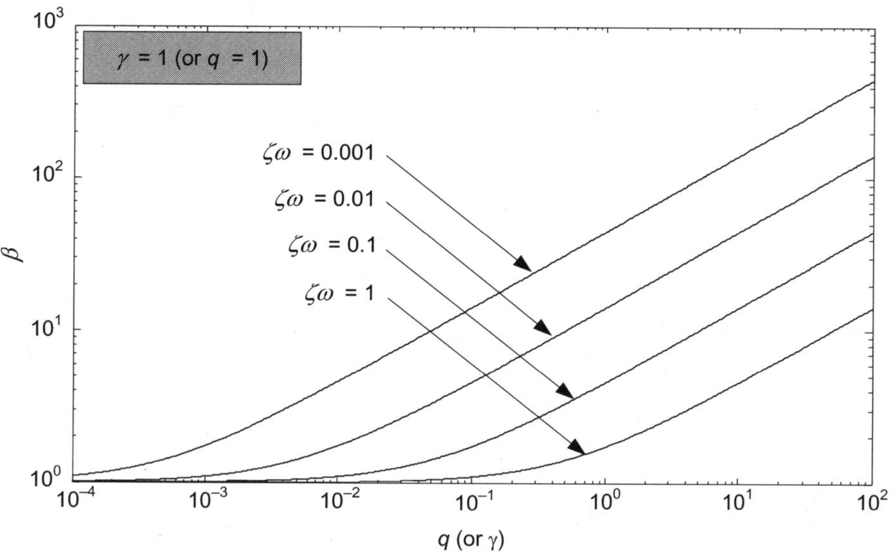

Figure 11.4. Coefficient β versus weight q for $\gamma = 1$, or versus the Hankel singular value γ for $q = 1$.

$$\mu_i = \sqrt{s_{ci} s_{ei}}, \qquad i = 1, \dots, n, \tag{11.49}$$

Matrix M is diagonally dominant, and the transformation R from the modal representation (A, B, C) to the LQG-balanced representation $(A_{lqg}, B_{lqg}, C_{lqg})$ is diagonally dominant as well, in the following form:

$$R \cong \operatorname{diag}(r_1 I_2,\ r_2 I_2, \dots, r_n I_2), \qquad r_i = \left(\frac{s_{ei}}{s_{ci}} \right)^{1/4}. \tag{11.50}$$

Since the state matrix A is diagonally dominant, the transformation scales only the state input and output matrices, while the state matrix A remains unchanged, i.e., $(A_{lqg}, B_{lqg}, C_{lqg}) \cong (A, R^{-1}B, CR)$.

Next, we determine weights that make a structure approximately LQG balanced.

Property 11.4. ***Weights that Approximately LQG Balance the Modal Representation.*** *If the system is in the almost-balanced modal representation, and the weights Q and V are equal and diagonal, $Q = V = \operatorname{diag}(q_i)$, the solutions of the Riccati equations are almost identical*

$$s_{ci} \cong s_{ei}, \qquad i = 1, \dots, n, \tag{11.51}$$

and the open-loop and LQG-balanced representations approximately coincide, i.e.,

$$(A_{lqg}, B_{lqg}, C_{lqg}) \cong (A, B, C). \tag{11.52}$$

Proof. Introducing $q_i = v_i$ and $w_{ci} = w_{oi}$ to (11.31) and (11.33) we find that $s_{ci} \cong s_{ei}$, $i = 1, ..., n$. In this case $R \cong I$ (from (11.50)); hence, the open-loop and LQG-balanced representations are approximately identical. ▣

11.8 Three Ways to Compute LQG Singular Values

From the above analysis we can use one of three ways to compute LQG singular values:

1. From the algorithm in Section 11.3. This algorithm gives the exact LQG singular values. However, the relationship between the LQG singular value and the corresponding natural mode it represents is neither explicit nor obvious.
2. From (11.48). This is an approximate value that gives a connection between the LQG singular values and the corresponding modes.
3. From (11.35). This is an approximate value related to a specific mode. It is a closed-form equation that gives an explicit relationship between structural parameters and the singular value.

11.9 The Tracking LQG Controller

Previously considered LQG controllers were designed for vibration suppression purposes, where the commanding signal was zero. A more complex task includes a tracking controller, where a structure must follow a command. It requires tracking performance in addition to vibration suppression properties. This is the case of controllers for radar and microwave antennas, such as the NASA Deep Space Network antennas. This kind of controller should assure zero steady-state tracking error, which is achieved by adding an integral of the plant position to the plant state-space representation, as reported in [4], [36], [39], [42], [80], [118], and [142]. The closed-loop system configuration of the tracking LQG controller is shown in Fig. 11.5. In this figure (A, B, C) is the plant state-space triple, x is the state, $\hat{x}$ is the estimated state, $\hat{x}_f$ is the estimated state of a flexible part, r is the command, u is the control input, y is the output, $\hat{y}$ is the estimated output, $e = r - \hat{y}$ is the servo error, e_i is the integral of servo error, v is the process noise of intensity V, and the measurement noise w is of intensity W. Both v and w are uncorrelated: $E(vw^T) = 0$, $V = E(vv^T)$, $W = E(ww^T) = I$, $E(v) = 0$, and $E(w) = 0$.

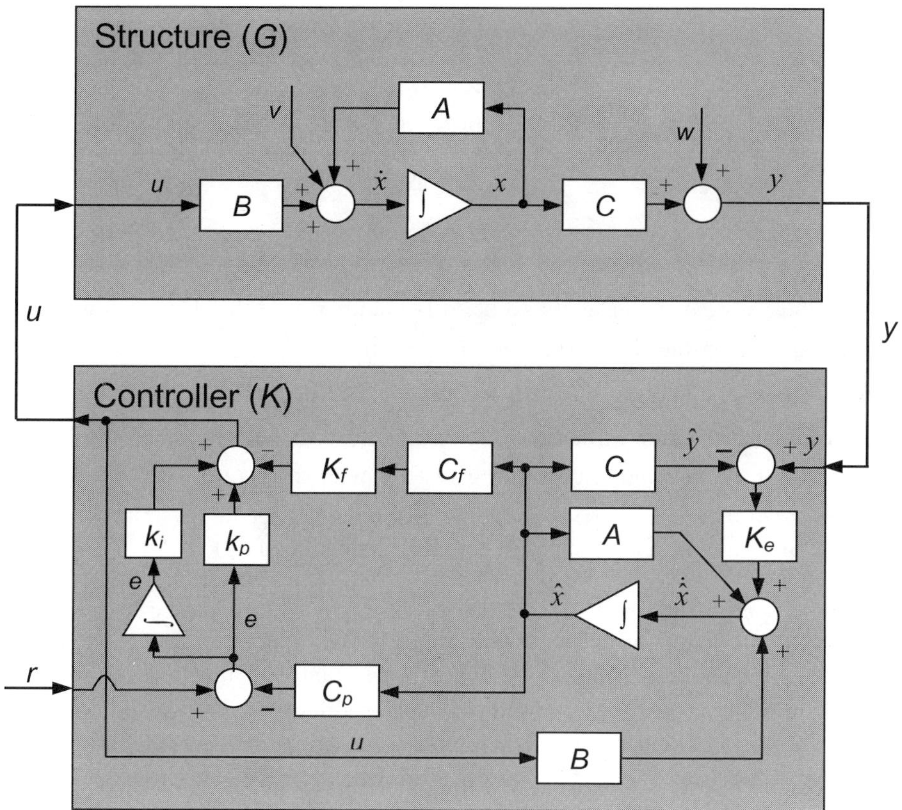

Figure 11.5. The tracking LQG controller with an integral upgrade.

For the open-loop state-space representation (A,B,C) of a flexible structure the state vector x is divided into the tracking, x_t, and flexible, x_f, parts, i.e.,

$$x = \begin{pmatrix} x_t \\ x_f \end{pmatrix}. \tag{11.53}$$

The tracking part includes the structural position, and its integral, while the flexible mode part includes modes of deformation. For this division the system triple can be presented as follows (see [59]):

$$A = \begin{bmatrix} A_t & A_{tf} \\ 0 & A_f \end{bmatrix}, \qquad B = \begin{bmatrix} B_t \\ B_f \end{bmatrix}, \qquad C = \begin{bmatrix} C_t & 0 \end{bmatrix}. \tag{11.54}$$

The gain, K_c, the weight, Q, and solution of CARE, S_c, are divided similarly to x,

$$K_c = \begin{bmatrix} K_{ct} & K_{cf} \end{bmatrix},$$
$$Q = \begin{bmatrix} Q_t & 0 \\ 0 & Q_f \end{bmatrix}, \qquad (11.55)$$
$$S_c = \begin{bmatrix} S_{ct} & S_{ctf} \\ S_{ctf}^T & S_{cf} \end{bmatrix}.$$

The tracking system is considered to be of low authority, if the flexible weights are much smaller than the tracking ones, i.e., such that $\|Q_t\| \gg \|Q_f\|$. It was shown by Collins, Haddad, and Ying [20] that for $Q_f = 0$ one obtains $S_{cf} = 0$ and $S_{ctf} = 0$. This means that the gain of the tracking part, K_{ct}, does not depend on the flexible part. And, for the low-authority tracking system (with small Q_f), one obtains weak dependence of the tracking gains on the flexible weights, due to the continuity of the solution. Similar conclusions apply to the FARE equation (11.8).

This property can be validated by observation of the closed-loop transfer functions for different weights. Consider the transfer function of the Deep Space Network antenna, as in Fig. 11.6. Denote by I_n and 0_n the identity and zero matrices of order n, then the magnitude of the closed-loop transfer function (azimuth angle to azimuth command) for $Q_t = I_2$ and $Q_f = 0_{10}$ is shown as a solid line, for $Q_t = I_2$ and $Q_f = 5 \times I_{10}$ as a dashed line, and for $Q_t = 8 \times I_2$ and $Q_f = 0_{10}$ as a dot–dashed line in Fig. 11.6. It follows from the plots that variations in Q_f changed the properties of the flexible subsystem only, while variations in Q_t changed the properties of both subsystems.

Note, however, that the larger Q_f increases dependency of the gains on the flexible system; only quasi-independence in the final stage of controller design is observed, while separation in the initial stages of controller design is still strong. The design consists therefore of the initial choice of weights for the tracking subsystem, and determination of the controller gains of the flexible subsystem. It is followed by the adjustment of weights of the tracking subsystem, and a final tuning of the flexible weights, if necessary.

11.10 Frequency Weighting

The LQG controller can be designed to meet tracking requirements and, at the same time, maintain the disturbance rejection properties. In order to achieve this, the problem should be appropriately defined in quantitative terms. For this purpose we use the frequency shaping filters to define tracking requirements, or disturbance rejection performance of the closed-loop system. Although these filters are used only in the controller design stage, they add to the complexity of the problem. This is because in the process of design the number of system equations varies and their parameters are modified. As stated by Voth et al. [132, p. 55], "the selection of the controller gains and filters as well as the controller architecture is an iterative, and often tedious, process that relies heavily on the designers' experience."

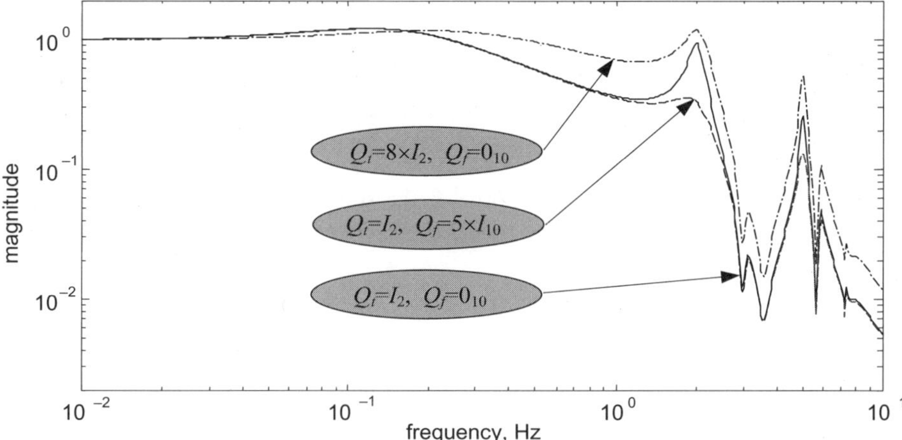

Figure 11.6. Magnitudes of the transfer function of a closed-loop system for different LQG weights: Q_f impacts the flexible modes (higher frequencies), while Q_t impacts the low and high frequencies.

We will show in this section that this comparatively complex task can be simplified in the case of flexible structure control. Structures have special properties that allow for a simple incorporation of filters. Namely, for the system in modal representation, the addition of a filter is equivalent to the multiplication of each row of the input matrix (or input gains) by a constant. The ith constant is the filter gain at the ith natural frequency of the structure. In this way each natural mode is weighted separately. This approach addresses the system performance at the mode level, which simplifies an otherwise ad hoc and tedious process.

Let (A, B, C) be the modal state-space representation of a structure transfer function G, with s inputs and r outputs, and let (A_i, B_i, C_i) and G_i be the state-space representation of the ith mode and its transfer function, respectively. Introduce the following transfer function:

$$\overline{G} = \sum_{i=1}^{n} \overline{G}_i, \tag{11.56}$$

where

$$\overline{G}_i = C_i (j\omega I - A_i)^{-1} \overline{B}_i \tag{11.57}$$

and

$$\bar{B}_i = B_i \alpha_i, \qquad \alpha_i = |F(\omega_i)|. \tag{11.58}$$

Above, α_i is the magnitude of the filter transfer function at the ith natural frequency, and $\bar{G}_i$ is a transfer function G_i with the scaled input matrix B_i.

For the LQG controller the H_2 norm of the transfer function GF is used as a system performance measure. In the modal representation this norm is approximated as follows:

Property 11.5(a). *LQG Input Filtering.* *The H_2 norm of a structure with a smooth input filter is approximately equal to the H_2 norm of a structure with a scaled input matrix B,*

$$\|GF\|_2 \cong \|\bar{G}\|_2 . \tag{11.59}$$

Proof. Using Property 5.7 one obtains

$$\|GF\|_2^2 \cong \sum_{i=1}^{n} \|G_i \alpha_i\|_2^2 = \sum_{i=1}^{n} \|\bar{G}_i\|_2^2 = \|\bar{G}\|_2^2 .$$ ▣

Equation (11.59) shows that the application of the input filter for the H_2 performance modeling is equivalent to the scaling of the $2 \times s$ modal input matrix B_i with α_i.

Similar results are obtained for the output filter:

Property 11.5(b). *LQG Output Filtering.* *The H_2 norm of a structure with a smooth output filter is approximately equal to the H_2 norm of a structure with a scaled output matrix C,*

$$\|FG\|_2 \cong \|\bar{G}\|_2 , \tag{11.60}$$

where

$$\bar{G} = \sum_{i=1}^{n} \bar{G}_i, \qquad \bar{G}_i = \bar{C}_i (j\omega I - A_i)^{-1} B_i, \tag{11.61}$$

and

$$\bar{C}_i = \alpha_i C_i. \tag{11.62}$$

Proof. Similar to the one of Property 11.5(a). ▣

In (11.62) α_i is the magnitude of the filter transfer function at the ith natural frequency. Note that Property 11.5 preserves the order of the system state-space model, as well as physical (modal) interpretation of the transfer function, and the corresponding state variables. This simplifies the controller design process, since the relationship between filter gains and system performance is readily available.

11.11 The Reduced-Order LQG Controller

We see from the previous analysis of the LQG controller that the size of the controller is equal to the size of the plant. However, the size of the plant is often large so that the corresponding controller size is too large to be acceptable for implementation. It is crucial to obtain a controller of the smallest possible order that preserves the stability and performance of the full-order controller. In order to do so the plant model should not be reduced excessively in advance, to assure the quality of the closed-loop system design. Therefore, controller reduction is a part of controller design. The modal LQG design procedure provides this opportunity.

11.11.1 The Reduction Index

In order to perform controller reduction successfully, we introduce an index of the importance of each controller mode. In the open-loop case, modal norms served as reduction indices. In the closed-loop case, Jonckheere and Silverman [82] used the LQG singular values as reduction indices for symmetric and passive systems. Unfortunately, they can produce unstable controllers. This we illustrate later in the simple structure example in this chapter, where the most important controller mode has the lowest LQG singular value.

In this chapter we evaluate the effectiveness of the closed-loop system using the degree of suppression of flexible motion of the structure. The suppression, in turn, depends on the pole mobility into the left-hand side of the complex plane. Therefore, if a particular pair of poles is moved "easily" (i.e., when a small amount of weight is required to move the poles), the respective states are easy to control and estimate. On the contrary, if a particular pair of poles is difficult to move (i.e., even large weights move the poles insignificantly), the respective states (or modes) are difficult to control and estimate, and the action of the controller is irrelevant. Therefore, the states, which are difficult to control and estimate, are reduced. This demonstrates heuristically the rationale of the choice of the pole mobility as an indicator of the importance of controller states.

We perform the reduction of the LQG system in modal coordinates. First, we define the reduction index σ_i as a product of a Hankel singular value and the LQG singular value of the system

$$\sigma_i = \gamma_i \mu_i. \tag{11.63}$$

This combines the observability and controllability properties of the open-loop system and the controller performance. This choice is a result of the fact that σ_i is a measure of the ith pole mobility. Indeed, note from (11.35) and (11.63) that σ_i is the geometric mean of the plant and the estimator pole mobility indexes, i.e.,

$$\sigma_i \cong 0.5\sqrt{(\beta_{ci}-1)(\beta_{ei}-1)}. \tag{11.64}$$

This equation reveals, for example, that for $\beta_{ci}=1$ (no shift of the ith controller pole) σ_i is equal to zero. Similarly, for $\beta_{ei}=1$ (no shift of the ith estimator pole) σ_i is equal to zero, too. However, for a shifted pole one obtains $\beta_{ci}>1$, $\beta_{ei}>1$; hence, the index is also "shifted," that is, $\sigma_i>0$.

We can find an alternative interpretation of σ_i. Denote by σ_{oi}^2 the variance of the open-loop response to white noise, and by σ_{ci}^2 the variance of the closed-loop response to white noise, and note that $w_{oci}=\sigma_{oi}^2$ and $w_{cci}=\sigma_{ci}^2$, where w_{oci} and w_{cci} are the diagonal entries of the open- and closed-loop controllability grammians. Denote by $\Delta\sigma_i^2=\sigma_{oi}^2-\sigma_{ci}^2$ the change of the response of the open- and closed-loop systems due to white noise. Then a useful interpretation of the reduction index follows from (11.43):

$$\sigma_i \cong \frac{\Delta\sigma_i^2}{2\sigma_{ci}^2}. \tag{11.65}$$

This equation shows that the reduction index is proportional to the relative change of the response of the open- and closed-loop systems due to white noise.

Having defined σ_i as the controller performance evaluation tool, we develop the reduction technique. The reduction index is determined from (11.63). But, in order to find the reduction index we need to find the Hankel singular values and LQG singular values. They are found as follows. In modal coordinates the Hankel singular values are approximately equal to the geometric mean of the corresponding controllability and observability grammians, i.e.,

$$\gamma_i \cong \sqrt{w_{cii} w_{oii}}, \tag{11.66}$$

where w_{cii} and w_{oii} are the ith diagonal entries of the controllability and observability grammians, respectively. Similarly, in modal coordinates the solutions of the CARE and FARE equations are approximately equal to the geometric mean of the corresponding CARE and FARE solutions

$$\mu_i \cong \sqrt{s_{cii} s_{eii}}, \tag{11.67}$$

where s_{cii} and s_{eii} are the ith diagonal entries of the CARE and FARE solutions, respectively. Thus, combining the last two equations and (11.63) we obtain

$$\sigma_i \cong \sqrt{w_{cii} w_{oii} s_{cii} s_{eii}}. \tag{11.68}$$

Thus, in modal coordinates the reduction index is obtained from the diagonal entries of the grammians and the diagonal entries of the CARE and FARE solutions.

11.11.2 The Reduction Technique

In order to introduce the controller reduction technique, we define the matrix Σ of the reduction indices as $\Sigma = \operatorname{diag}(\sigma_1, \sigma_2, \ldots, \sigma_N)$, and from (11.63) it follows that

$$\Sigma = \Gamma \mathrm{M}. \tag{11.69}$$

Next, we arrange the diagonal entries σ_i in Σ in descending order, i.e., $\sigma_i > 0$, $\sigma_i \geq \sigma_{i+1}$, $i = 1, \ldots, N$, and divide the matrix as follows:

$$\Sigma = \begin{bmatrix} \Sigma_r & 0 \\ 0 & \Sigma_t \end{bmatrix}, \tag{11.70}$$

where Σ_r consists of the first k entries of Σ, and then Σ_t the remaining ones. If the entries of Σ_t are small in comparison with the entries of Σ_r, the controller can be reduced by truncating its last $N-k$ states.

Note that the value of the index σ_i depends on the weight q_i, so that the reduction depends on the weight choice. For example, if for a given weight a particular resonant peak is too large to be accepted (or a pair of poles is too close to the imaginary axis, or the amplitudes of vibrations at this resonance frequency are unacceptably high), the weighting of this particular mode should be increased to suppress this mode. The growth of weight increases the value of σ_i, which can save this particular mode from reduction.

11.11.3 Stability of the Reduced-Order Controller

The question of stability of the closed-loop system with the reduced-order controller should be answered before implementation of the controller. In order to answer this question, consider the closed-loop system as in Fig. 11.2. Denote the state of the closed-loop system as

$$x_o = \begin{Bmatrix} x \\ \varepsilon \end{Bmatrix}, \tag{11.71}$$

and let $\varepsilon = x - \hat{x}$ be the estimation error. For this state we obtain the following closed-loop equations:

$$\begin{aligned} \dot{x}_o &= A_o x_o + B_o r + B_v v + B_w w, \\ y &= C_o x_o + w, \end{aligned} \tag{11.72}$$

where

$$A_o = \begin{bmatrix} A - BK_c & BK_c \\ 0 & A \quad K_e C \end{bmatrix},$$

$$B_o = \begin{bmatrix} B \\ 0 \end{bmatrix}, \qquad B_v = \begin{bmatrix} I \\ I \end{bmatrix}, \qquad B_w = \begin{bmatrix} 0 \\ -K_e \end{bmatrix}, \qquad C_o = \begin{bmatrix} C & 0 \end{bmatrix}.$$

Let the matrices A, B, C of the estimator be partitioned conformingly to Σ in (11.70),

$$A = \begin{bmatrix} A_r & 0 \\ 0 & A_t \end{bmatrix}, \qquad B = \begin{bmatrix} B_r \\ B_t \end{bmatrix}, \qquad C = \begin{bmatrix} C_r & C_t \end{bmatrix}, \tag{11.73}$$

then the reduced controller representation is (A_r, B_r, C_r). The controller gains are divided similarly

$$K_c = \begin{bmatrix} K_{cr} & K_{ct} \end{bmatrix},$$
$$K_e = \begin{bmatrix} K_{er} \\ K_{et} \end{bmatrix}, \tag{11.74}$$

and the resulting reduced closed-loop system is as follows:

$$A_{or} = \begin{bmatrix} A - BK_p & BK_{pr} \\ 0 & A_r - K_{er}C_r \end{bmatrix}, \tag{11.75}$$

$$B_{or} = \begin{bmatrix} B \\ 0 \end{bmatrix}, \quad B_{vr} = \begin{bmatrix} I \\ I_r \end{bmatrix}, \quad B_w = \begin{bmatrix} 0 \\ -K_{er} \end{bmatrix}, \quad C_o = \begin{bmatrix} C & 0_r \end{bmatrix}. \tag{11.76}$$

Define the stability margin of matrix A_o as follows:

$$m(A_o) = \min_i [-\mathrm{Re}(\lambda_i(A_o))], \tag{11.77}$$

where Re(.) denotes a real part of a complex variable, and $\lambda_i(.)$ is the ith eigenvalue of a matrix, then the following property is valid:

Property 11.6. ***Stability of the Reduced-Order Controller.*** *For $\|\Sigma_t\| \ll \|\Sigma_r\|$, one obtains $m(A_o) \cong m(A_{or})$, where A_{or} is a closed-loop matrix of a system with the reduced controller, and A_t is the state matrix of the truncated part. Hence, the reduced-order controller is stable.*

Proof. Introduce (11.73), (11.74), (11.75) to (11.76) to obtain

$$A_o = \begin{bmatrix} A_{or} & A_{o1} \\ A_{o2} & A_t - K_{et}C_t \end{bmatrix},$$

where

$$A_{o1} = \begin{bmatrix} B_r K_{ct} \\ B_t K_{ct} \\ -K_{er}C_t \end{bmatrix}, \qquad A_{o2} = \begin{bmatrix} 0 & 0 & -K_{et}C_r \end{bmatrix}.$$

The matrix A_o is divided into four blocks, with the upper left block A_{or}. Thus, in order to prove that $m(A_o) \cong m(A_{or})$, it is sufficient to show that: (a) in the lower left block $\|K_{et}C_r\| \cong 0$; and (b) $m(A_t) \geq m(A_o)$; i.e., that in the lower right block $\|K_{et}C_t\| \cong 0$. But for (a) from (11.9), for the LQG-balanced system, one obtains

$\|K_{et}C_r\| = \|M_t C_t^T C_r\| \ll \|M_t C_t^T C_t\| \cong \|\Sigma_t\| \cong 0$; similarly, for (b) $\|K_{et}C_t\| = \|M_t C_t^T C_t\|$ $\cong \|\Sigma_t\| \cong 0$. ▣

11.11.4 Performance of the Reduced-Order Controller

In addition to the stability evaluation, we assess the performance of the reduced-order controller. Denote by $\varepsilon^T = \begin{bmatrix} \varepsilon_r^T & \varepsilon_t^T \end{bmatrix}$ the estimation error of the full-order controller and by ε_{rr} the estimation error of the reduced-order controller, and then we obtain the following property:

Property 11.7. ***Performance of the Reduced-Order Controller.*** *If the states with small reduction indices are truncated, then one obtains*

$$\varepsilon_r \cong \varepsilon_{rr} \qquad \text{and} \qquad \varepsilon_t \cong 0. \tag{11.78}$$

Proof. Note that for A_o as in the previous proof the estimation error is

$$\dot{\varepsilon}_r = (A_r - K_{er}C_r)\varepsilon_r - K_{er}C_t\varepsilon_t$$

and that

$$\dot{\varepsilon}_t = K_{et}C_r\varepsilon_r + (A_t - K_{et}C_t)\varepsilon_t.$$

But, from (11.75), the error of the reduced-order controller is

$$\dot{\varepsilon}_{rr} = (A_r - K_{er}C_r)\varepsilon_{rr}.$$

As shown previously $\|K_{er}C_t\| \cong 0$, and $\|K_{et}C_r\| \cong 0$ for small σ_i; thus, $\varepsilon_r \cong \varepsilon_{rr}$. Additionally, we obtain $\dot{\varepsilon}_t \cong A_t\varepsilon_t$, imposing that for stable A_t the truncation error vanishes ($\varepsilon_t \to 0$) with elapsing time ($t \to \infty$). ▣

The above property implies that for $\|\Sigma_t\| \ll \|\Sigma_r\|$ the performance of the reduced- and full-order controllers is approximately the same. We will show this in the design examples section, where we compare the performance of the full- and reduced-order controllers.

11.11.5 Weights of Special Interest

Here we discuss weights that produce a special form of the CARE/FARE solutions and closed-loop response. First, for a fully controllable system, consider the weights Q and V as follows:

$$Q = 2W_c^{-1}BB^TW_c^{-1},$$
$$V = 2W_o^{-1}C^TCW_o^{-1}. \tag{11.79}$$

In this case we obtain the inverses of the controllability and observability grammians as the CARE and FARE solutions, i.e.,

$$S_c = W_c^{-1},$$
$$S_e = W_o^{-1}. \tag{11.80}$$

We prove this by the introduction of (11.79) and (11.80) into CARE, which gives

$$A^TS_c + S_cA + S_cBB^TS_c = 0. \tag{11.81}$$

Introducing $S_c = W_c^{-1}$ gives the Lyapunov equations (4.5). A similar proof can be shown for the solution of FARE.

The weights as in (11.79) penalize each state reciprocally to its degree of controllability and observability. Particularly, when the weights Q and V are determined in the modal representation, we obtain from (11.79) that the system LQG singular values are $\mathrm{M} \cong \sqrt{S_cS_e} = \sqrt{W_c^{-1}W_o^{-1}} = \Gamma^{-1}$. In this case, the reduction index Σ from (11.69) is $\Sigma = \mathrm{M}\Gamma = \Gamma^{-1}\Gamma = I$, i.e., that all modes are equally important and no reduction is allowed.

Consider another set of weights of a fully controllable system, namely,

$$Q = C^TC + W_oBB^TW_o,$$
$$V = BB^T + W_cC^TCW_c, \tag{11.82}$$

then we obtain the observability and controllability grammians as solutions of the CARE and FARE equations

$$S_c = W_o,$$
$$S_e = W_c. \tag{11.83}$$

We prove this by the introduction of (11.83) to the CARE and FARE equations. For the system in the modal representation the LQG singular values are equal to the Hankel singular values, since $M \cong \sqrt{S_c S_e} = \sqrt{W_c W_o} = \Gamma$. In this case, the reduction index Σ from (11.69) is $\Sigma = M\Gamma = \Gamma^2$, i.e., that the closed-loop reduction can be performed as an open-loop reduction, using Hankel singular values.

11.12 Controller Design Procedure

The following steps help to design an LQG controller:

1. Put the structural model into modal coordinates 1 or 2.
2. Define the performance criteria, such as bandwidth, settling time, overshoot, etc.
3. Assign initial values of weighting matrices Q and V (remember: these matrices are diagonal).
4. Solve the Riccati equations (11.8) and (11.10), find controller gains from (11.7) and (11.9), and simulate the closed-loop performance. Check if the performance satisfies the performance criteria. If not, continue.
5. Check which modes do not satisfy the criteria. Change corresponding weights q_i and v_i, and return to p. 4.
6. If the criteria are not fully satisfied, consider the addition of a filter to achieve the goal. Use the procedure of Section 11.10, by appropriately scaling the input (B) or output (C) matrices in modal coordinates.
7. When the goal is achieved, perform controller reduction. Determine the reduction index as in (11.63) or (11.68), and eliminate the controller states with the small reduction indexes. Simulate the closed-loop system with the reduced-order controller. If the performance of the system with the reduced-order controller is close to the performance of the system with the full-order controller, accept the reduced-order controller; or you may consider further reduction. If the performance of the reduced-order controller departs significantly from the performance of the full-order controller, increase the order of the reduced-order controller, until its performance is satisfactory.

In the above design procedure we show that we can achieve the design goals due to two facts: First, the modes are almost independent, therefore by changing a single weight (or rather a single pair of weights) we change the properties of a single mode, leaving other modes almost unchanged. Second, we know approximately from (11.41) how much weight we need to add in order to damp the vibrations of a selected mode.

11.13 Controller Design Examples

Here we present examples of the design of a modal LQG controller for a simple structure, for the 3D truss structure, and for the Deep Space Network antenna.

11.13.1 A Simple Structure

The Matlab code for this example is in Appendix B. Design the LQG controller for the system shown in Fig. 1.1. The system masses are $m_1 = m_2 = m_3 = 1$, stiffness $k_1 = 10$, $k_2 = 3$, $k_3 = 4$, and $k_4 = 3$, and a damping matrix $D = 0.004K + 0.001M$, where K, M are the stiffness and mass matrices, respectively. The input force is applied to mass m_3, the output is the displacement of the same mass, and the poles of the open-loop system are

$$\lambda_{o1,o2} = -0.0024 \pm j0.9851,$$
$$\lambda_{o3,o4} = -0.0175 \pm j2.9197,$$
$$\lambda_{o5,o6} = -0.0295 \pm j3.8084.$$

The system Hankel singular values for each mode are $\gamma_1 = 64.60$, $\gamma_2 = 1.71$, and $\gamma_3 = 0.063$, hence,

(a) $$\Gamma = \operatorname{diag}(64.60,\ 64.60,\ 1.71,\ 1.71,\ 0.063,\ 0.063).$$

We select the following weight matrix Q and the covariance matrix V: $Q = V = \operatorname{diag}(0.5, 0.5, 1, 1, 2.5, 2.5)$. For these matrices the solution S_c of CARE and the solution S_e of FARE are diagonally dominant,

$$S_c \cong \operatorname{diag}(1.83,\ 1.91,\ 4.45,\ 4.35,\ 20.01,\ 19.81),$$
$$S_e \cong \operatorname{diag}(1.12,\ 0.94,\ 3.75,\ 3.95,\ 20.56,\ 20.87).$$

Next, we obtain the approximate LQG singular values from (11.48) as a geometric mean of the CARE/FARE solutions

(b) $$M_{a1} \cong (S_c S_e)^{1/2} \cong \operatorname{diag}(1.34,\ 1.43,\ 4.09,\ 4.15,\ 20.31,\ 20.33),$$

and the exact LQG singular values obtained from the algorithm in Section 11.3 are

$$M = \operatorname{diag}(1.09,\ 1.60,\ 3.93,\ 4.17,\ 20.17,\ 20.37).$$

The LQG singular values are plotted in Fig. 11.7(a).

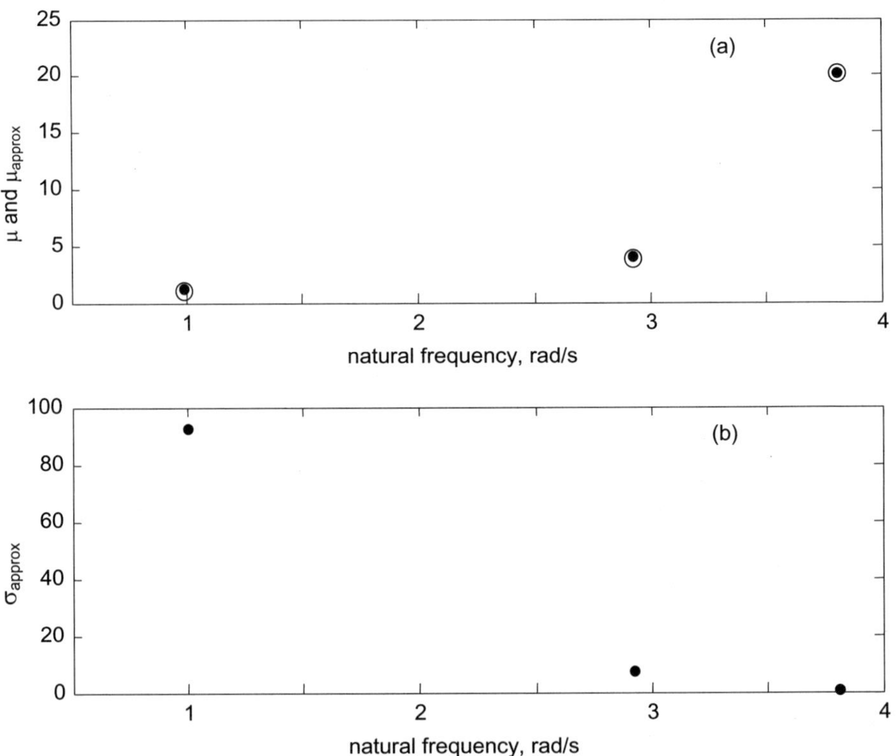

Figure 11.7. A simple system: (a) The exact (○) and approximate (•) LQG singular values coincide; and (b) the controller reduction index shows that the third mode is redundant.

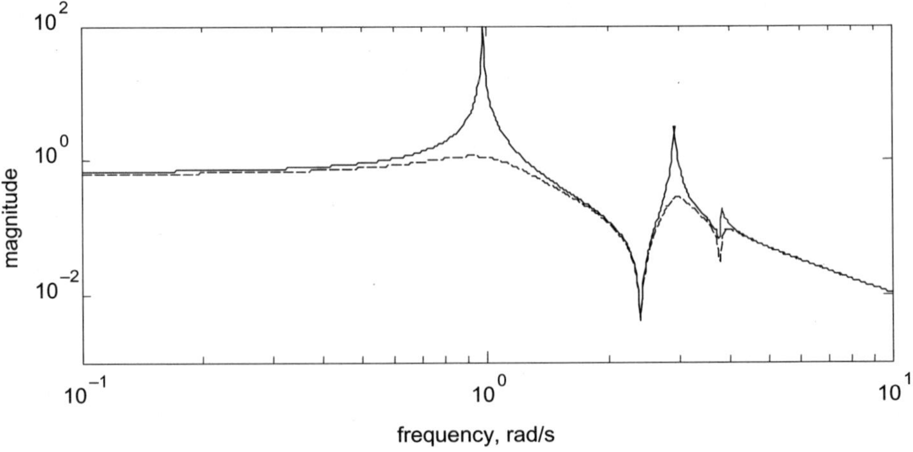

Figure 11.8. Magnitudes of the transfer function of the open-loop (solid line) and closed-loop (dashed line) simple structures: Closed-loop damping increased since resonant peaks are flattened.

The magnitudes of the open- and closed-loop transfer functions are shown in Fig. 11.8. The weights Q and V shift the poles to the right, causing the peaks of the open-loop transfer function (solid line in Fig. 11.8) to flatten, see the closed-loop transfer function (dashed line in Fig. 11.8).

Controller reduction. We obtain the controller reduction matrix Σ as $\Sigma = \mathrm{M}\Gamma$ from the approximate values of M and Γ, using equations (a) and (b); obtaining

$$\Sigma = \mathrm{M}\Gamma = \ \mathrm{diag}(86.56,\ 92.38,\ 6.99,\ 7.10,\ 1.28,\ 1.28).$$

Their plots are shown in Fig. 11.7(b). We truncate the third mode, which has the smallest reduction index ($\sigma_3 = 1.28$), and the reduced LQG controller with the two-mode estimator is applied. Note that a mode with the largest LQG singular value is truncated. The closed-loop transfer functions, with full- and reduced-order controllers coincident, are shown in Fig. 11.9, solid and dashed lines, respectively.

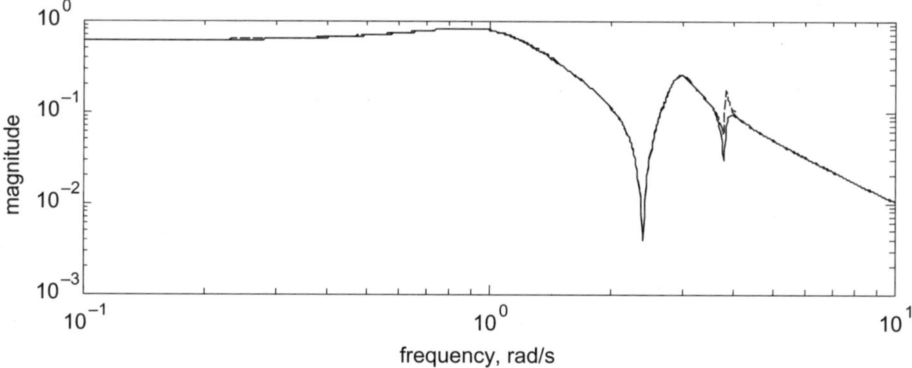

Figure 11.9. Coincident magnitudes of the transfer function of the closed-loop simple structure with full-order (solid line) and reduced-order (dashed line) LQG controllers.

11.13.2 The 3D Truss

We design the LQG controller for the 3D truss as presented in Fig. 1.3. A vertical control force is applied simultaneously at nodes 18 and 24 (the first input), and a horizontal force is applied simultaneously at nodes 6 and 18 (the second input). The combined vertical displacement at nodes 6 and 12 is the first output, and the combined horizontal displacement at nodes 5 and 17 is the second output. The system is in modal almost-balanced representation, and it has (after reduction) 34 states (or 17 modes). We assume the weight (Q) and the covariance (V) matrices equal and diagonal, i.e., $Q = V = \mathrm{diag}(q_1, q_1, q_2, q_2, \ldots, q_{17}, q_{17})$ where $q_1 = q_2 = 400$, $q_3 = q_4 = 4000$, $q_5 = q_6 = 40000$, $q_7 = \cdots = q_{17} = 400$. In this case the CARE and FARE solutions are approximately equal and diagonally dominant, as stated in Section 11.5. In Fig. 11.10(a) we show the exact LQG singular values (from the algorithm, Section 11.3), and the approximate singular values (from (11.35)); they confirm satisfactory coincidence. Poles of the open- as well as the closed-loop system and the estimator are shown in Fig. 11.11. For the modal almost-balanced controller the poles of the closed-loop system and the estimator overlap.

In Fig. 11.12(a) we compare the open-loop (solid line) and closed-loop (dashed line) impulse responses from the first input to the first output. They show that the closed-loop system has increased damping. Comparing the open-loop transfer function (solid line in Fig. 11.12(b)) and the closed-loop transfer function (dashed line) we see that the oscillatory motion of the structure is damped out.

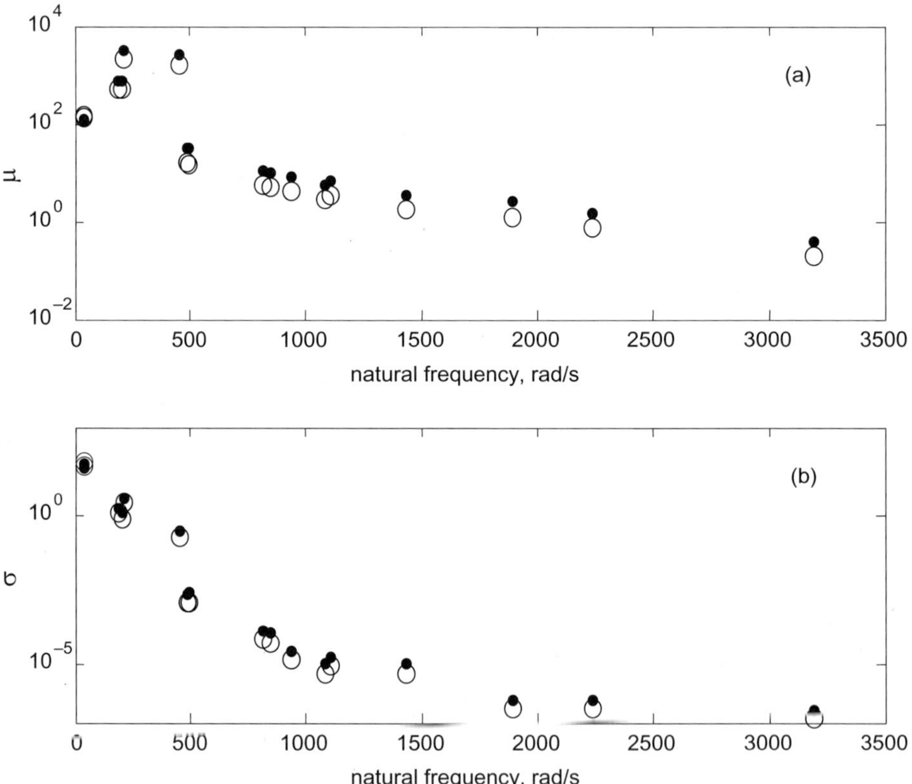

Figure 11.10. The 3D truss: (a) The coinciding exact (○) and approximate (•) LQG singular values; and (b) the coinciding exact (○) and approximate (•) controller reduction indices.

The diagonal entries of the reduction matrix Σ are shown in Fig. 11.10(b). We obtained them from (11.69) using exact and approximate values of Γ and M. We reduced the order of the controller by truncating 18 states that are associated with the small reduction indices (i.e., such that $\sigma_i < 0.01$). The resulting reduced-order controller has 16 states. The reduction did not impact the closed-loop dynamics, since the magnitude of the transfer function of the full-order controller (solid line) and reduced-order controller (dashed line) overlap; see the illustration in Fig. 11.13.

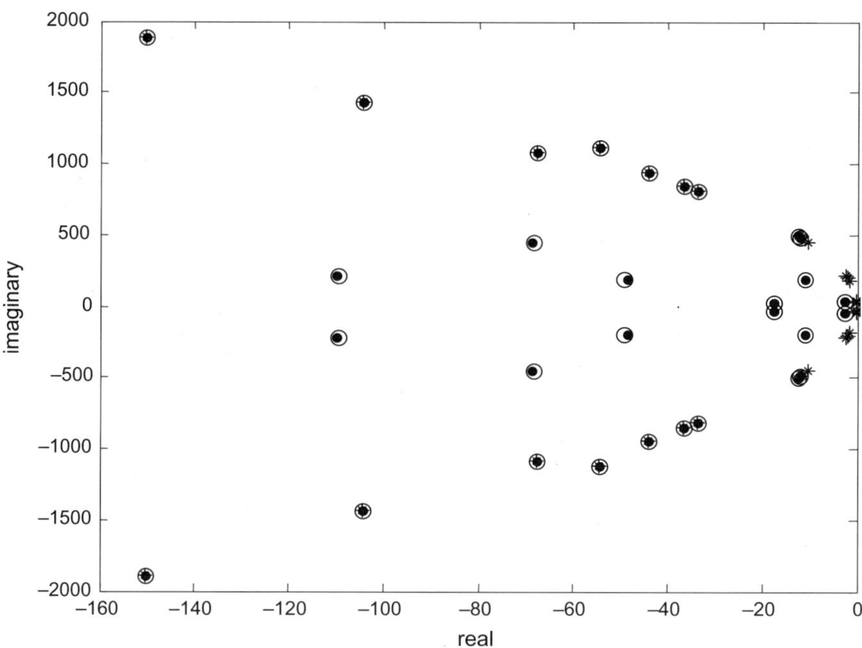

Figure 11.11. Poles of the open-loop truss (∗), of the closed-loop truss (•), and of the estimator (○).

11.13.3 The 3D Truss with Input Filter

We design the LQG controller for the steel truss as in Fig. 1.3. The disturbance is applied at node 7 in the z-direction, the performance is measured at node 21, in the same direction. The input u is applied at node 20 in the z-direction, and the output y is a displacement of node 28, in the same direction. The open-loop transfer function from the disturbance to the performance is shown in Fig. 11.14 (solid line). The disturbance input is filtered with a low-pass filter, of transfer function $F(s) = 1/(1+0.011s)$. The magnitude of its transfer function is shown in the same figure by a dot–dashed line. The resulting transfer function of the structure and filter is represented by the dotted line.

We obtained the equivalent structure with the filter by scaling the disturbance input according to (11.57) and (11.58). The magnitude of its transfer function is shown in Fig. 11.14 (dashed line). It is clear from that figure that the structure with the filter, and the structure with the scaled disturbance input, have similar frequency characteristics. In order to compare how close they are, we calculated their H_2 norms, obtaining $\|G\|_2 = 2.6895$ for the structure with the filter, and $\|G\|_2 = 2.6911$ for the structure with the scaled disturbance input.

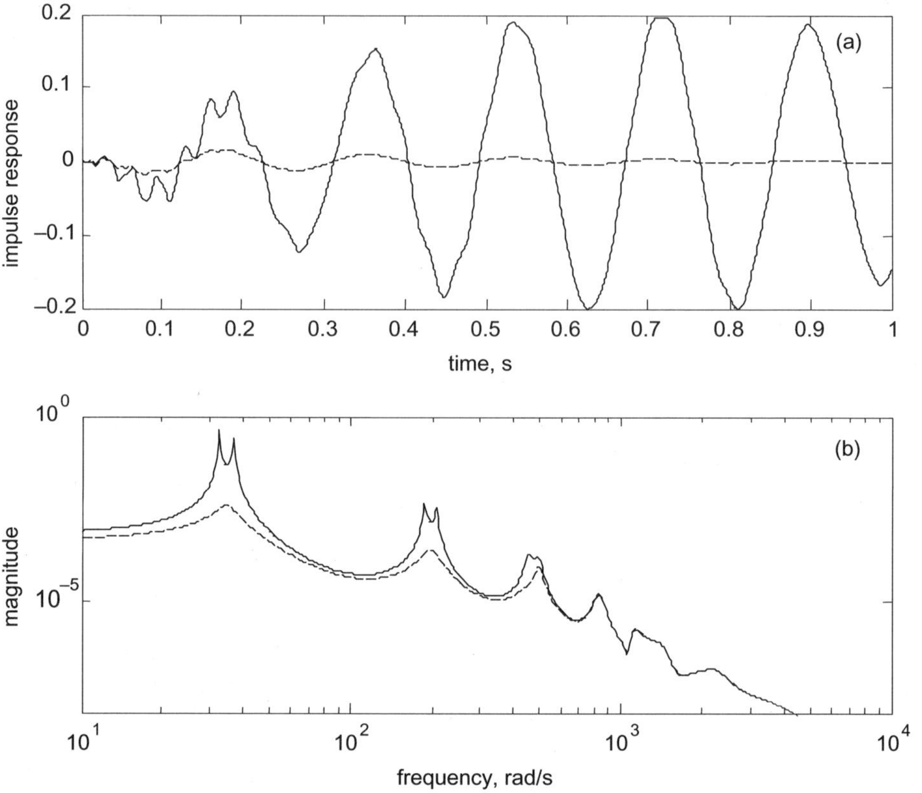

Figure 11.12. The 3D truss: (a) Impulse responses; and (b) magnitudes of the transfer function of the open-loop (solid line) and closed-loop (dashed line) trusses, from the first input to the first output. Closed-loop responses show increased damping.

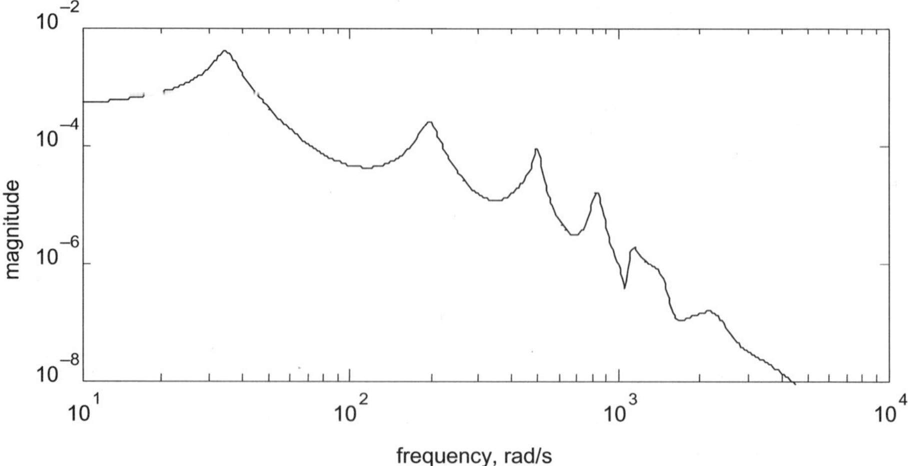

Figure 11.13. Overlapped magnitudes of the transfer function of the closed-loop truss with a full-order (solid line), and reduced-order (dashed line) LQG controller.

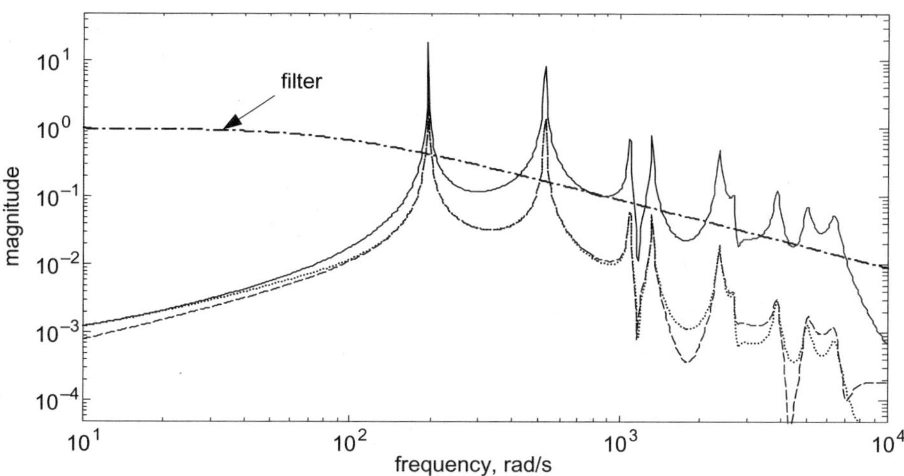

Figure 11.14. Magnitudes of the transfer function of the truss (solid line), filter (dash–dot line), truss with filter (dotted line), and truss with scaled disturbance input (dashed line): Scaling has a similar effect as filter application.

We designed two frequency weighted LQG controllers for this structure. The first one is based on the structure with a filter, while the second is based on the structure with the scaled input matrix. The magnitudes of the closed-loop transfer function of a structure with a filter, and with the scaled input matrix are shown in Fig. 11.15. The plot shows that both systems have almost identical performance. Indeed, the closed-loop H_2 norms are as follows: $\|G_{cl}\|_2 = 0.4153$ for the structure with the filter, and $\|G_{cl}\|_2 = 0.4348$ for the structure with the scaled disturbance input.

11.13.4 The Deep Space Network Antenna

We illustrate the design of a modal LQG controller for the azimuth axis of the Deep Space Network antenna. For this design we use the 18-state reduced antenna modal model obtained in Example 6.9. We assume the weight, Q, and plant noise covariance, V, equal and diagonal.

In the first step, we upgrade the Deep Space Network antenna model with the integral-of-the-position state. After upgrade the model consists of two tracking states (azimuth angle and its integral), a state with the real pole that corresponds to the drive dynamics, and eight flexible modes (consisting of 16 states). For the tracking subsystem (consisting of the angle y and its integral y_i) we assumed the preliminary weights of $q_1 = q_2 = 1$, and the drive state weight of $q_3 = 33$. We chose the weights for the flexible subsystem such that the flexible modes show increased damping; this was obtained for the following weights: $q_4 = \cdots = q_7 = 33$ and for $q_8 = \cdots = q_{19} = 10$. Next, we calculate the step response of the closed-loop system, which is shown in

Fig. 11.16(a). It shows 8 s settling time. We also obtain the closed-loop transfer function, and it is shown in Fig. 11.16(b). It has a bandwidth of 0.2 Hz.

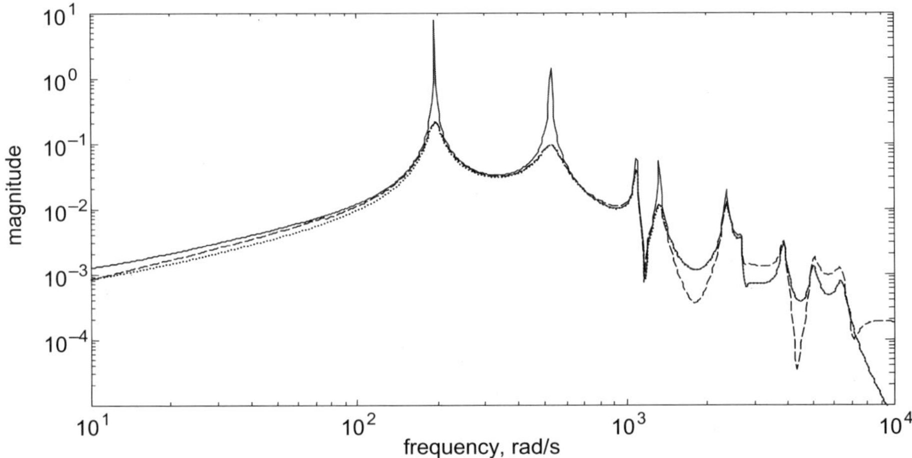

Figure 11.15. Open–loop transfer function (solid line) and closed-loop transfer function, for a structure with scaled input matrix (dashed line) and for a structure with a filter (dotted line): Scaling has a similar effect as filter application.

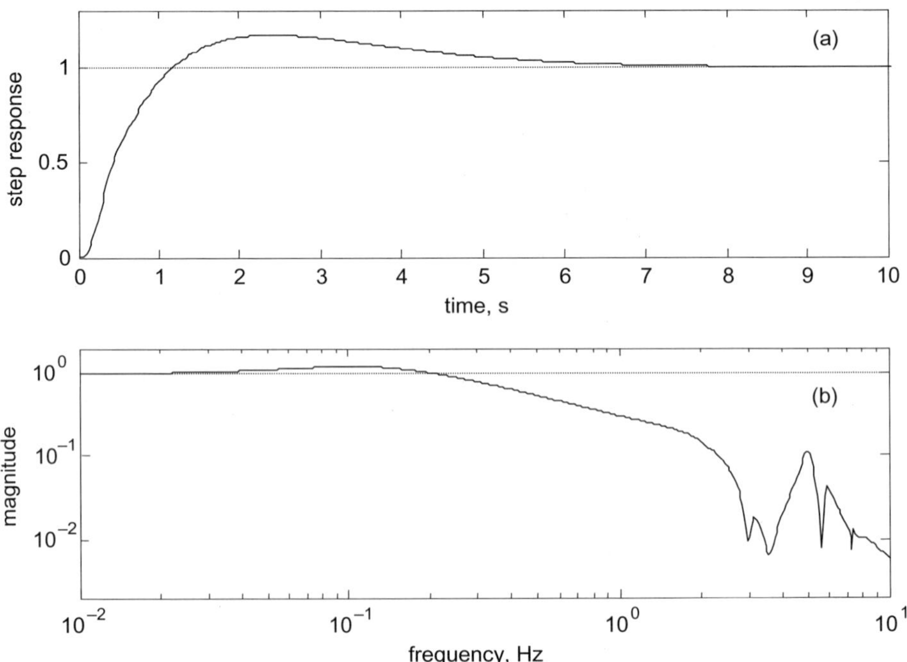

Figure 11.16. The initial design of the antenna LQG controller shows a large response time and low bandwidth: (a) Closed-loop step response; and (b) closed-loop magnitude of the transfer function.

In the next step we improve the tracking properties of the system by the weight adjustment of the tracking subsystem. By increasing the proportional and integral weights to $q_1 = q_2 = 100$ the tracking properties are improved, see the step response in Fig. 11.17(a) (small overshoot and settling time is 3 s) and in the magnitude of the transfer function Fig. 11.17(b) (the bandwidth is extended up to 2 Hz).

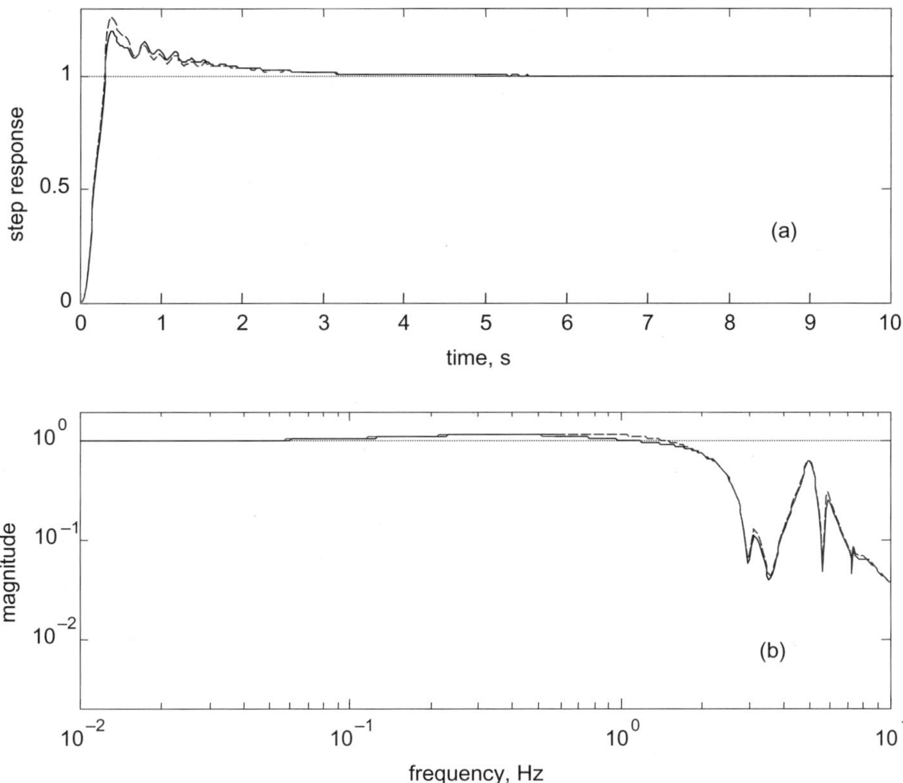

Figure 11.17. Final design of the antenna LQG controller, full-order (solid line), reduced-order (dashed line): (a) Closed-loop step response; and (b) closed-loop magnitude of the transfer function. The design features fast step response and increased bandwidth.

Next we obtain the reduced-order controller through the evaluation of controller reduction indices σ_i. The plot of σ_i is shown in Fig. 11.18. Reducing the order of the estimator to 10 states (preserving the tracking states, and the eight flexible mode states) yields a stable and accurate closed-loop system. Indeed, the reduced-order controller shows satisfactory accuracy, when compared with the full-order controller in the step response plots in Fig. 11.17(a) and with the transfer function plots in Fig. 11.17(b). For more on the LQG controller for the antennas, see [42] and http://tmo.jpl.nasa.gov/tmo/progress_report/42-112/112J.PDF.

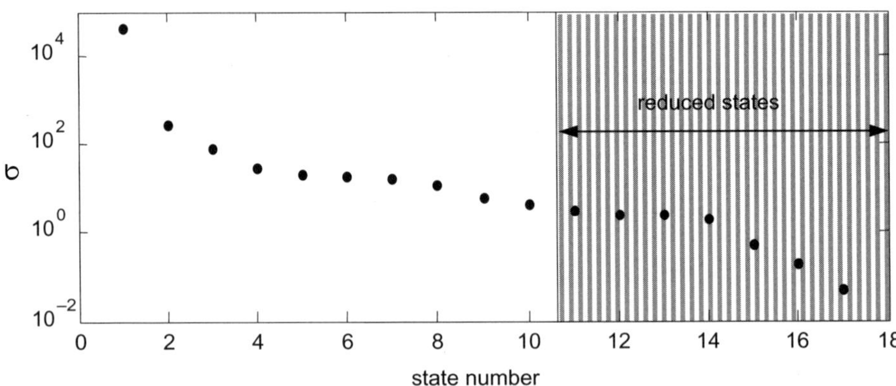

Figure 11.8. Controller reduction index for the Deep Space Network antenna

12
H_∞ and H_2 Controllers

how to control a generalized structure

Black holes are
where God divided by zero.
—*Steven Wright*

In the LQG controller design we assumed that the control inputs were collocated with disturbances, and that the control outputs were collocated with the performance. This assumption imposes significant limits on the LQG controller possibilities and applications. The locations of control inputs do not always coincide with the disturbance locations, and the locations of controlled outputs are not necessarily collocated with the location where the system performance is evaluated. This was discussed earlier, when the generalized structure was introduced. The H_2 and H_∞ controllers address the controller design problem in its general configuration of non-collocated disturbance and control inputs, and noncollocated performance and control outputs. Many books and papers have been published addressing different aspects of H_∞ controller design, and [12], [30], [94], [99], [100], [104], [122], and [129] explain the basic issues of the method. The H_∞ method addresses a wide range of the control problems, combining the frequency- and time-domain approaches. The design is an optimal one in the sense of minimization of the H_∞ norm of the closed-loop transfer function. The H_∞ model includes colored measurement and process noise. It also addresses the issues of robustness due to model uncertainties, and is applicable to the single-input–single-output systems as well as to the multiple-input–multiple-output systems.

In this chapter we present the H_∞ controller design for flexible structures. We chose the modal approach to H_∞ controller design, which allows for the determination of a stable reduced-order H_∞ controller with performance close to the full-order controller.

12.1 Definition and Gains

The closed-loop system architecture is shown in Fig. 12.1. In this figure G is the transfer function of a plant (or structure), K is the transfer function of a controller, w is the exogenous input (such as commands, disturbances), u is the actuator input, z is the regulated output (at which performance is evaluated), and y is the sensed (or controlled) output. This system is different from the LQG control system as in Fig. 11.1: besides the actuator input and controlled output it has disturbance input and the regulated output. Needless to say, it represents a broader class of systems than the LQG control system.

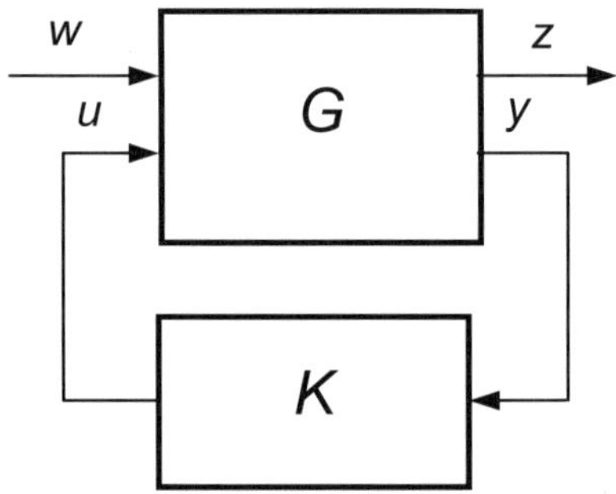

Figure 12.1. The H_∞ closed-loop system configuration: G—plant, K—controller, u—actuator input, w—exogenous input, y—sensed output, and z—regulated output.

For a closed-loop system as in Fig. 12.1 the plant transfer function $G(s)$ and the controller transfer function $K(s)$ are such that

$$\begin{pmatrix} z(s) \\ y(s) \end{pmatrix} = G(s) \begin{pmatrix} w(s) \\ u(s) \end{pmatrix},$$
$$u(s) - K(s)y(s), \tag{12.1}$$

where u, w are control and exogenous inputs and y, z are measured and controlled outputs, respectively. The related state-space equations of a structureare as follows:

$$\begin{aligned} \dot{x} &= Ax + B_1 w + B_2 u, \\ z &= C_1 x + D_{12} u, \\ y &= C_2 x + D_{21} w. \end{aligned} \tag{12.2}$$

Hence, the state-space representation in the H_∞ controller description consists of the quintuple (A, B_1, B_2, C_1, C_2). For this representation (A, B_2) is stabilizable and (A, C_2) is detectable, and the conditions

$$D_{12}^T \begin{bmatrix} C_1 & D_{12} \end{bmatrix} = \begin{bmatrix} 0 & I \end{bmatrix},$$
$$D_{21} \begin{bmatrix} B_1^T & D_{21}^T \end{bmatrix} = \begin{bmatrix} 0 & I \end{bmatrix} \tag{12.3}$$

are satisfied. When the latter conditions are satisfied the H_∞ controller is called the central H_∞ controller. These are quite common assumptions, and in the H_2 control they are interpreted as the absence of cross terms in the cost function $(D_{12}^T C_1 = 0)$, and the process noise and measurement noise are uncorrelated $(B_1 D_{21}^T = 0)$.

The H_∞ control problem consists of determining controller K such that the H_∞ norm of the closed-loop transfer function G_{wz} from w to z is minimized over all realizable controllers K, that is, one needs to find a realizable K such that

$$\|G_{wz}(K)\|_\infty \tag{12.4}$$

is minimal. Note that the LQG control system depends on y and u rather than on w and z, as above.

The solution says that there exists an admissible controller such that $\|G_{wz}\|_\infty < \rho$, where ρ is the smallest number such that the following four conditions hold:

1. $S_{\infty c} \geq 0$ solves the following central H_∞ controller algebraic Riccati equation (HCARE),

$$S_{\infty c} A + A^T S_{\infty c} + C_1^T C_1 - S_{\infty c}(B_2 B_2^T - \rho^{-2} B_1 B_1^T) S_{\infty c} = 0. \tag{12.5}$$

2. $S_{\infty e} \geq 0$ solves the following central H_∞ filter (or estimator) algebraic Riccati equation (HFARE),

$$S_{\infty e} A^T + A S_{\infty e} + B_1 B_1^T - S_{\infty e}(C_2^T C_2 - \rho^{-2} C_1^T C_1 S_{\infty e} = 0. \tag{12.6}$$

3.

$$\lambda_{\max}(S_{\infty c} S_{\infty e}) < \rho^2, \tag{12.7}$$

where $\lambda_{\max}(X)$ is the largest eigenvalue of X.

4. The Hamiltonian matrices

$$\begin{bmatrix} A & \rho^{-2}B_1B_1^T - B_2B_2^T \\ -C_1^TC_1 & -A^T \end{bmatrix},$$
$$\begin{bmatrix} A^T & \rho^{-2}C_1^TC_1 - C_2^TC_2 \\ -B_1B_1^T & -A \end{bmatrix}, \tag{12.8}$$

do not have eigenvalues on the $j\omega$-axis.

With the above conditions satisfied the optimal closed-loop system is presented in Fig. 12.2, and the controller state-space equations, from the input y to the output u, are obtained from the block-diagram in Fig. 12.2,

$$\begin{aligned} \dot{\hat{x}} &= (A + \rho^{-2}B_1B_1^TS_{\infty c} - B_2K_c - K_eC_2)\hat{x} + K_ey, \\ u &= -K_c\hat{x}. \end{aligned} \tag{12.9}$$

According to the above equations the H_∞ controller state-space representation $(A_\infty, B_\infty, C_\infty)$ is as follows:

$$\begin{aligned} A_\infty &= A + \rho^{-2}B_1B_1^TS_{\infty c} - B_2K_c - K_eC_2, \\ B_\infty &= K_e, \\ C_\infty &= -K_c, \end{aligned} \tag{12.10}$$

where

$$K_c = B_2^TS_{\infty c} \tag{12.11}$$

and

$$\begin{aligned} K_e &= S_oS_{\infty e}C_2^T, \\ S_o &= (I - \rho^{-2}S_{\infty e}S_{\infty c})^{-1}. \end{aligned} \tag{12.12}$$

The gain K_c is called the controller gain, while K_e is the filter (estimator) gain. The order of the controller state-space representation is equal to the order of the plant. Note that the form of the H_∞ solution is similar to the LQG solution. However, the LQG gains are determined independently, while the H_∞ gains are coupled through the inequality (12.7), and through the component S_o in (12.12).

How is the H_∞ norm of the closed-loop transfer system w to z minimized? Through the gains that depend on the solution of the Riccati equations (12.5) and (12.6), which in turn depend of the w input matrix B_1, and the z output matrix C_1.

12.2 The Closed-Loop System

We derive the closed-loop equations starting from the state-space equations of the open-loop system, see (12.2),

$$\begin{aligned} \dot{x} &= Ax + B_1 w + B_2 u, \\ z &= C_1 x + D_{12} u, \\ y &= C_2 x + D_{21} w. \end{aligned} \tag{12.13}$$

Next, we obtain the state-space equations of the central H_∞ controller from the block-diagram in Fig. 12.1, or (12.9),

$$\begin{aligned} \dot{\hat{x}} &= (A - K_e C_2 + \rho^{-2} B_1 B_1^T S_{\infty c} - B_2 K_c)\hat{x} + K_e y, \\ u &= -K_c \hat{x}. \end{aligned} \tag{12.14}$$

Defining a new state variable

$$x_o = \begin{Bmatrix} x \\ \varepsilon \end{Bmatrix}, \tag{12.15}$$

where $\varepsilon = x - \hat{x}$, we obtain the closed-loop state-space equations in the following form:

$$\begin{aligned} \dot{x}_o &= A_o x_o + B_o w, \\ z &= C_o x_o, \end{aligned} \tag{12.16}$$

where

$$\begin{aligned} A_o &= \begin{bmatrix} A - B_2 K_c & B_2 K_c \\ -\rho^{-2} B_1 B_1^T S_{\infty c} & A - K_e C_2 + \rho^{-2} B_1 B_1^T S_{\infty c} \end{bmatrix}, \\ B_o &= \begin{bmatrix} B_1 \\ B_1 - K_c D_{21} \end{bmatrix}, \\ C_o &= \begin{bmatrix} C_1 + D_{12} K_c & -D_{12} K_c \end{bmatrix}. \end{aligned} \tag{12.17}$$

The block diagram of the closed-loop system is shown in Fig. 12.2.

Assuming $\rho^{-1} = 0$ in (12.16) and (12.17), one obtains the H_2 system, which has structure identical to the LQG controller as in Fig. 11.2.

12.3 The Balanced H_∞ Controller

The balanced H_∞ controller helps to reduce the controller size. An H_∞ controller is balanced if the related HCARE and HFARE solutions are equal and diagonal, see [110] and [56], i.e., if

$$S_{\infty c} = S_{\infty e} = M_\infty,$$
$$M_\infty = \text{diag}(\mu_{\infty 1}, \mu_{\infty 2}, ..., \mu_{\infty N}), \tag{12.18}$$

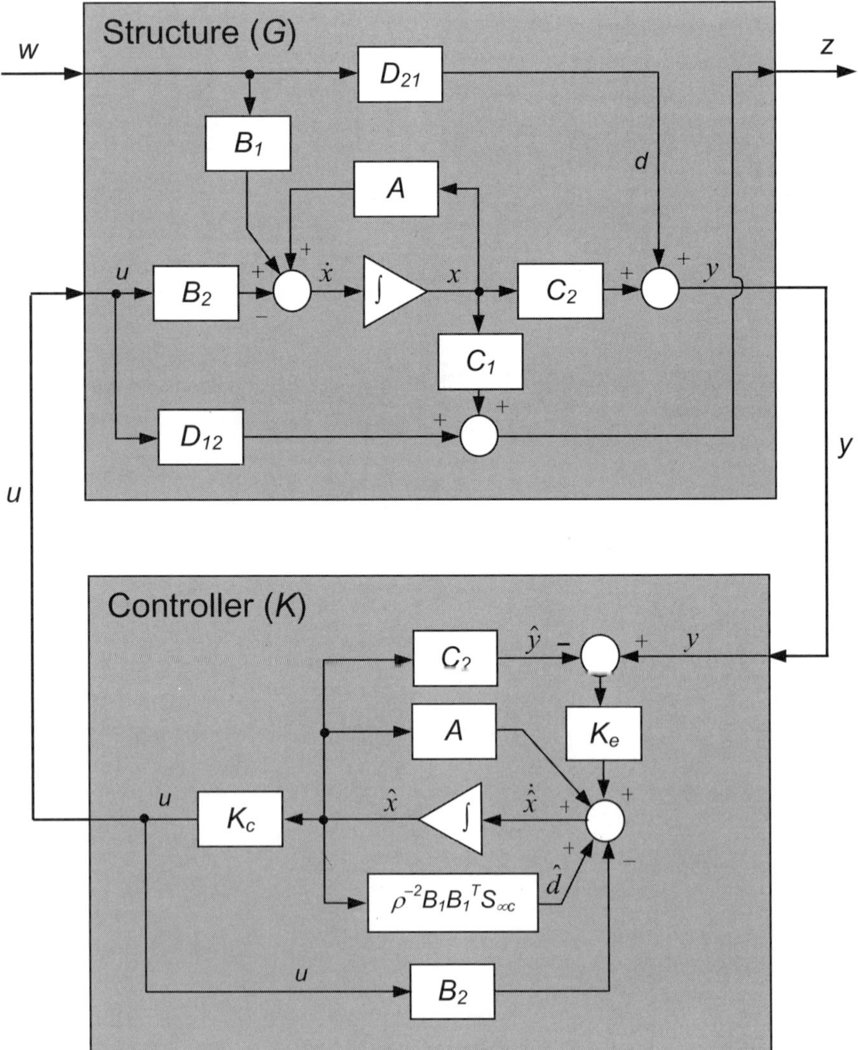

Figure 12.2. An H_∞ closed-loop system.

$\mu_{\infty 1} \geq \mu_{\infty 2} \geq \cdots \geq \mu_{\infty N} > 0$, where $\mu_{\infty i}$ is the ith H_∞ singular (or characteristic) value.

The transformation R to the H_∞-balanced representation is determined as follows:

- Find the square roots $P_{\infty c}$ and $P_{\infty e}$, of the HCARE and HFARE solutions

$$P_{\infty c}^T P_{\infty c} = S_{\infty c}, \qquad P_{\infty e} P_{\infty e}^T = S_{\infty e}. \tag{12.19}$$

- Denote $N_\infty = P_{\infty c} P_{\infty e}$ and find the singular value decomposition of N_∞,

$$N_\infty = V_\infty M_\infty U_\infty^T. \tag{12.20}$$

- Obtain the transformation R in the following form:

$$R = P_{\infty e} U_\infty \mathrm{M}_\infty^{-1/2} \qquad \text{or} \qquad R = P_{\infty c}^{-1} V_\infty \mathrm{M}_\infty^{1/2}. \tag{12.21}$$

The state $\bar{x}$, such that $\bar{x} = Rx$, is H_∞ balanced and the state-space representation is $(R^{-1}AR, R^{-1}B_1, R^{-1}B_2, C_1R, C_2R)$.

In order to prove this, note that the solutions of HCARE and HFARE in new coordinates are $\bar{S}_{\infty c} = R^T S_{\infty c} R$, $\bar{S}_{\infty e} = R^{-1} S_{\infty e} R^{-T}$. Introducing *R*, as in (12.21), we obtain the balanced HCARE and HFARE solutions.

The Matlab function *bal_H_inf.m* in Appendix A.13 transforms a representation (A, B_1, B_2, C_1, C_2) to the H_∞-balanced representation $(A_b, B_{b1}, B_{b2}, C_{b1}, C_{b2})$.

For the H_∞-balanced solution, the condition in (12.7) simplifies to

$$\mu_{\infty 1} < \rho \qquad \text{and} \qquad \mu_{\infty n} > 0. \tag{12.22}$$

In the following we establish the relationship between H_∞ singular values and open-loop (or Hankel) singular values. Let the matrix inequalities be defined as follows: $X_1 > X_2$ if the matrix $X_1 - X_2$ is positive definite, and by $X_1 \geq X_2$ if the matrix $X_1 - X_2$ is positive semidefinite. For asymptotically stable A, and for $V > 0$, consider two Riccati equations:

$$A^T S_1 + S_1 A - S_1 W_1 S_1 + V = 0, \qquad A^T S_2 + S_2 A - S_2 W_2 S_2 + V = 0. \tag{12.23}$$

If $W_2 > W_1 > 0$, we obtain

$$S_1 \geq S_2 \geq 0, \tag{12.24}$$

see [25].

Property 12.1. H_∞ and Hankel Singular Values. *Let Γ_1 be a matrix of Hankel singular values of the state representation (A, B_1, C_1), and let M_∞ be a matrix of H_∞ singular values defined in* (12.18). *Then, for an asymptotically stable A, and for $B_2B_2^T - \rho^{-2}B_1B_1^T \geq 0$, $C_2^TC_2 - \rho^{-2}C_1^TC_1 \geq 0$, we obtain*

$$M_\infty \leq \Gamma_1 \quad or \quad \mu_{\infty i} \leq \gamma_{1i}, \qquad i = 1,\ldots,N. \tag{12.25}$$

Proof. Note that (12.25) is a consequence of the property given by (12.24). This property is applied to (12.5), and to the Lyapunov equation $W_oA + A^TW_o + C_1^TC_1 = 0$. It is also a consequence of property (12.24) applied to (12.6), and to the Lyapunov equation $W_cA^T + AW_c + B_1B_1^T = 0$. In this way, we obtain $W_{c1} \geq S_{\infty e}$ and $W_{o1} \geq S_{\infty c}$. From the latter inequalities it follows that $\lambda_i(W_{c1}) \geq \lambda_i(S_{\infty e})$ and $\lambda_i(W_{o1}) \geq \lambda_i(S_{\infty c})$ (see [73, p. 471]); thus, $\lambda_i(W_{c1}W_{o1}) \geq \lambda_i(S_{c1}S_{\infty e})$ or $M_\infty \leq \Gamma_1$. ▣

12.4 The H_2 Controller

The H_2 controller is a special case of the H_∞ controller but, at the same time, it is a generalization of the LQG controller. It minimizes the H_2 norm similarly to the LQG index, but its two-input–two-output structure (disturbance and control inputs are not collocated and performance and sensor outputs are not collocated either) is similar to the H_∞ controller.

12.4.1 Gains

The open-loop state-space representation for the H_2 controller is given by (12.2). It is the same as for the H_∞ system, and we define its matrices $A, B_1, B_2, C_1, C_2, D_{21}$, and D_{12} in the following, based on [12].

The controlled system consists of state x, control input u, measured output y, exogenous input $w^T = \begin{bmatrix} v_u^T & v_y^T \end{bmatrix}$, and regulated variable $z = C_1x + D_{12}u$, where v_u and v_y are process and measurement noises, respectively. The noises v_u and v_y are uncorrelated, and have constant power spectral density matrices V_u and V_y, respectively. For the positive-semidefinite matrix V_u, matrix B_1 has the following form:

$$B_1 = \begin{bmatrix} V_u^{1/2} & 0 \end{bmatrix}. \qquad (12.26)$$

The task is to determine the controller gain (K_c) and the estimator gain (K_e), such that the performance index (J) as in (11.5) is minimal, where R is a positive-definite input weight matrix and Q is a positive-semidefinite state weight matrix.

Matrix C_1 is defined through the weight Q,

$$C_1 = \begin{bmatrix} 0 \\ Q^{1/2} \end{bmatrix} \qquad (12.27)$$

and, without loss of generality, we assume $R = I$ and $V_v = I$, obtaining

$$D_{12} = \begin{bmatrix} I \\ 0 \end{bmatrix}, \quad D_{21} = \begin{bmatrix} 0 & I \end{bmatrix}. \qquad (12.28)$$

The minimum of J is achieved for the feedback with gain matrices (K_c and K_e) as follows:

$$K_c = B_2^T S_{2c}, \quad K_e = S_{2e} C_2^T, \qquad (12.29)$$

where S_{2c} and S_{2e} are solutions of the controller algebraic Riccati equation (CARE) and the estimator algebraic Riccati equation (FARE), respectively, which in this case are as follows:

$$S_{2c}A + A^T S_{2c} + C_1^T C_1 - S_{2c} B_2 B_2^T S_{2c} = 0, \quad S_{2e}A^T + AS_{2e} + B_1 B_1^T - S_{2e} C_2^T C_2 S_{2e} = 0. \qquad (12.30)$$

Note by comparing (12.5), (12.6), and (12.30) that the H_2 solution is a special case of the H_∞ solution by assuming $\rho^{-1} = 0$, for which the inequality (12.7) is unconditionally satisfied.

12.4.2 The Balanced H_2 Controller

An H_2 controller is balanced if the related CARE and FARE solutions are equal and diagonal. We derive the relationship between the H_∞ and H_2 characteristic values as follows:

Property 12.2. ***The Relationship Between H_∞, H_2, and Hankel Singular Values.***

$$\mathrm{M}_2 \leq \mathrm{M}_\infty \qquad \text{or} \qquad \mu_{2i} \leq \mu_{\infty i}, \tag{12.31}$$

$$\mathrm{M}_2 \leq \Gamma_1 \qquad \text{or} \qquad \mu_{2i} \leq \gamma_{1i}, \qquad i = 1, \ldots, N. \tag{12.32}$$

Proof. Let $\beta = \inf\{\rho : \mathrm{M}_\infty(\rho) \geq 0\}$. Then on the segment $(\beta, +\infty)$ all H_∞ characteristic values are smooth nonincreasing functions of ρ, and the maximal characteristic value $\mu_{\infty 1}$ is a nonincreasing convex function of ρ; see [95]. As a consequence, for $\rho \to \infty$ one obtains $\mathrm{M}_\infty \to \mathrm{M}_2$. However, $\mu_{\infty i}$ are increasing functions of ρ, and $\mu_{\infty i} \to \mu_{2i}$ as $\rho \to \infty$, thus $\mu_{2i} \leq \mu_{\infty i}$. The second part is a direct consequence of (12.31) and Property 12.1. ▣

12.5 The Low-Authority H_∞ Controller

We extend the properties of flexible structures to H_∞ control design. These properties are valid for a low-authority controller of moderate action. In this case flexible structure properties are reflected in the properties of the H_∞ controller. Let (A, B_1, B_2, C_1, C_2) be the open-loop modal representation of a flexible structure (in the modal form 1 or 2), and let $A_{c1} = A - B_2 B_2^T S_{\infty c}$, $A_{c2} = A - S_o S_{\infty e} C_2 C_2$ be the closed-loop matrices, where $S_{\infty c}$ and $S_{\infty e}$ are the solutions of the HCARE and HFARE equations, respectively, and $S_o = (I - \rho^{-2} S_{\infty e} S_{\infty c})^{-1}$. Denote by b_i the ith row of B. The H_∞ controller is of low authority if for the closed-loop matrix A_{c1} we obtain $\mathrm{eig}(A_{c1}) \cong \mathrm{eig}(A - \mathrm{diag}(BB^T S_c)$. In other words, for the low-authority controller we can replace BB^T with its diagonal terms. Similarly, for the low-authority H_∞ controller we can replace $C^T C$ with its diagonal terms, obtaining $\mathrm{eig}(A_{c2}) \cong \mathrm{eig}(A - S_o S_e \,\mathrm{diag}(C^T C))$.

We can find a positive scalar s_o such that for $\|S_c\|_2 \leq s_o$ and $\|S_e\|_2 \leq s_o$ the H_∞ controller is of low authority. For a low-authority controller the following property holds:

Property 12.3. ***HCARE and HFARE in Modal Coordinates.*** *For the low-authority H_∞ controller the solutions of HCARE and HFARE in modal coordinates are diagonally dominant*

$$S_{\infty c} \cong \operatorname{diag}(s_{\infty ci} I_2),$$
$$S_{\infty e} \cong \operatorname{diag}(s_{\infty ei} I_2), \qquad i = 1, \ldots, n, \tag{12.33}$$

and the H_∞ singular values are obtained as follows:

$$\mu_{\infty i} = \sqrt{s_{\infty ci} s_{\infty ei}}, \qquad i = 1, \ldots, n. \tag{12.34}$$

Furthermore, if A is in the modal form 2, one can use (11.25) *as replacements for BB^T (or $C^T C$).*

Proof. The proof is similar to the proof of Property 11.1. The second part follows from the diagonally dominant solutions of HCARE and HFARE. ▣

We obtained the diagonal solutions of HCARE and HFARE under low-authority assumption. Often, for low values of the parameter ρ, some modes do not satisfy the low-authority conditions. The HCARE and HFARE solutions for these modes are no longer diagonal, and the total solution is in the block-diagonal form, as in Fig. 12.3. However, this block-diagonal form is equally useful in applications, since it remains diagonally dominant for those modes that preserve the low-authority properties. These modes are subjected to truncation in the controller reduction process. They are weakly correlated with the remaining modes, and their reduction index is small, which makes the truncation stable and the truncation error small.

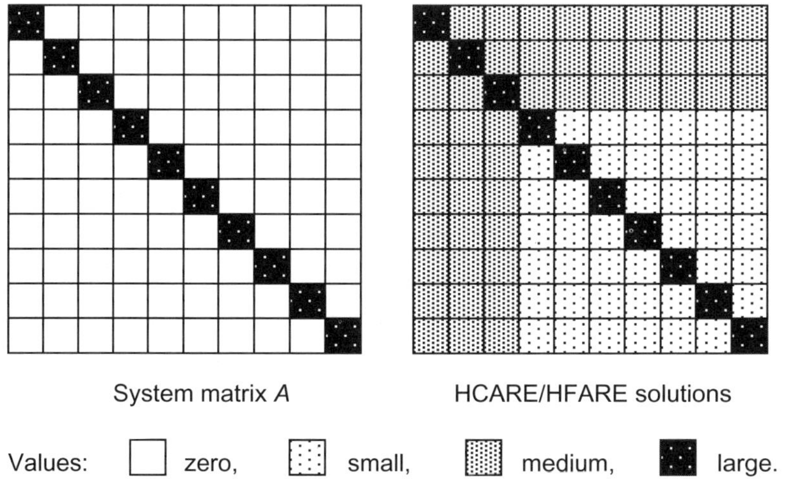

Figure 12.3. Modal matrix A and HCARE/HFARE solutions for the partially low-authority H_∞ controller: The solution for the flexible mode part is diagonally dominant.

12.6 Approximate Solutions of HCARE and HFARE

We obtain the approximate solutions of HCARE and HFARE in closed-form, in order to understand the relationship between structural parameters and the closed-loop system performance. For flexible structures in modal coordinates, we use Properties 12.3 and 11.1 to obtain the Riccati equations (12.5), (12.6) in the following form:

$$\begin{aligned} &\kappa_{ci} s_{c\infty i}^2 + s_{c\infty i} - w_{o1i} \cong 0, \\ &\kappa_{ei} s_{e\infty i}^2 + s_{e\infty i} - w_{c1i} \cong 0, \qquad i = 1, \ldots, n, \end{aligned} \tag{12.35}$$

where

$$\begin{aligned} \kappa_{ci} &= w_{c2i} - \frac{w_{c1i}}{\rho^2}, \\ \kappa_{ei} &= w_{o2i} - \frac{w_{o1i}}{\rho^2}. \end{aligned} \tag{12.36}$$

The solutions of the ith equation are

$$\begin{aligned} s_{c\infty i} &\cong \frac{\beta_{ci} - 1}{2\kappa_{ci}}, \\ s_{e\infty i} &\cong \frac{\beta_{ei} - 1}{2\kappa_{ei}}, \end{aligned} \tag{12.37}$$

where

$$\begin{aligned} \beta_{ci} &= \sqrt{1 + 4 w_{o1i} \kappa_{ci}} = \sqrt{1 + 4\gamma_{21i}^2 - 4\rho^{-2}\gamma_{11i}^2}, \\ \beta_{ei} &= \sqrt{1 + 4 w_{c1i} \kappa_{ei}} = \sqrt{1 + 4\gamma_{12i}^2 - 4\rho^{-2}\gamma_{11i}^2}, \end{aligned} \tag{12.38}$$

and γ_{jki} is the ith Hankel singular value between the jth input and the kth output. The H_∞ singular values are real and positive for $\kappa_{ci} > 0$ and $\kappa_{ei} > 0$.

From (12.35), we obtain $\kappa_{ci} s_{c\infty i}^2 + s_{c\infty i} \cong w_{o1i}$ and $\kappa_{ei} s_{e\infty i}^2 + s_{e\infty i} \cong w_{c1i}$. Thus,

$$s_{c\infty i} \le w_{o1i} \qquad \text{for} \qquad \kappa_{ci} \ge 0, \tag{12.39}$$

and

$$s_{e\infty i} \le w_{c1i} \qquad \text{for} \qquad \kappa_{ei} \ge 0. \tag{12.40}$$

Introducing (12.37) to (12.34) we obtain the approximate H_∞ singular values as follows:

$$\mu_{\infty i} \cong \frac{\sqrt{(\beta_{ci}-1)(\beta_{ei}-1)}}{2\kappa_i}, \tag{12.41}$$

where

$$\kappa_i = \sqrt{\gamma_{22i}^2 - \rho^{-2}\gamma_{12i}^2 - \rho^{-2}\gamma_{21i}^2 + \rho^{-4}\gamma_{11i}^2}. \tag{12.42}$$

Consider a special case of the equal cross-coupling between the two inputs and two outputs, i.e., $\gamma_{12} = \gamma_{21}$. For this case, $\beta_{ci} = \beta_{ei} = \beta$ and $\gamma_{12}^2 = \gamma_{21}^2 = \gamma_{11}\gamma_{22}$; therefore,

$$\mu_{\infty i} \cong \frac{\beta - 1}{2\kappa_i}, \qquad \kappa_i = \gamma_{22} - \rho^{-2}\gamma_{11}. \tag{12.43}$$

Setting $\rho^{-1} = 0$ specifies the above results for the H_2 systems. Thus, for the H_2 controller $\kappa_{ci} = w_{c2i}$, and for $\kappa_{ei} = w_{o2i}$ from (12.37) and (12.38), it follows that

$$\begin{aligned} s_{c2i} &= \frac{\beta_{2ci} - 1}{2w_{c2i}}, \\ s_{e2i} &= \frac{\beta_{2ei} - 1}{2w_{o2i}}, \end{aligned} \tag{12.44}$$

are the approximate solutions of the modal H_2 Riccati equations, and

$$\begin{aligned} \beta_{2ci} &= \sqrt{1 + 4\gamma_{21i}^2}, \\ \beta_{2ei} &= \sqrt{1 + 4\gamma_{12i}^2}. \end{aligned} \tag{12.45}$$

Thus, $\mu_{2i} = \sqrt{s_{c2i}s_{e2i}}$ is the ith characteristic value of an H_2 system, obtained from (12.39) and (12.40) for $\rho^{-1} = 0$,

$$\mu_{2i} \cong \frac{\sqrt{(\beta_{2ci}-1)(\beta_{2ei}-1)}}{\gamma_{22i}}. \tag{12.46}$$

Also, from (12.43) one obtains

$$\mu_{2i} \le \mu_{\infty i} \le \gamma_{1i}^2, \\ \mu_{21} \le \rho \le \gamma_{11}^2, \tag{12.47}$$

for $\kappa_{ci} > 0$ and $\kappa_{ei} > 0$.

12.7 Almost H$_\infty$-Balanced Modal Representation

For the diagonally dominant solutions of HCARE and HFARE in modal coordinates, see (12.33), we find the approximately balanced solution M_∞ of HCARE and HFARE, which is also diagonally dominant, i.e.,

$$\mathrm{M}_\infty \cong \mathrm{diag}(\mu_{\infty i} I_2), \\ \mu_{\infty i} = \sqrt{s_{\infty ci} s_{\infty ei}}, \qquad i = 1, \ldots, n. \tag{12.48}$$

The modal representation for which the solutions of HCARE and HFARE are approximately equal is called the almost H$_\infty$-balanced representation. The transformation R from the modal representation (A, B_1, B_2, C_1, C_2) to the H$_\infty$ almost-balanced representation $(A_{abh}, B_{abh1}, B_{abh2}, C_{abh1}, C_{abh2})$ is diagonal

$$R = \mathrm{diag}(r_1 I_2, r_2 I_2, \ldots, r_n I_2), \\ r_i = \left(\frac{s_{\infty ei}}{s_{\infty ci}} \right)^{1/4}, \tag{12.49}$$

and

$$(A_{abh}, B_{abh1}, B_{abh2}, C_{abh1}, C_{abh2}) = (A, R^{-1} B_1, R^{-1} B_2, C_1 R, C_2 R). \tag{12.50}$$

Note that this transformation requires only a rescaling of the input and output matrices.

Indeed, the modal representation $(A, R^{-1}B_1, R^{-1}B_2, C_1 R, C_2 R)$ is almost H$_\infty$ balanced, and the HCARE, HFARE solution M_∞ is diagonally dominant in the modal almost-balanced coordinates. This we can prove by noting that the solutions of HCARE and HFARE are $S_{\infty ch} = R^T S_{\infty c} R$ and $S_{\infty eh} = R^{-1} S_{\infty e} R^{-T}$ and introducing R, as in (12.49), we obtain the balanced solution as in (12.50). Note that the values of $s_{\infty ci}$ and $s_{\infty ei}$ depend on the choice of coordinates, but their product does not.

12.8 Three Ways to Compute H_∞ Singular Values

The above analysis allows us to compute the H_∞ singular values in three different ways:

1. From the algorithm in Section 12.3. This algorithm gives the exact H_∞ singular values. However, the relationship between the H_∞ singular value and the corresponding natural mode it represents is not explicit.
2. From (12.33), (12.34). These approximate values give an explicit connection between H_∞ singular values and natural modes.
3. From (12.41). This is an approximate value related to a single mode. The largest singular values may be inaccurate, but the closed-form equation gives an explicit relationship between structural parameters and the singular value.

12.9 The Tracking H_∞ Controller

The tracking control problem differs from the regulation problem because controller performance depends not only on the plant parameters, but also on the tracking command profile (its rate, acceleration, etc.). It is useful to formulate the tracking problem such that the requirements are met by definition. One important requirement for tracking systems is to maintain zero steady-state error for constant-rate command. Upgrading the plant with an integrator can satisfy this requirement, as was already discussed in the LQG controller design in Chapter 11. An H_∞ tracking controller with an integral upgrade is presented in Fig. 12.4. For this configuration the design approach is similar to the LQG tracking controller design presented earlier, see Section 11.9.

12.10 Frequency Weighting

In order to meet the specified performance requirements we need smooth pre- and post-compensating filters. Typically, filters are smooth, i.e., their transfer function satisfies conditions (5.28), and for smooth filters Property 5.8 is valid. This property says that the H_∞ norm of a smooth filter in series with a flexible structure is approximately equal to the norm of a structure alone with the input (output) matrices scaled by the filter gains at natural frequencies.

Denote by $\bar{G}_i$ a transfer function of the ith mode G_i with the scaled input matrix B_i; see (11.58). We show that the H_∞ norms of both transfer functions are approximately equal.

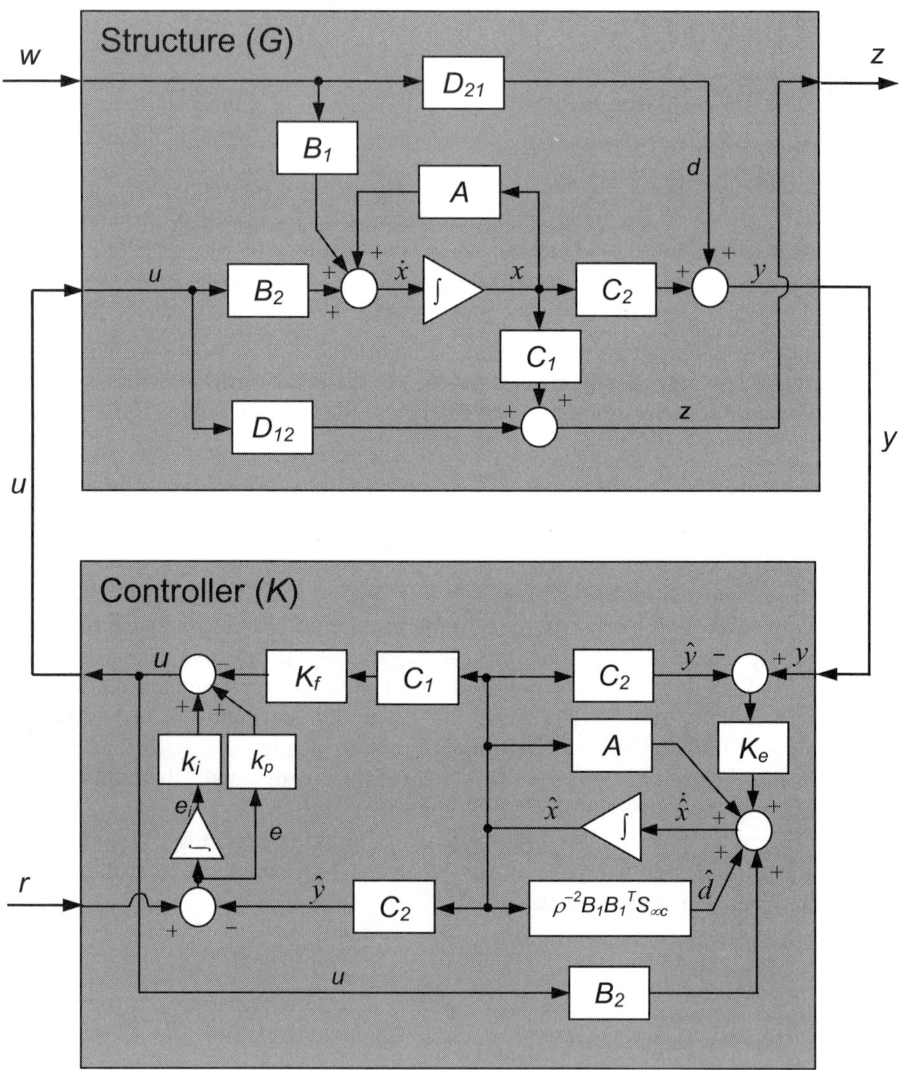

Figure 12.4. An H_∞ tracking controller with an integral upgrade.

Property 12.4(a). H_∞ *Input Filtering.* *The H_∞ norm of a structure with a smooth input filter is approximately equal to the H_∞ norm of a structure with a scaled input matrix B,*

$$\|GF\|_\infty \cong \|\bar{G}\|_\infty, \tag{12.51}$$

where

$$\bar{G} = \sum_{i=1}^{n} \bar{G}_i, \qquad (12.52)$$
$$\bar{G}_i = C_i (j\omega I - A_i)^{-1} \bar{B}_i,$$

and

$$\bar{B}_i = B_i \alpha_i, \qquad (12.53)$$
$$\alpha_i = |F(\omega_i)|.$$

Proof. From Property 5.8 we obtain

$$\|GF\|_\infty \cong \max_i \|G_i \alpha_i\|_\infty = \max_i \|\bar{G}_i\|_\infty \cong \|\bar{G}\|_\infty, \qquad i = 1, \ldots, n. \qquad ■$$

Equation (12.51) shows that the application of the input filter for the H_∞ performance modeling is equivalent to the scaling of the $2 \times n$ input matrix B_i with α_i, where α_i is the magnitude of the filter transfer function at the resonant frequency ω_i, $\alpha_i = |F(\omega_i)|$; see (5.27).

Property 12.4(b). H_∞ *Output Filtering.* *The H_∞ norm of a structure with a smooth output filter is approximately equal to the H_∞ norm of a structure with a scaled output matrix C,*

$$\|FG\|_\infty \cong \|\bar{G}\|_\infty, \qquad (12.54)$$

where

$$\bar{G} = \sum_{i=1}^{n} \bar{G}_i, \qquad (12.55)$$
$$\bar{G}_i = \bar{C}_i (j\omega I - A_i)^{-1} B_i,$$

and

$$\bar{C}_i = \alpha_i C_i. \qquad (12.56)$$

Proof. Similar to Property 12.4(a). ■

Equation (12.54) shows that the application of the output filter for the H_∞ performance modeling is equivalent to the scaling of the $2 \times n$ output matrix C_i

with α_i, where α_i is the magnitude of the filter transfer function at the resonant frequency.

12.11 The Reduced-Order H_∞ Controller

The order of the state-space representation of the H_∞ controller is equal to the order of the plant, which is often too large for implementation. Order reduction is therefore a design issue worth consideration. The reduction of a generic H_∞ controller is not a straightforward task; however, an H_∞ controller for flexible structures inherits special properties that are used for controller reduction purposes.

12.11.1 The Reduction Index

We introduce the following reduction index for the H_∞ controller:

$$\sigma_{\infty i} = \gamma_{22i}\mu_{\infty i}. \tag{12.57}$$

In this index γ_{22i} is the ith Hankel singular value of (A, B_2, C_2), and $\mu_{\infty i}$ is the ith H_∞ singular value. The index $\sigma_{\infty i}$ serves as an indicator of importance of the ith mode of the H_∞ controller. If $\sigma_{\infty i}$ is small, the ith mode is considered negligible and can be truncated.

When $\rho^{-1} = 0$ the H_∞ controller becomes the H_2 controller. Indeed, for $\rho^{-1} = 0$, we get $\sigma_{\infty i} = \sigma_{2i}$,

$$\sigma_{2i} = \gamma_{22i}\mu_{2i}, \tag{12.58}$$

i.e., the H_2 controller reduction index.

The choice of reduction index as in (12.57) is justified by the properties of the closed-loop system, presented below.

12.11.2 Closed-Loop Poles

Let $(A_\infty, B_\infty, C_\infty)$ be the state-space representation of the central H_∞ controller as in (12.10). Defining the closed-loop state variable as in (12.15), we obtain the closed-loop modal state-space equations as in (12.16). Divide A_o into submatrices

$$A_o = \begin{bmatrix} A_{11} & A_{12} \\ A_{21} & A_{22} \end{bmatrix}, \tag{12.59}$$

where

$$\begin{aligned} A_{11} &= A - B_2 K_c, \\ A_{12} &= B_2 K_c, \\ A_{21} &= -\rho^{-2} B_1 B_1^T \mathrm{M}_\infty, \\ A_{22} &= A - K_e C_2 + \rho^{-2} B_1 B_1^T \mathrm{M}_\infty, \end{aligned} \tag{12.60}$$

to prove the following property:

Property 12.5. ***Closed-Loop Poles.*** *If*

$$\sigma_{\infty i} \ll \sigma_{\infty 1}, \qquad \text{for} \qquad i = k+1, \ldots, n, \tag{12.61}$$

then the ith pole is shifted approximately by $2\sigma_{\infty i}$ with respect to the open-loop location, i.e.,

$$A_{22i} \cong A_i - 2\sigma_{\infty i} I_2. \tag{12.62}$$

Proof. In modal coordinates, A is diagonal and the following components are diagonally dominant:

$$\begin{aligned} B_2 K_c = B_2 B_2^T \mathrm{M}_\infty &\cong \operatorname{diag}(2\zeta_i \omega_i w_{c2i} \mu_{\infty i}), \\ K_e C_2 = \mathrm{M}_\infty C_2^T C_2 &\cong \operatorname{diag}(2\zeta_i \omega_i w_{o2i} \mu_{\infty i}), \\ \rho^{-2} B_1 B_1^T \mathrm{M}_\infty &\cong \operatorname{diag}\left(\frac{2\zeta_i \omega_i w_{c1i} \mu_{\infty i}}{\rho^2} \right); \end{aligned} \tag{12.63}$$

thus, each of the four blocks of A_o is diagonally dominant. If $\sigma_{\infty i} \ll \sigma_{\infty 1}$ for $i = k+1, \ldots, n$, then the ith diagonal components of A_{12} and A_{21} are small for $i = k+1, \ldots, n$. Thus for those components the separation principle is valid and gains k_{ci} and k_{ei} are independent. Furthermore, the ith diagonal block A_{22i} of the matrix A_{22} is as follows:

$$A_{22i} \cong A_i - s_{oi} \mu_{\infty i} C_{2i}^T C_{2i} - \rho^{-2} B_{1i} B_{1i}^T \mu_{\infty i}, \tag{12.64}$$

where A_i is given by (2.53). Note, in addition, that $s_{oi} \cong 1$ for $\sigma_{\infty i} \ll \sigma_{\infty 1}$, thus $\mu_{\infty i} C_{2i}^T C_{2i} \cong 2\zeta_i \omega_i w_{o2i} \mu_{\infty i} I_2$, and also that $\rho^{-2} B_{1i} B_{1i}^T \mu_{\infty i} \cong 2\zeta_i \omega_i \rho^{-2} w_{c1i} \mu_{\infty i} I_2$. Consequently, (12.64) now becomes $A_{22i} \cong A_i - 2\sigma_{\infty i} I_2$ or (12.62). ▣

12.11.3 Controller Performance

Let the error vector ε be partitioned as follows:

$$\varepsilon = \begin{Bmatrix} \varepsilon_r \\ \varepsilon_t \end{Bmatrix} \tag{12.65}$$

with ε_r of dimension n_r, ε_t of dimension n_t, such that $n_r + n_t = n$. Let the matrix of the reduction indices be arranged in decreasing order, $\Sigma_\infty = \operatorname{diag}(\sigma_{\infty 1} I_2, ..., \sigma_{\infty n} I_2)$, $\sigma_{\infty i} \geq \sigma_{\infty i+1}$, and be divided consistently by ε,

$$\Sigma_\infty = \begin{bmatrix} \Sigma_{\infty r} & 0 \\ 0 & \Sigma_{\infty t} \end{bmatrix}, \tag{12.66}$$

where $\Sigma_{\infty r} = \operatorname{diag}(\sigma_{\infty 1} I_2, ..., \sigma_{\infty k} I_2)$, $\Sigma_{\infty t} = \operatorname{diag}(\sigma_{\infty k+1} I_2, ..., \sigma_{\infty n} I_2)$. Divide the matrix M_∞ accordingly, $\mathrm{M}_\infty = \operatorname{diag}(\mathrm{M}_{\infty r}, \mathrm{M}_{\infty t})$. The closed-loop system representation (A_o, B_o, C_o) is rearranged according to the division of ε, i.e.,

$$A_o = \begin{bmatrix} A_{or} & A_{ort} \\ A_{otr} & A_{ot} \end{bmatrix},$$
$$B_o = \begin{bmatrix} B_{or} \\ B_{ot} \end{bmatrix}, \tag{12.67}$$
$$C_o = \begin{bmatrix} C_{or} & C_{ot} \end{bmatrix}.$$

Hence, the closed-loop states of the reduced-order system are now

$$x_o = \begin{Bmatrix} x_r \\ \varepsilon_t \end{Bmatrix}, \qquad x_r = \begin{Bmatrix} x \\ \varepsilon_r \end{Bmatrix}. \tag{12.68}$$

The reduced-order controller representation is (A_{or}, B_{or}, C_{or}), and let the closed-loop system state be denoted by $\bar{x}_r$.

If condition (12.61) is satisfied, the performance of the closed-loop system with the reduced-order controller is almost identical to the full-order controller in the sense that $\|x_r - \bar{x}_r\|_2 \cong 0$. It follows from (12.63) that for $\sigma_{\infty i} \ll \sigma_{\infty 1}$

($i = k+1,\ldots,n$) we obtain $\|A_{otr}\| \cong \|A_{ort}\| \cong 0$, and the closed-loop block A_{ot} is almost identical to the open-loop block A_t, i.e., $A_{ot} \cong A_t$. In this case, from (12.67) and (12.68), we obtain

$$\dot{x}_r = A_{or} x_r + A_{ort} \varepsilon_t + B_{or} w \cong A_{or} x_r + B_{or} w = \dot{\bar{x}}_r, \tag{12.69}$$

or, thus, $x_r \cong \bar{x}_r$.

The above approximations hold for low-authority controllers, i.e., for the controllers that modify only moderately the system natural frequencies. Typically, the modes with largest H_∞ singular values do not fall under this category, but the modes with the smallest H_∞ singular values are under low-authority control. Thus the latter modes are the ones that are the most suitable for reduction. Therefore the presented reduction procedure is applicable in this case.

12.12 Controller Design Procedure

The following steps help to design an H_∞ controller:

1. Put the structural model into modal coordinates 1 or 2.
2. Define the performance criteria, such as bandwidth, settling time, overshoot, etc.
3. Assign the initial values of the disturbance matrix B_1 and performance matrix C_1 (these matrices are known to a certain degree).
4. Solve the Riccati equations (12.5), (12.6), and (12.7), find controller gains from (12.11) and (12.12), and simulate the closed-loop performance. Check if the performance satisfies the performance criteria. If not, continue.
5. Check which modes do not satisfy the criteria. Scale the corresponding components of B_1 and/or C_1, and return to p. 4.
6. If the criteria are not fully satisfied, consider the addition of a filter to achieve the goal. Use the procedure of Section 12.10, by appropriately scaling the input (B_1) or output (C_1) matrices in modal coordinates.
7. When the goal is achieved, perform controller reduction. Determine the reduction index as in (12.57) and eliminate the controller states with the small reduction indexes. Simulate the closed-loop system with the reduced-order controller. If the performance of the system with the reduced-order controller is close to the performance of the system with the full-order controller, accept the reduced-order controller; or you may consider further reduction. If the performance of the reduced-order controller departs significantly from the performance of the full-order controller, increase the order of the reduced-order controller, until its performance is satisfactory.

The above design procedure achieves the design goals because the modes are almost independent; therefore by scaling a single entry of the disturbance matrix B_1 or performance matrix C_1 we change the properties of a single mode, leaving other modes almost unchanged.

12.13 Controller Design Examples

We illustrate the H_∞ design method using a simple structure, the truss structure, and the Deep Space Network antenna.

12.13.1 A Simple Structure

We design an H_∞ controller for a system as in Fig. 1.1. The system parameters are as follows: $m_1 = 3$, $m_2 = 1$, $m_3 = 2$, $k_1 = 30$, $k_2 = k_3 = k_4 = 6$, $D = 0.004K + 0.001M$, where M, K, and D are mass, stiffness, and damping matrices, respectively. The control input (u) acts at mass 2 and mass 3 in opposite directions. The first disturbance (w_1) acts at mass 2 and mass 3 in opposite directions, with an amplification factor of 3, the second disturbance (w_2) acts at mass 2, and the third disturbance (w_3) is the output noise. The output (y) is a displacement of mass 2, and the controlled outputs (z_1, z_2, and z_3) are the displacement of mass 2 with an amplification factor of 3, a rate of mass 3, and an input u. Thus, the corresponding input and output matrices are as follows:

$$B_1 = \begin{bmatrix} 0 & 0 & 0 \\ 0 & 0 & 0 \\ 0 & 0 & 0 \\ 0 & 0 & 0 \\ 1 & 3 & 0 \\ 0 & -1.5 & 0 \end{bmatrix}, \qquad B_2 = \begin{bmatrix} 0 \\ 0 \\ 0 \\ 0 \\ 1 \\ -0.5 \end{bmatrix},$$

$$C_1 = \begin{bmatrix} 0 & 0 & 1 & 0 & 0 & 0 \\ 0 & 0 & 0 & 0 & 3 & 0 \\ 0 & 0 & 0 & 0 & 0 & 0 \end{bmatrix}, \qquad C_2 = \begin{bmatrix} 0 & 0 & 0 & 0 & 1 & 0 \end{bmatrix},$$

and

$$D_{12}^T = D_{21} = \begin{bmatrix} 0 & 0 & 1 \end{bmatrix}.$$

First we find the parameter ρ such that the condition (12.7) is satisfied, obtaining $\rho = 7.55$. Next we determine the H_∞ singular values: the exact ones obtained from (12.18) and the approximate ones obtained from (12.41). They are shown in Fig. 12.5. The figure shows good coincidence for the two smallest values. Next we calculated the open- and closed-loop impulse responses, and show them in

Fig. 12.6(a) (from the first input to the first output). We also calculated the magnitudes of the transfer function of the open- and closed-loop systems, and they are compared in Fig. 12.6(b), showing significant vibration suppression.

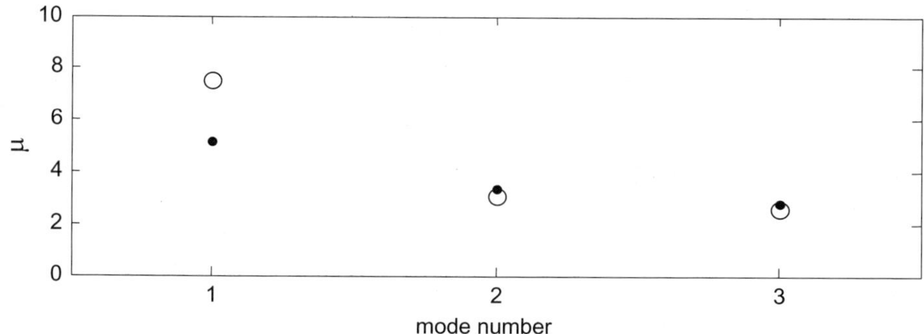

Figure 12.5. Exact (○) and approximate (•) H_∞ singular values of a simple system: Good coincidence for modes 2 and 3.

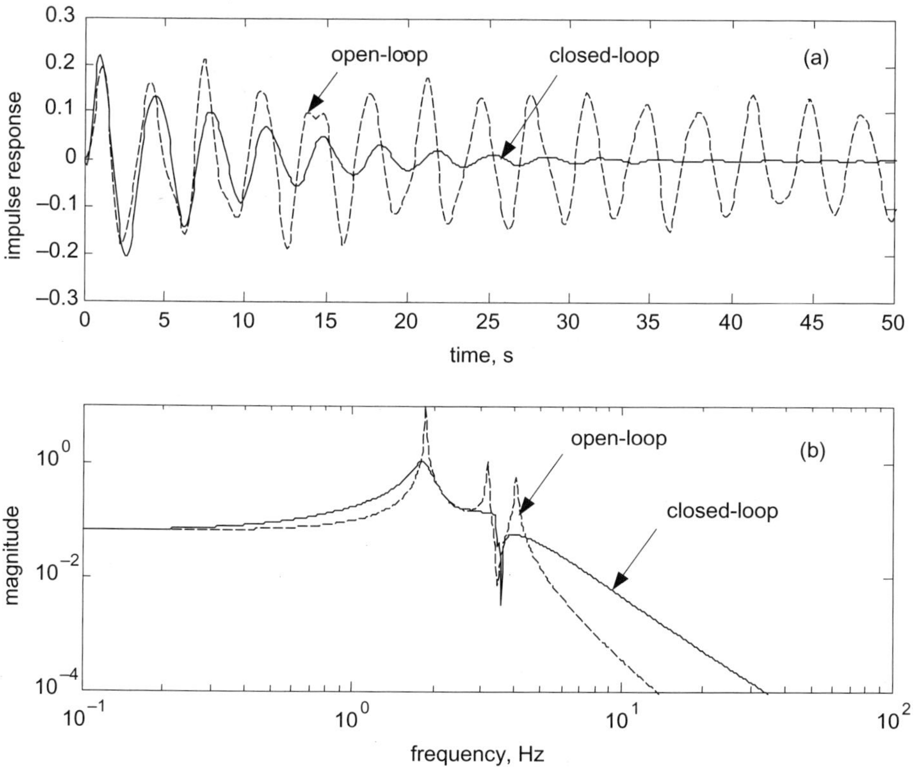

Figure 12.6. A simple system: (a) Open- and closed-loop impulse responses; and (b) magnitudes of the open- and closed-loop transfer functions. Damping is added to the closed-loop system.

Next we reduce the controller using the reduction indices. They are shown in Fig. 12.7. The index of the third mode is small and is truncated; hence the controller order is reduced from six to four states. The obtained closed-loop system is stable, with comparable performance. This is confirmed with the impulse responses of the full and reduced controller in Fig. 12.8 (from the first input to the first output).

12.13.2 The 2D Truss

The Matlab code for this example is in Appendix B. We present the design of the H_∞ controller for the 2D truss structure, as shown in Fig. 1.2. The structural model has 16 modes, or 32 states. The control input, u, is applied to node 4, in the vertical direction, the controller and the output y is collocated with u. The disturbances act at the input u with an amplification factor of 90, and at node 10 (the horizontal direction). The performance output z is measured at output y, and at node 9, in the horizontal direction.

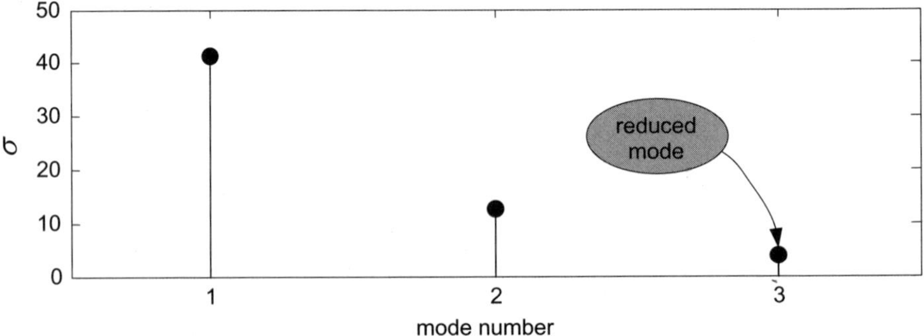

Figure 12.7. Reduction index of the simple system.

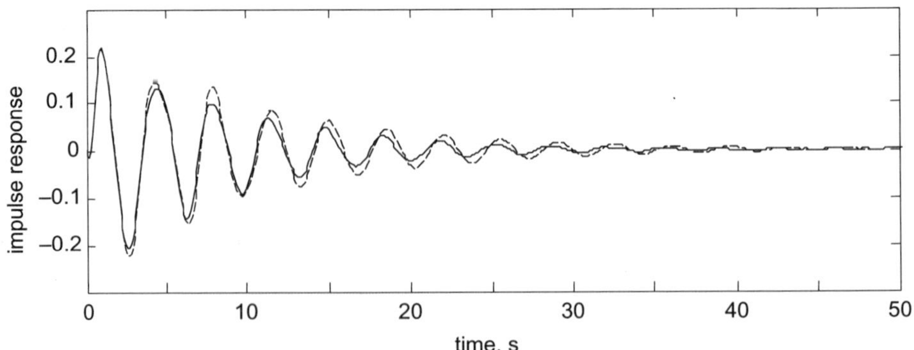

Figure 12.8. Almost identical impulse responses of the full (solid line) and reduced (dashed line) H_∞ closed-loop system.

First, we obtain the system H_∞ singular values and compare them in Fig. 12.9 with the approximate ones, obtained from (12.41). Similarly to the previous example, the small values show good coincidence while the large values diverge.

This property is explained by the fact that for the largest singular values the closed-loop modal damping is large enough to diverge from the low-authority conditions. Nevertheless, this is not a significant obstacle, since only small H_∞ singular values are used to evaluate the modes subjected to reduction.

Next, we compare the H_∞ singular values (○), the H_2 singular values (•), and the Hankel singular values γ_{11i} (◊) in Fig. 12.10, showing that Properties 12.1 and 12.2 hold. Namely, the Hankel singular values dominate the H_∞ singular values, and the latter dominate the H_2 singular values. The critical value of ρ is ρ=125.1.

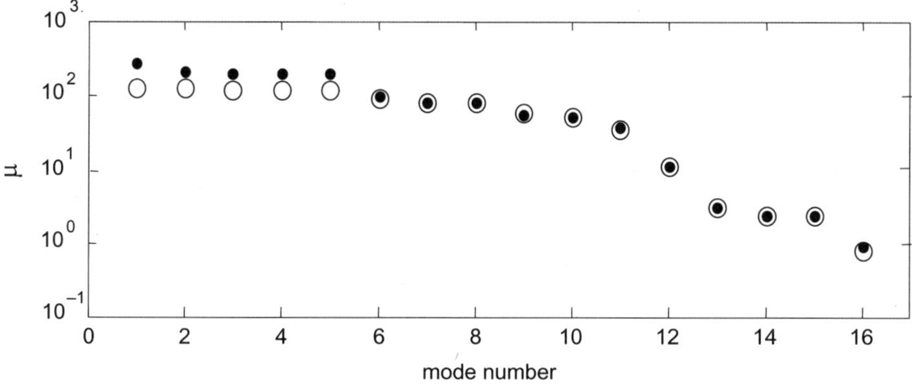

Figure 12.9. The exact (○) and approximate (•) H_∞ singular values of the 2D truss are almost equal for higher modes.

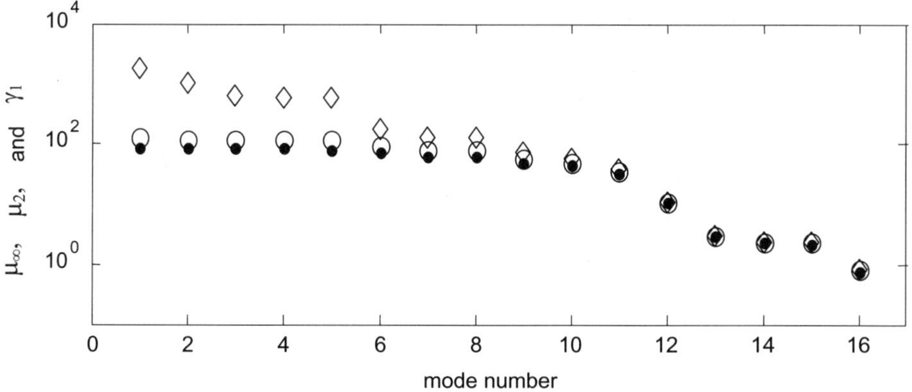

Figure 12.10. H_∞ (○), H_2 (•), and Hankel singular values (◊) of the 2D truss satisfy (12.47).

We also compare the open- and closed-loop impulse responses and magnitudes of the transfer functions in Fig. 12.11, showing that the closed-loop performance is improved when compared to the open-loop performance.

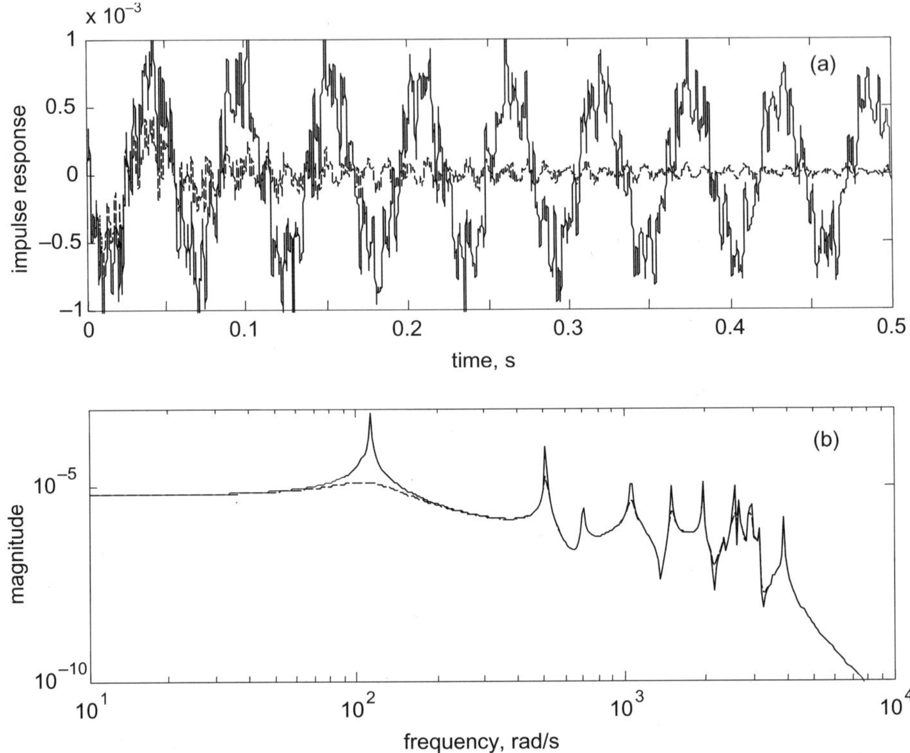

Figure 12.11. The 2D truss: (a) Open-loop (solid line) and closed-loop (dashed line) impulse responses; and (b) magnitudes of the open-loop (solid line) and closed-loop (dashed line) transfer functions.

Next we compute the H_∞ reduction indices, and show them in Fig. 12.12. The H_∞ reduction index satisfies the condition in (12.61) for $k = 6, ..., 16$, i.e., $\sigma_{\infty k} \ll \sigma_{\infty 1}$. Hence, the reduced controller contains five modes, or 10 states. Indeed, the controller with five modes (of order 10) is stable, and its performance is almost identical to the full-order controller, since the closed-loop impulse responses of the full-order (see Fig. 12.11(a)) and reduced-order controllers overlap.

12.13.3 Filter Implementation Example

Consider the 3D truss with a filter as in Subsection 11.13.3. The magnitude of the transfer function of the truss with a filter is shown in Fig. 12.13(a) (solid line). We obtained an equivalent structure with filter by scaling the disturbance input, according to (12.51), and the magnitude of its transfer function is shown in Fig. 12.13(a) (dashed line). It is clear from that figure that the structure with the filter, and the structure with the scaled disturbance input, have very similar frequency characteristics and their norms are as follows: $\|G\|_\infty = 2.6895$ for the structure with the filter and $\|G\|_\infty = 2.6911$ for the structure with the scaled disturbance input.

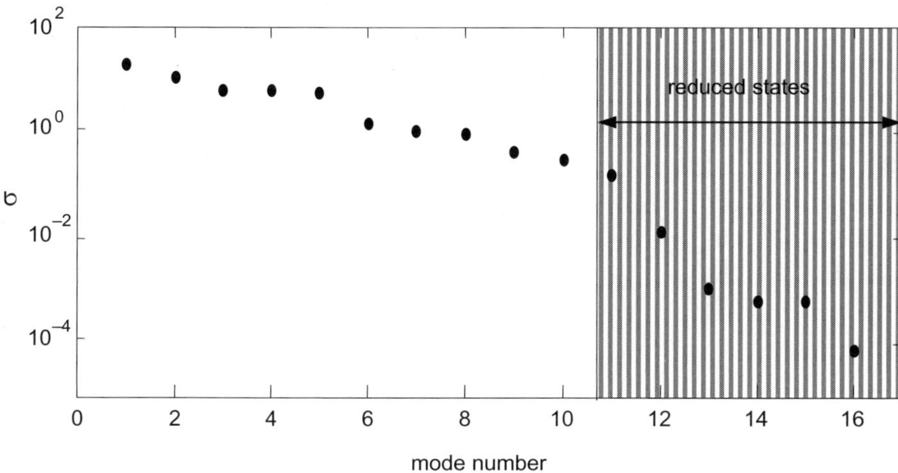

Figure 12.12. Reduction indices of the 2D truss.

We designed two frequency weighted H_∞ controllers for this structure. The first one is based on a structure with a filter, while the second one is based on a structure with a scaled input matrix. The closed-loop transfer functions are shown in Fig. 12.13(b). The closed-loop performance of the structure with the filter, and with the scaled input, are quite close. The closed-loop H_∞ norms are as follows: $\|G_{cl}\|_\infty = 0.4221$ for the structure with the filter and $\|G_{cl}\|_\infty = 0.2852$ for the structure with the scaled disturbance input. The scaled system has better performance because it had a minimum for a smaller value of ρ $(\rho = 5.7)$ than the system with a filter $(\rho = 7.0)$. If we use $\rho = 7.0$, the norm of the scaled system is $\|G_{cl}\|_\infty = 0.4034$.

12.13.4 The Deep Space Network Antenna with Wind Disturbance Rejection Properties

A significant portion of the antenna tracking error is generated by the antenna vibrations excited by wind gusts. The LQG controller designed in Subsection 11.13.4 improved its tracking, but we did not address directly the disturbance rejection properties in the design process, therefore they are rather moderate. The H_∞ controller allows for addressing simultaneously its tracking and disturbance rejection properties, as we show in the following.

In [48] the wind spectra were determined from the wind field data. Based on these spectra, and using the antenna model in the modal representation, we add the wind filter by an appropriate scaling of the input matrix B_1 of the antenna. The scaling factors are the filter gains at the natural frequencies of the antenna.

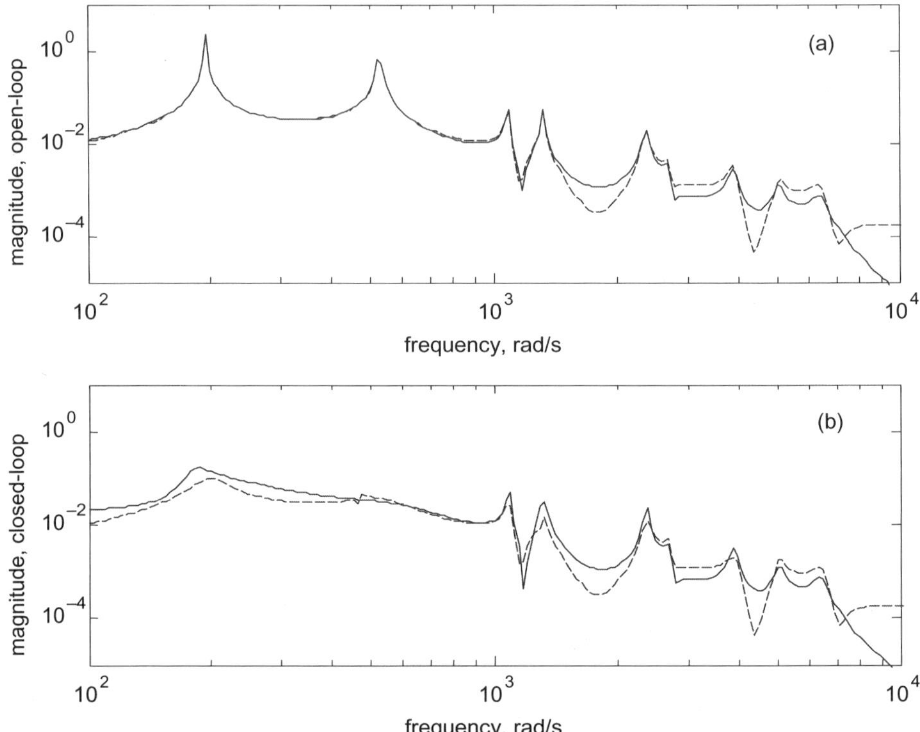

Figure 12.13. Magnitudes of the transfer function: (a) The open-loop system; and (b) the closed-loop system (with filter (dashed line) and with equivalent weights (dotted line)).

We obtained and simulated the H_∞ controller for the azimuth axis, and compared its tracking performances with the LQG controller performance. First, we compare the transfer functions from the command input to the encoder output; see Fig. 12.14(a). The plot shows improved tracking performance of the H_∞ controller (the bandwidth is 2.2 Hz for the H_∞ controller and 1.2 Hz for the LQG controller). The wind disturbance rejection properties are represented by the transfer functions from the wind disturbance input to the encoder output, Fig. 12.14(b), and by the simulated wind gusts action on the antenna in Fig. 12.15, where the tracking errors of the H_∞ and LQG controllers are plotted. In Fig. 12.14(b) the H_∞ controller disturbance transfer function is about a decade lower than the LQG controller, showing improved disturbance rejection properties of the H_∞ controller. This is confirmed by the plot of the tracking error in a 50 km/h wind, see Fig. 12.15. The rms encoder error of the LQG controller is 0.70 mdeg, while the error of the H_∞ controller is 0.12 mdeg, showing an almost six-fold improvement. For more on the antenna controllers and its practical limitations, see [42] and

http://tmo.jpl.nasa.gov/tmo/progress_report/42-127/127G.pdf.

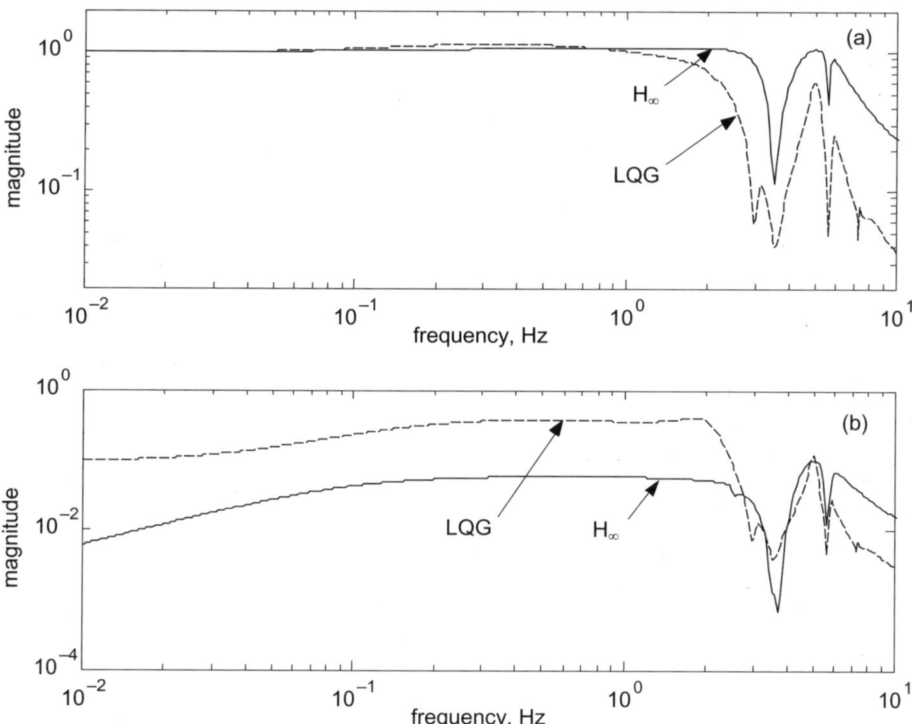

Figure 12.14. Magnitudes of the azimuth transfer functions of the H_∞ (solid line) and LQG (dashed line) controllers: (a) From the command input to the encoder output; and (b) from the wind disturbance input to the encoder output. The H_∞ controller shows a wider bandwidth and improved disturbance rejection properties.

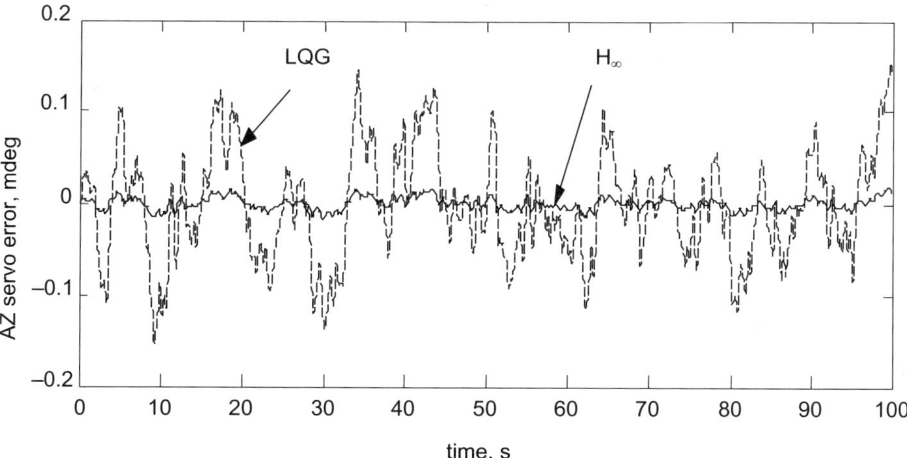

Figure 12.15. The azimuth tracking error due to wind gusts of the H_∞ (solid line) and LQG (dashed line) controllers: The H_∞ controller shows improved disturbance rejection properties.

Appendices

Matlab functions, Matlab examples, and structural parameters

Fast cars, fast women, fast algorithms ...
what more could a man want?
—*Joe Mattis*

A
Matlab Functions

The following Matlab® functions are given in this appendix:

- *modal1*, for the determination of the modal 1 state-space representation from a generic state-space representation;
- *modal2*, for the determination of the modal 2 state-space representation from a generic state-space representation;
- *modal1m*, for the determination of the modal 1 state-space representation from natural frequencies, modal damping, modal mass, modal matrix, etc.;
- *modal2m*, for the determination of the modal 2 state-space representation from natural frequencies, modal damping, modal mass, modal matrix, etc.;
- *modal1n*, for the determination of the modal 1 state-space representation from mass, stiffness, damping matrices, etc.;
- *modal2n*, for the determination of the modal 2 state-space representation from mass, stiffness, damping matrices, etc.;
- *modal_time_fr,* for the determination of modal representation in limited time and frequency ranges;
- *balan2*, for the determination of the open-loop balanced representation;
- *norm_H2*, for the determination of modal H_2 norms;
- *norm_Hinf*, for the determination of modal H_∞ norms;
- *norm_Hankel*, for the determination of modal Hankel norms;
- *bal_LQG*, for the determination of the LQG-balanced representation; and
- *bal_H_inf*, for the determination of the H_∞-balanced representation.

These functions use the following standard Matlab routines: *are, cdf2rdf, inv, lqe, lqr, lyap, norm, size, sqrt, svd.*

A.1 Transformation from an Arbitrary State-Space Representation to the Modal 1 State-Space Representation

The *modal1* state-space representation is obtained by the initial transformation of an arbitrary representation (A,B,C) to the modal representation (A_n, B_n, C_n); the system matrix A_n is complex and diagonal. Its diagonal entries are the eigenvalues of A,

$$\begin{aligned} a_n(i,i) &= -\zeta_i\omega_i + j\omega_i\sqrt{1-\zeta_i^2}, \\ a_n(i+1,i+1) &= -\zeta_i\omega_i - j\omega_i\sqrt{1-\zeta_i^2}. \end{aligned} \tag{A.1}$$

This transformation, denoted V, is obtained using the Matlab command eig(A). In the next step the representation (A_n, B_n, C_n) is turned into the modal form 1 (A_m, B_m, C_m) by applying the following transformation:

$$\begin{aligned} T &= \operatorname{diag}(t_i), \\ t_i &= \begin{bmatrix} \zeta_i - j\sqrt{1-\zeta_i^2} & 1 \\ \zeta_i + j\sqrt{1-\zeta_i^2} & 1 \end{bmatrix}. \end{aligned} \tag{A.2}$$

The system matrix A_m in the obtained representation has the block-diagonal form as in (2.47), and its 2×2 blocks are as in (2.52). However, the input and output matrices B_m and C_m are not in the form as in (2.52). The first entry of B_{mi} is nonzero, and the first and second entries of C_{mi} do not correspond to the displacement and rate sensors. This happens because the representation with block-diagonal A_m is not unique. Indeed, define the transformation S as follows:

$$\begin{aligned} S &= \operatorname{diag}(s_i), \\ s_i &= \begin{bmatrix} \alpha_i + 2\zeta_i\beta_i & \beta_i \\ -\beta_i & \alpha_i \end{bmatrix}. \end{aligned} \tag{A.3}$$

This leaves A_m unchanged, although B_m and C_m have been changed.

We use the above property to obtain B_m and C_m as in (2.52). In order to do this for a single input system, note that the input matrix B_m is in the following form:

$$B_m = \begin{bmatrix} B_{m1} \\ B_{m2} \\ \vdots \\ B_{mn} \end{bmatrix},$$

where $B_{mi} = \begin{bmatrix} b_{mi1} \\ b_{mi2} \end{bmatrix}$ has both entries nonzero (unlike (2.52)). By choosing parameters α_i and β_i in the transformation (A.3), such that

$$\alpha_i = -\beta_i \frac{b_{mi2}}{b_{mi1}}, \tag{A.4}$$

we obtain (after transformation) $b_{mi1} = 0$.

Care should be taken in the permutation of variables in the state-space representation: the modal displacement will be placed before the modal velocity. As a result, the total transformation from the given representation (A, B, C) to modal representation (A_m, B_m, C_m) is $R=VTS$, such that $A_m = R^{-1}AR$, $B_m = R^{-1}B$, and $C_m = CR$.

```
function [r,am,bm,cm] = modal1(a,b,c)
% this function determines the modal representation 1 (am,bm,cm)
% given a generic state-space representation (a,b,c)
% and the transformation r to the modal representation
% such that am=inv(r)*a*r, bm=inv(r)*b, and cm=c*r

% transformation to complex-diagonal form:
[v,an]=eig(a);
bn=inv(v)*b;
cn=c*v;

% transformation to modal form 1:
i = find(imag(diag(an))');
index = i(1:2:length(i));
   j = sqrt(-1);
   t = eye(length(an));

if isempty(index)
    am=an;bm=bn;cm=cn;

else
```

```
    for i=index
    om=abs(an(i,i));
    z(i)=-real(an(i,i))/abs(an(i,i));
    t(i:i+1,i:i+1)=[z(i)-j*sqrt(1-z(i)^2) 1;z(i)+j*sqrt(1-z(i)^2) 1];
    end

% modal form 1:
    am=real(inv(t)*an*t);
    bm=real(inv(t)*bn);
    cm=real(cn*t);

    beta=1;
    for i=index
     alpha=-beta*bm(i+1,1)/bm(i);
     s(i,i)=alpha+2*z(i)*beta;
     s(i,i+1)=beta;
     s(i+1,i)=-beta;
     s(i+1,i+1)=alpha;
    end

am=inv(s)*am*s;
bm=inv(s)*bm;
cm=cm*s;

% the transformation:
  r=v*t*s;
end
```

A.2 Transformation from an Arbitrary State-Space Representation to the Modal 2 State-Space Representation

Note that unlike the modal form 1, a structure can be transformed to modal form 2 approximately (assuming small damping, terms with the squared damping coefficient ζ_i^2 are ignored). First, we transform (A,B,C) to the diagonal complex modal form (A_n, B_n, C_n), as before. Next, we apply the following transformation:

$$T = \text{diag}(t_i),$$
$$t_i = \begin{bmatrix} j & 1 \\ -j & 1 \end{bmatrix}, \tag{A.5}$$

to turn the (A_n, B_n, C_n) into modal form 2, and eventually correcting for nonzero terms in B_{mi} using transformations (A.3) and (A.4).

```
function [r,am,bm,cm] = modal2(a,b,c)
% this function determines the modal representation 2 (am,bm,cm)
% given a generic state-space representation (a,b,c)
% and the transformation r to the modal representation
% such that am=inv(r)*a*r, bm=inv(r)*b, and cm=c*r

% transformation to complex-diagonal form:
[v,an]=eig(a);
bn=inv(v)*b;
cn=c*v;

% transformation to modal form 2:
i = find(imag(diag(an))');
index = i(1:2:length(i));
   j = sqrt(-1);
   t = eye(length(an));

if isempty(index)
    am=an;bm=bn;cm=cn;

else
   for i=index
      t(i:i+1,i:i+1)=[j 1;-j 1];
   end

% modal form 2:
   am=real(inv(t)*an*t);
   bm=real(inv(t)*bn);
   cm=real(cn*t);

% the transformation
   r=v*t;
end
```

A.3 Transformation from Modal Parameters to the Modal 1 State-Space Representation

If *coord* = 1 this function determines the modal state-space representation in form 2, as in (2.53), or if *coord* = 0 this determines the state-space representation in modal coordinates in form 1, as in (2.52). The input data include natural frequencies, modal damping, modal mass, a modal matrix, an input matrix, and displacement and rate output matrices.

```
function [am,bm,cm]=modal1m(om,z,mm,phi,b,cq,cv,coord)
% the determination of modal form 1 (am,bm,cm)
% from modal data
% n     - number of modes
% nd    - number of degrees of freedom
% om    - vector of natural frequencies (nx1)
% z     - vector of modal damping (nx1)
% mm    - vector of modal masses (nx1)
% phi   - modal matrix (ndxn)
% b     - input matrix (ndxs)
% cq    - displacement output matrix (rxnd)
% cv    - rate output matrix (rxnd)
% coord - if coord=0 -> state-space representation in modal
%                       coordinates
%         if coord=1 -> modal state-space representation

% arranging input data
mm=diag(mm);
om=diag(om);
z=diag(z);
c=[cq cv];

% modal input and output matrices:
bm=inv(mm)*phi'*b;
cmq=cq*phi;
cmv=cv*phi;

% representation in modal coordinates
Am=[0*om om;-om -2*z*om];
Bm=[0*bm;bm];
Cm=[cmq*inv(om) cmv];

if coord==0;
    % representation in modal coordinates:
```

```
    am=Am;
    bm=Bm;
    cm=Cm;
else
    % modal representation:
    n=max(size(Am))/2;
    for i=1:n; ind(2*i-1)=i;ind(2*i)=i+n;end
    am=Am(ind,ind);
    bm=Bm(ind,:);
    cm=Cm(:,ind);
end
```

A.4 Transformation from Modal Parameters to the Modal 2 State-Space Representation

If *coord* = 1 this function determines the modal state-space representation in form 1, as in (2.52), or if *coord* = 0 this determines the state-space representation in modal coordinates in form 2, as in (2.53). The input data include natural frequencies, modal damping, modal mass, a modal matrix, an input matrix, and displacement and rate output matrices.

```
function [am,bm,cm]=modal2m(om,z,mm,phi,b,cq,cv,coord)
% the determination of modal form 2 (am,bm,cm)
% from modal data
% n     - number of modes
% nd    - number of degrees of freedom
% om    - vector of natural frequencies (nx1)
% z     - vector of modal damping (nx1)
% mm    - vector of modal masses (nx1)
% phi   - modal matrix (ndxn)
% b     - input matrix (ndxs)
% cq    - displacement output matrix (rxnd)
% cv    - rate output matrix (rxnd)
% coord - if coord=0 -> state-space representation in modal
%                        coordinates
%         if coord=1 -> modal state-space representation

% arranging input data
mm=diag(mm);
om=diag(om);
```

```
z=diag(z);
c=[cq cv];

% modal input and output matrices:
bm=inv(mm)*phi'*b;
cmq=cq*phi;
cmv=cv*phi;

% representation in modal coordinates
Am=[-z*om om;-om -z*om];
Bm=[0*bm;bm];
Cm=[cmq*inv(om)-cmv*z cmv];

if coord==0;
    % representation in modal coordinates:
    am=Am;
    bm=Bm;
    cm=Cm;
else
    % modal representation:
    n=max(size(Am))/2;
    for i=1:n; ind(2*i-1)=i;ind(2*i)=i+n;end
    am=Am(ind,ind);
    bm=Bm(ind,:);
    cm=Cm(:,ind);
end
```

A.5 Transformation from Nodal Parameters to the Modal 1 State-Space Representation

If *coord* = 1 this function determines the modal state-space representation in form 2, as in (2.53), or if *coord* = 0 this determines the state-space representation in modal coordinates in form 1, as in (2.52). The input data include mass, stiffness, and damping matrices, an input matrix, and displacement and rate output matrices.

```
function [am,bm,cm]=modal1n(m,damp,k,b,cq,cv,n,coord)
% the determination of modal form 1 (am,bm,cm)
% from nodal data
% n      - number of modes
% nd     - number of degrees of freedom
% m      - mass matrix (ndxnd)
```

```
% damp  - damping matrix (ndxnd)
% k     - stiffness matrix (ndxnd)
% b     - input matrix (ndxs)
% cq    - displacement output matrix (rxnd)
% cv    - rate output matrix (rxnd)
% coord - if coord=0 -> state-space representation in modal
%                       coordinates
%         if coord=1 -> modal state-space representation

% modal matrix:
[phi,om2]=eig(k,m);
nn=1:n;
phi=phi(:,nn);

% natural frequency matrix
om=sqrt(om2);

% modal mass, stiffness and damping matrices:
mm=phi'*m*phi;
km=phi'*k*phi;
dm=phi'*damp*phi;
z=0.5*inv(mm)*dm*inv(om);

% input and output matrices
c=[cq cv];
bm=inv(mm)*phi'*b;
cmq=cq*phi;
cmv=cv*phi;

% representation in modal coordinates
Am=[0*om om;-om -2*z*om];
Bm=[0*bm;bm];
Cm=[cmq*inv(om) cmv];

if coord==0;
    % representation in modal coordinates:
    am=Am;
    bm=Bm;
    cm=Cm;
else
    % modal representation:
    n=max(size(Am))/2;
    for i=1:n; ind(2*i-1)=i;ind(2*i)=i+n;end
```

```
    am=Am(ind,ind);
    bm=Bm(ind,:);
    cm=Cm(:,ind);
end
```

A.6 Transformation from Nodal Parameters to the Modal 2 State-Space Representation

If *coord* = 1 this function determines the modal state-space representation in form 1, as in (2.52), or if *coord* = 0 the state-space representation in modal coordinates in form 2, as in (2.53). The input data include mass, stiffness, and damping matrices, an input matrix, and displacement and rate output matrices.

```
function [am,bm,cm]=modal2n(m,damp,k,b,cq,cv,n,coord)
% the determination of modal form 2 (am,bm,cm)
% from nodal data
% n     - number of modes
% nd    - number of degrees of freedom
% m     - mass matrix (ndxnd)
% damp  - damping matrix (ndxnd)
% k     - stiffness matrix (ndxnd)
% b     - input matrix (ndxs)
% cq    - displacement output matrix (rxnd)
% cv    - rate output matrix (rxnd)
% coord - if coord=0 -> state-space representation in modal
%                       coordinates
%         if coord=1 -> modal state-space representation

% modal matrix:
[phi,om2]=eig(k,m);
nn=1:n;
phi=phi(:,nn);

% natural frequency matrix
om=sqrt(om2);

% modal mass, stiffness and damping matrices:
mm=phi'*m*phi;
km=phi'*k*phi;
dm=phi'*damp*phi;
```

```
z=0.5*inv(mm)*dm*inv(om);

% input and output matrices
c=[cq cv];
bm=inv(mm)*phi'*b;
cmq=cq*phi;
cmv=cv*phi;

% representation in modal coordinates
Am=[-z*om om;-om -z*om];
Bm=[0*bm;bm];
Cm=[cmq*inv(om)-cmv*z cmv];

if coord==0;
    % representation in modal coordinates:
    am=Am;
    bm=Bm;
    cm=Cm;
else
    % modal representation:
    n=max(size(Am))/2;
    for i=1:n; ind(2*i-1)=i;ind(2*i)=i+n;end
    am=Am(ind,ind);
    bm=Bm(ind,:);
    cm=Cm(:,ind);
end
```

A.7 Determination of the Modal 1 State-Space Representation and the Time- and Frequency-Limited Grammians

This function determines the modal state-space representation (form 1), and

- the time-limited grammians and Hankel singular values, for the time interval $T=[t_1,t_2]$, $t_2>t_1$;
- the frequency-limited grammians and Hankel singular values, for the frequency interval $\Omega=[\omega_1,\omega_2]$, $\omega_2>\omega_1$; and
- the time- and frequency-limited grammians and Hankel singular values, for the time interval $T=[t_1,t_2]$, $t_2>t_1$, and for the frequency interval $\Omega=[\omega_1,\omega_2]$, $\omega_2>\omega_1$.

The input data include the system state-space representation (a,b,c), lower (t_1) and upper (t_2) time interval limits, and lower (ω_1) and upper (ω_2) frequency interval limits.

For the time-only case, assume $\omega_1 = 0$ and $\omega_2 \gg \omega_b$, where ω_n is the highest natural frequency.

For the frequency-only case, assume $t_1 = 0$ and $t_2 \gg 2\pi / \omega_n$, where ω_n is the highest natural frequency.

```
function [am,bm,cm,g,r,wc,wo]=modal_time_fr(a,b,c,t1,t2,om1,om2);
%
% This function finds modal representation (am,bm,cm)
% and transformation r
% in limited-time interval [t1 t2],
% and limited-frequency interval  [om1, om2]
% It uses modal1.m function
%
% modal representation:
[r,a,b,c] = modal1(a,b,c);
am=a;
bm=b;
cm=c;

% finite-frequency transformation matrix sw,
% and finite-frequency grammians wcw and wow:
j=sqrt(-1);
[n1,n2]=size(a);
i=eye(n1);
x1=j*om1*i+a;
x2=inv(-j*om1*i+a);
s1=(j/2/pi)*logm(x1*x2);
x1=j*om2*i+a;
x2=inv(-j*om2*i+a);
s2=(j/2/pi)*logm(x1*x2);
sw=s2-s1;
%
% grammians:
wc=lyap(a,b*b');              % controllability grammian
wo=lyap(a',c'*c);             % observability grammian
%
% finite-frequency grammians:
wcw=wc*conj(sw)'+sw*wc;
wow=conj(sw)'*wo+wo*sw;
```

```
%
% finite-time transformation matrices st1, st2,
% and finite time and frequency grammians wcTW and woTW :

st1=-expm(a*t1);
st2=-expm(a*t2);

wct1W=st1*wcw*st1';
wct2W=st2*wcw*st2';
wcTW=wct1W-wct2W;

wot1W=st1'*wow*st1;
wot2W=st2'*wow*st2;
woTW=wot1W-wot2W;

% sorting in descending order of the Hankel singular values:
wc=real(wcTW);wo=real(woTW);
g=sqrt(abs(diag(wc*wo)))
[g,ind]=sort(-g);
g=-g;
am=am(ind,ind);
bm=bm(ind,:);
cm=cm(:,ind);
```

A.8 Open-Loop Balanced Representation

```
% function [Ab,Bb,Cb,Gamma,R]=balan2(A,B,C);
% This function finds the open-loop balanced representation
% (Ab,Bb,Cb) so that controllability (Wc) and observability (Wo)
% grammians are equal and diagonal:
%               Wc=Wo=Gamma
%
% Input parameters:
% (A,B,C)            - system state-space representation
%
% Output parameters:
% (Ab,Bb,Cb)         - balanced representation
% R                  - transformation to the balanced representation
% Gamma              - Hankel singular values
%
```

```
function [Ab,Bb,Cb,Gamma,R]=balan2(A,B,C);
%
Wc=lyap(A,B*B');                % controllability grammian
Wo=lyap(A',C'*C);               % observability grammian
[Uc,Sc,Vc]=svd(Wc);             % SVD of the controllability grammian
[Uo,So,Vo]=svd(Wo);             % SVD of the observability grammian
Sc=sqrt(Sc);
So=sqrt(So);
P=Uc*Sc;
Q=So*Vo';
H=Q*P;                          % Hankel matrix
[V,Gamma,U]=svd(H);             % SVD of the Hankel matrix
G1=sqrt(Gamma1);
R=P*U*inv(G1);                  % transformation matrix R
Rinv=inv(G1)*V'*Q;              % inverse of R
Ab=Rinv*A*R;
Bb=Rinv*B;
Cb=C*R;                         % balanced representation (Ab,Bb,Cb)
```

A.9 H_2 Norm of a Mode

```
% function norm=norm_H2(om,z,bm,cmq,cmr,cma);
%
% This function finds an approximate H2 norm
% for each mode of a structure with displacement, rate,
% and acceleration sensors
%
% Input parameters:
% om    - vector of natural frequencies
% z     - vector of modal damping
% bm    - modal matrix of actuator location
% cmq   - modal matrix of displacement sensor location
% cmr   - modal matrix of rate sensor location
% cma   - modal matrix of accelerometer location
%
% Output parameter:
% norm  - H2 norm
%
function norm=norm_H2(om,z,bm,cmq,cmr,cma);
%
om2=diag(om.*om);
```

```
bb=diag(bm*bm');
cc=diag(cma'*cma*om2+cmr'*cmr+cmq'*cmq*inv(om2));
h=sqrt(bb.*cc)/2;
h=h./sqrt(z);
norm=h./sqrt(om);
```

A.10 H_∞ Norm of a Mode

```
% function norm=norm_Hinf(om,z,bm,cmq,cmr,cma);
%
% This function finds an approximate H∞ norm
% for each mode of a structure with displacement, rate, and
% acceleration sensors
%
% Input parameters:
% om    - vector of natural frequencies
% z     - vector of modal damping
% bm    - modal matrix of actuator location
% cmq   - modal matrix of displacement sensor location
% cmr   - modal matrix of rate sensor location
% cma   - modal matrix of accelerometer location
%
% Output parameter:
% norm  - H∞ norm
%
function norm=norm_Hinf(om,z,bm,cmq,cmr,cma);
%
om2=diag(om.*om);
bb=diag(bm*bm');
cc=diag(cma'*cma*om2+cmr'*cmr+cmq'*cmq*inv(om2));
h=sqrt(bb.*cc)/2;
h=h./z;
norm=h./om;
```

A.11 Hankel Norm of a Mode

```
% function norm=norm_Hankel(om,z,bm,cmq,cmr,cma);
%
% This function finds an approximate Hankel norm
```

```
% for each mode of a structure with displacement, rate,
% and acceleration sensors
%
% Input parameters:
% om    - vector of natural frequencies
% z     - vector of modal damping
% bm    - modal matrix of actuator location
% cmq   - modal matrix of displacement sensor location
% cmr   - modal matrix of rate sensor location
% cma   - modal matrix of accelerometer location
%
% Output parameter:
% norm  - Hankel norm
%
function norm=norm_Hankel(om,z,bm,cmq,cmr,cma);
%
om2=diag(om.*om);
bb=diag(bm*bm');
cc=diag(cma'*cma*om2+cmr'*cmr+cmq'*cmq*inv(om2));
h=sqrt(bb.*cc)/4;
h=h./z;
norm=h./om;
```

A.12 LQG-Balanced Representation

```
% function [Ab,Bb,Cb,Mu,Kpb,Keb,Qb,Vb,R]=bal_LQG(A,B,C,Q,R,V,W)
%
% This function finds the LQG-balanced representation (Ab,Bb,Cb)
% so that CARE (Sc) and FARE (Se) solutions are equal and diagonal:
%             Sc=Se=Mu
%
% Input parameters:
% (A,B,C)      - system state-space representation,
% Q            - state weight matrix,
% R            - input weight matrix,
% V            - process noise covariance matrix,
% W            - measurement noise covariance matrix.
%
% Output parameters:
% (Ab,Bb,Cb)  - LQG-balanced representation,
% R            - LQG-balanced transformation,
```

```
% Mu          - balanced CARE, FARE solutions
% Qb          - balanced weight matrix,
% Vb          - balanced process noise covariance matrix,
% Kpb, Keb    - balanced gains.
%
%
function [Ab,Bb,Cb,Mu,Kpb,Keb,Qb,Vb,R]=bal_LQG(A,B,C,Q,R,V,W)
%
V1=V;
R1=R;
[n1,n2]=size(A);
[Kp,Sc,ec]=lqr(A,B,Q,R);              % solution of CARE
[Ke,Se,ee]=lqe(A,eye(n1),C,V,W);      % solution of FARE
[Uc,Ssc,Vc]=svd(Sc);
Pc=sqrt(Ssc)*Vc';                     % Pc
[Ue,Sse,Ve]=svd(Se);
Pe=Ue*sqrt(Sse);                      % Pe
H=Pc*Pe;                              % H
[V,Mu,U]=svd(H);                      % SVD of H
mu=sqrt(Mu);
R=Pe*U*inv(mu);                       % transformation R
Rinv=inv(mu)*V'*Pc;                   % inverse of R
Ab=Rinv*A*R;
Bb=Rinv*B;
Cb=C*R;                               % LQG balanced representation
Qb=R'*Q*R;                            % balanced weight matrix
Vb=Rinv*V1*Rinv';                     % balan. process noise cov.
matrix
[Kpb,Scb,ecb]=lqr(Ab,Bb,Qb,R1);
[Keb,Seb,eeb]=lqe(Ab,eye(n1),Cb,Vb,W);  % balanced gains
```

A.13 H_∞-Balanced Representation

```
% function [Ab,Bb1,Bb2,Cb1,Cb2,Mu_inf,R]=bal_H_inf(A,B1,B2,C1,C2,ro)
%
% This function finds the H_inf-balanced representation
%          (Ab,Bb1,Bb2,Cb1,Cb2)
% so that HCARE (Sc) and HFARE (Se) solutions are equal and diagonal
%              Sc=Se=Mu_inf
%
```

```
% Input parameters:
% (A,B1,B2,C1,C2)        - system state-space representation
% ro                     - parameter in HCARE and HFARE
%
% Output parameters:
% (Ab,Bb1,Bb2,Cb1,Cb2)   - H_inf balanced representation,
% R                      - H_inf balanced transformation,
% Mu                     - balanced HCARE, HFARE solutions,
%
%
function [Ab,Bb1,Bb2,Cb1,Cb2,Mu_inf,R]=bal_H_inf(A,B1,B2,C1,C2,ro)
%
[n1,n2]=size(A);
Qc=C1'*C1;
gi=1/(ro*ro);
Rc=B2*B2'-gi*B1*B1';
[Sc,sc1,sc2,wellposedc]=are(A,Qc,Rc,'eigen');      % HCARE  solution
Qe=B1*B1';
Re=C2'*C2-gi*C1'*C1;
[Se,se1,se2,wellposef]=are(A',Qe,Re,'eigen');      % HFARE solution
%
if(norm(imag(Se))>1e-6 | norm(imag(Sc))>1e-6) ...
disp('nonpositive solution');end
%
[Uc,Ssc,Vc]=svd(Sc);
Pc=sqrt(Ssc)*Vc';
   % Pc
[Ue,Sse,Ve]=svd(Se);
Pe=Ue*sqrt(Sse);
   % Pe
N=Pc*Pe;
[V,Mu_inf,U]=svd(N);
mu_inf=sqrt(Mu_inf);
R=Pe*U*inv(mu_inf);
Rinv=inv(mu_inf)*V'*Pc;
Ab=Rinv*A*R;
Bb1=Rinv*B1;
Bb2=Rinv*B2;
Cb1=C1*R;
Cb2=C2*R;                                     % H∞-balanced representation
```

B
Matlab Examples

B.1 Example 2.5

```
% impulse responses and transfer functions
clear
% stiffness matrix:
k1=3;k2=3;k3=3;k4=0;
k=[k1+k2 -k2 0;
  -k2 k2+k3 -k3;
   0 -k3 k3+k4];
% mass matrix:
m1=1;m2=1;m3=1;
m=[m1 m2 m3];
m=diag(m);
% damping matrix:
damp=.01*k;
% state matrix:
```

```
k=inv(m)*k;
damp=inv(m)*damp;
a=[0*eye(3) eye(3);-k -damp];
% input matrix:
b=[0 0 1]';
b=inv(m)*b;
b=[0*b;b];
% output matrix:
c=[0 0 0 1 0 0];
% feed-through matrix:
d=[0];
[va,am,bm,cm]=modal2(a,b,c);

% impulse response:
dt=.1
t=0:dt:1000-dt;
y=impulse(a,b,c,d,1,t);

figure(1)
subplot(211);
plot(t,y)
axis([0 300 -.8 .8])
xlabel('time, s')
ylabel('velocity, mass 1')

% spectrum of the impulse response:
nn=max(size(t));
n=nn/4;
p=spectrum(y,n);
pp=sqrt(p(:,1));
fs=1/dt;
f=fs*(0:n/2-1)/n;              % frequency range
om=2*pi*f;
nf=max(size(f));
nnf=1:nf;
subplot(212)
plot(om,pp(nnf,1))
```

```
axis([0 5 0 1.8])
xlabel('frequency, rad/s')
ylabel('spectrum of v_1')

% transfer function:
figure(2)
subplot(211);
w=logspace(-1,1,900);
[ma,ph]=bode(a,b,c,d,1,w);
loglog(w,ma);
axis([.1 10 .01 100])
xlabel('frequency, rad/s')
ylabel('magnitude')

subplot(212)
semilogx(w,ph);
xlabel('frequency, rad/s')
ylabel('phase, deg')

% single mode analysis
% mode 1:
nn=1:2;
am1=am(nn,nn);
bm1=bm(nn,:);
cm1=cm(:,nn);
ym1=impulse(am1,bm1,cm1,d,1,t);
[ma1,ph1]=bode(am1,bm1,cm1,d,1,w);

% mode 2:
nn=3:4;
am2=am(nn,nn);
bm2=bm(nn,:);
cm2=cm(:,nn);
ym2=impulse(am2,bm2,cm2,d,1,t);
[ma2,ph2]=bode(am2,bm2,cm2,d,1,w);

% mode 3:
```

```
nn=5:6;
am3=am(nn,nn);
bm3=bm(nn,:);
cm3=cm(:,nn);
ym3=impulse(am3,bm3,cm3,d,1,t);
[ma3,ph3]=bode(am3,bm3,cm3,d,1,w);

% impulse responses of the single modes:
figure(3)
subplot(311);
plot(t,ym1)
axis([0 200 -.4 .4])
xlabel('time, s')
ylabel('mode 1')

subplot(312);
plot(t,ym2)
axis([0 200 -.4 .4])
xlabel('time, s')
ylabel('mode 2')

subplot(313);
plot(t,ym3)
axis([0 200 -.4 .4])
xlabel('time, s')
ylabel('mode 3')

% transfer functions of the single modes:
figure(4)
subplot(211)
loglog(w,ma,w,ma1,'--',w,ma2,'--',w,ma3,'--');
xlabel('frequency, rad/s')
ylabel('magnitude')
subplot(212)
semilogx(w,ph,w,ph1,'--',w,ph2,'--',w,ph3,'--')
xlabel('frequency, rad/s')
ylabel('phase, deg')
```

B.2 Example 3.3

```
% impulse response and transfer function
% with acceleration output
clear
% stiffness matrix:
k1=3;k2=3;k3=3;k4=0;
k=[k1+k2 -k2 0;
  -k2 k2+k3 -k3;
   0 -k3 k3+k4];
% mass matrix:
m1=1;m2=1;m3=1;
m=[m1 m2 m3];
m=diag(m);
% damping matrix:
damp=.01*k;
mi=inv(m);
% state matrix:
a=[0*eye(3) eye(3);
  -mi*k -mi*damp];
% input matrix:
bo=[0 0 1]';
bo=inv(m)*bo;
b=[0*bo;
   bo];
% acceleration output:
ca=[ 1 0 0];
d=ca*mi*bo;
c=[-ca*mi*k -ca*mi*d];

% impulse response:
dt=.1
t=0:dt:200;
y=impulse(a,b,c,d,1,t);

figure(1)
subplot(211);
```

```
plot(t,y)
axis([0 200 -1.6 1.6])
xlabel('time, s')
ylabel('impulse response')

% transfer function:
w=logspace(-1,1,900);
[ma,ph]=bode(a,b,c,d,1,w);
figure(2)
subplot(211);
loglog(w,ma);
axis([.1 10 .01 100])
xlabel('frequency, rad/s')
ylabel('magnitude')
subplot(212)
semilogx(w,ph);
axis([.1 10 -400 200])
xlabel('frequency, rad/s')
ylabel('magnitude')
```

B.3 Example 4.11

```
% grammians in limited time and frequency intervals
clear
% stiffness matrix:
k1=10;k2=50;k3=50;k4=10;
k=[k1+k2 -k2 0;
  -k2 k2+k3 -k3;
   0 -k3 k3+k4];
% mass matrix:
m1=1;m2=1;m3=1;
m=[m1 m2 m3];
m=diag(m);
% damping matrix:
damp=.005*k+.1*m;
```

```
% state-space representation:
% matrix A:
k=inv(m)*k;
damp=inv(m)*damp;
a=[0*eye(3) eye(3);
   -k -damp];
% matrix B:
b=[0 0 1]';
b=inv(m)*b;
b=[0*b;
    b];
% matrix C:
c=[0 0 0 0 0 1];

% computing grammians in limited time and frequency intervals:
gam=[];
for iw=1:200,iw                    % frequency iterations
   for it=1:200;                   % time iterations
      t(it)=(it-1)*.125;
   om(iw)=(iw-1)*.1;
   [ab,bb,cb,g]=modal_time_fr(a,b,c,0,t(it),0,om(iw));

   g1(iw,it)=g(1,1);               % Hankel singular value, first mode
   g2(iw,it)=g(3,1);               % Hankel singular value, second mode
   g3(iw,it)=g(5,1);               % Hankel singular value, third mode
end
end

% plotting:
figure(1)
waterfall(t,om,g1)
xlabel('time, s')
ylabel('frequency, rad/s')
zlabel('c/o grammian, mode 1')

figure(2)
waterfall(t,om,g2)
```

```
xlabel('time, s')
ylabel('frequency, rad/s')
zlabel('c/o grammian, mode 2')

figure(3)
waterfall(t,om,g3)
xlabel('time, s')
ylabel('frequency, rad/s')
zlabel('c/o grammian, mode 3')
```

B.4 Example 5.3

```
% H2 and H∞ norms of modes
clear
% mass (m) and stiffness (k) matrices of the truss (see Appendix C):
load c:\truss_2D
% damping matrix
damp=.00001*k;

% input matrix, bo:
nd=max(size(k));
bo=zeros(nd,2);
bo(14,1)=1;
bo(16,2)=1;

% output matrices, coq and cov:
coq=zeros(2,nd);
cov=zeros(2,nd);
cov(1,5)=1;
cov(2,7)=1;

% second-order model:
ko=inv(m)*k;
dampo=inv(m)*damp;
[phi,om2]=eig(k,m);          % modal matrix, phi
```

```
om=sqrt(om2);                  % matrix of natural frequencies, om
mm=phi'*m*phi;                 % modal mass matrix
km=phi'*k*phi;                 % modal stiffness matrix
dm=phi'*damp*phi;
z=0.5*inv(om)*inv(mm)*dm;      % modal damping
bm=inv(mm)*phi'*bo;            % modal input matrix
cmq=coq*phi;                   % modal output matrix (displacement)
cmv=cov*phi;                   % modal output matrix (velocity)

% almost-balanced model:
for i=1:nd;
nq=norm(cmq(:,i),'fro');
nv=norm(cmv(:,i),'fro');
nc(i)=sqrt(nq*nq/om2(i,i)+nv*nv);  % output gain
nb(i)=norm(bm(i,:),'fro');         % input gain
end

% Hinf norm:
hinf=diag(nb)*diag(nc)*inv(om)*inv(z)/2;
% H2 norm:
h2=diag(nb)*diag(nc)*(sqrt(inv(om)*inv(z)))/2;

omega=diag(om);
hinf=diag(hinf);
h2=diag(h2);
f=omega/2/pi;
semilogy(omega,hinf,'o');
hold on
semilogy(omega,h2,'.','markersize',18);
axis([0 4000 1e-4 10])
xlabel('natural frequency, rad/s')
ylabel('H_\infty  and  H_2 norms')
for i=1:16;
    plot([1 1]*omega(i),[.0001,max(hinf(i),h2(i))]);
end

% Hankel singular values:
```

```
% sort in descending order:
[gam,ix]=sort(-hinf);
gam=-gam/2;
omega=omega(ix);

% state-space representation:
a=[zeros(nd,nd) eye(nd);
    -om2      -2*z*om];
b=[zeros(nd,2);bm];
c=[cmq 0*cmq;
   0*cmv cmv];

% balanced state-space representation:
[ab,bb,cb,g,r]=balan2(a,b,c);

% Hankel singular values:
gamma=diag(g);
% Hankel singular values for each mode:
gamma=gamma(1:2:32);

% plot exact Hankel singular values (gamma):
% and approximate ones (gam):
figure(2)
n=1:16;
semilogy(n,gamma,'o',n,gam,'.')
hold on
for i=1:16;
plot([i i],[1e-5,max(gamma(i),gam(i))])
end
axis([0 17 .00001 1])
xlabel('mode number')
ylabel('Hankel singular values')
```

B.5 Example 6.7

```
% Advanced Supersonic Transport
clear
% state matrix A:
a=[-0.0127 -0.0136 -0.036 0 0 0 0 0;
   -0.0969 -0.401 0 0.961 19.59 -0.1185 -9.2 -0.1326;
    0 0 0 1 0 0 0 0;
   -0.229 1.726 0 -0.722 -12.021 -0.342 1.8422 0.881;
    0 0 0 0 0 1 0 0;
    0 0.1204 0 0.0496 -44 -1.2741 -4.0301 -0.508;
    0 0 0 0 0 0 0 1;
    0 0.1473 0 0.301 -7.4901 -0.1257 -21.7 -0.803];
% input matrix B:
b=[0 -0.215 0 -1.097 0 -0.64 0 -1.882;
   0.0194 0 0 0 0 0 0 0;
   0 -0.004 0 0.366 0 0.1625 0 0.472;
   0 -1.786 0 -0.0569 0 -0.037 0 -0.0145]';
% output matrix C:
c=[0 1 0 0 0 0 0 0];

% time interval T=[t1,t2]:
t1=0;
t2=3.5;
% modal coordinates and Hankel singular values in time interval T:
[ab,bb,cb,g]=modal_time_fr(a,b,c,t1,t2,0,1e6);

% reduction:
nn=[1:4];
ar=ab(nn,nn);br=bb(nn,:);cr=cb(:,nn);

% step response:
t=0:.01:5;
% full model:
y=step(ab,bb,cb,zeros(1,4),1,t);
% reduced model:
yr=step(ar,br,cr,zeros(1,4),1,t);
```

```
subplot(211)
plot(t,y,t,yr,'--');
axis([0 5 -7 1])
hold on
plot([1 1]*3.5,[-7 1])
fill([3.5 3.5 5 5 3.5],[-7 1 1 -7 -7],'y')
plot(t,y,t,yr,'--')
xlabel('time, s')
ylabel('step response')
```

B.6 Example 7.2

```
% sensor placement for a clamped beam
% using Hinf and H2 norms
clear
% beam mass and stiffness matrices (see Appendix C):
load c:/beam

z=.00001*k;                              % damping

% modal parameters:
[phi,om2]=eig(k,m);
z=phi'*z*phi;
z=diag(z);
w=diag(om2);
w=sqrt(w);
ms=phi;

% shaker location:
nshaker=17;                              % node 6, dir y
b0=[zeros(nshaker-1,1);1;zeros(42-nshaker,1)];
bm=(ms)'*b0;

cm=zeros(1,42);
```

```
mode_no1=1;                          % first mode under consideration
mode_no2=2;                          % second mode under consideration
mode_no3=3;                          % third mode under consideration
mode_no4=4;                          % fourth mode under consideration

% location of y-direction sensors:
nsens=[2:3:42];

% determination of Hinf and H2 norms:
for i=1:14;
   cm1=cm;
   cm1(1,nsens(i))=1;                % first sensor location
   c=cm1*ms;                         % both sensors, modal coordinates
hi=norm_inf(w,z,bm,1*c,0*c,0*c);    % Hinf norm, both sensors
hi1(i)=hi(mode_no1,1);          % Hinf norm, both sensors, first mode
hi2(i)=hi(mode_no2,1);          % Hinf norm, both sensors, second mode
hi3(i)=hi(mode_no3,1);          % Hinf norm, both sensors, third mode
hi4(i)=hi(mode_no4,1);          % Hinf norm, both sensors, fourth mode
h2=norm_h2(w,z,bm,1*c,0*c,0*c);   % H2 norm, both sensors
h21(i)=h2(mode_no1,1);          % H2 norm, both sensors, first mode
h22(i)=h2(mode_no2,1);          % H2 norm, both sensors, second mode
h23(i)=h2(mode_no3,1);          % H2 norm, both sensors, third mode
h24(i)=h2(mode_no4,1);          % H2 norm, both sensors, fourth mode
end;

% add zero norm at the beam fixed ends:
hi1=[0 hi1 0];
hi2=[0 hi2 0];
hi3=[0 hi3 0];
hi4=[0 hi4 0];

h21=[0 h21 0];
h22=[0 h22 0];
h23=[0 h23 0];
h24=[0 h24 0];

% normalization:
```

```
ho=max(max(hi1));
hi1=hi1/ho;
ho=max(max(hi2));
hi2=hi2/ho;
ho=max(max(hi3));
hi3=hi3/ho;
ho=max(max(hi4));
hi4=hi4/ho;

ho=max(max(h21));
h21=h21/ho;
ho=max(max(h22));
h22=h22/ho;
ho=max(max(h23));
h23=h23/ho;
ho=max(max(h24));
h24=h24/ho;

% computing Hinf and H2 norms for two, three, and four modes:
for i=1:16;
        hi12(i)=max(hi1(i),hi2(i));
        hi123(i)=max(hi12(i),hi3(i));
        hi1234(i)=max(hi123(i),hi4(i));

        h212(i)=sqrt((h21(i))^2+(h22(i))^2);
        h2123(i)=sqrt((h212(i))^2+(h23(i))^2);
        h21234(i)=sqrt((h2123(i))^2+(h24(i))^2);
end

% plotting:

node=0:15;          % node number

figure(1);

subplot(221)
plot(node,hi1)
```

```
xlabel('node number');
ylabel('H_\infty index');
axis([0 15 0 1.3])

subplot(222)
plot(node,hi2,'r')
xlabel('node number');
ylabel('H_\infty index');
axis([0 15 0 1.3])

subplot(223)
plot(node,hi3,'r')
xlabel('node number');
ylabel('H_\infty index');
axis([0 15 0 1.3])

subplot(224)
plot(node,hi4,'r')
xlabel('node number');
ylabel('H_\infty index');
axis([0 15 0 1.3])

figure(2);

subplot(221)
plot(node,hi1)
xlabel('node number');
ylabel('H_\infty index');
axis([0 15 0 1.3])

subplot(222)
plot(node,hi1,node,hi2,node,hi12,'r')
xlabel('node number');
ylabel('H_\infty index');
axis([0 15 0 1.3])

subplot(223)
```

```
plot(node,hi1,node,hi2,node,hi3,node,hi123,'r')
xlabel('node number');
ylabel('H_\infty index');
axis([0 15 0 1.3])

subplot(224)
plot(node,hi1,node,hi2,node,hi3,node,hi4,node,hi1234,'r')
xlabel('node number');
ylabel('H_\infty index');
axis([0 15 0 1.3])

figure(3)

subplot(221)
plot(node,h21,'r')
xlabel('node number');
ylabel('H_2 index');
axis([0 15 0 1.8])

subplot(222)
plot(node,h21,node,h22,node,h212,'r')
xlabel('node number');
ylabel('H_2 index');
axis([0 15 0 1.8])

subplot(223)
plot(node,h21,node,h22,node,h23,node,h2123,'r')
xlabel('node number');
ylabel('H_2 index');
axis([0 15 0 1.8])

subplot(224)
plot(node,h21,node,h22,node,h23,node,h24,node,h21234,'r')
xlabel('node number');
ylabel('H_2 index');
axis([0 15 0 1.8])
```

B.7 Example 8.1

```
% modal actuator
clear
% beam mass and stiffness matrices (see Appendix C)
load c:/beam
% damping matrix:
damp=.00001*k;

nd=max(size(k));
% sensor locations:
ns=1;                           % number of sensors
coq=zeros(ns,nd);
% displacement sensors at each node 6, y-direction:
coq(1,17)=1;

% no velocity sensors:
cov=0*coq;

% second-order model:
[phi,om2]=eig(k,m);
om2=diag(om2);
[om2,ind]=sort(om2);
phi=phi(:,ind);
nm=9;
nn=1:nm;                        % number of modes considered
om2=om2(nn);
phi=phi(:,nn);

om=diag(sqrt(om2));
mm=phi'*m*phi;
km=phi'*k*phi;
dm=phi'*damp*phi;
z=0.5*inv(om)*inv(mm)*dm;
cmq=coq*phi;
cmv=cov*phi;
om=diag(om);
```

```
z=diag(z);

% state-space model:
a=[zeros(42,42) eye(42);-inv(m)*k -inv(m)*damp];
c=[coq cov];

phi2=phi(:,1:nm);              % nm modes avaliable

na=1;                          % number of actuators
bm_assumed=.01*[0 1 0 0 0 0 0 0]';

% weight such that the resonances are equal:
normb=diag(bm_assumed*bm_assumed');
normb=sqrt(normb);
normc=diag(cmq'*cmq);
normc=sqrt(normc);
h=ones(nm,1);                  % assumed height of the resonances
xx=2*z.*om.*om;
weight=(xx./normc);
for i=1:nm;
    nb(i,1)=norm(bm_assumed(i,:),'fro');
end

weight=diag(weight.*h);
bm_assumed=weight*bm_assumed;
r=inv(mm)*phi2';
bo2=pinv(r)*bm_assumed;
b=[zeros(42,na);inv(m)*bo2];     % input matrix

% system state-space representation:
sys=ss(a,b,c,zeros(ns,na));

% Bode plot:
w=logspace(0,4,1000)*2*pi;
f=w/2/pi;
[magn,ph]=bode(sys,w);
figure(1)
```

```
loglog(f,squeeze(magn(1,1,:)),'k');
axis([1 10000 1e-9 1e-2])
xlabel('frequency, Hz')
ylabel('magnitude')

% impulse response:
t=0:.0001:.2;
y=impulse(sys,t);
figure(2);
subplot(311)
plot(t,y,'k');
hold on
plot([0 0.2],[0 0],': k')
hold off
axis([0 0.2 -0.005 0.005]);
xlabel('time, s')
ylabel('displacement, node 6')

% nodal displacement:
cc=[eye(42) 0*eye(42)];
sysc=ss(a,b,cc,zeros(42,na));
t=(0:10)*0.0002;
yc=impulse(sysc,t);
q=yc(:,2:3:42);
q=[zeros(11,1),q,zeros(11,1)];
subplot(312)
nn=0:15;
plot(nn,q(1:10,:)','k');
axis([0 15 -.0022 .0022]);
xlabel('node number')
ylabel('displacement')

% actuator gain:
bx=inv(m)*bo2;
bx=bx(2:3:42);
bx=[0;bx;0];
subplot(313)
```

```
stem(nn,bx,'k');
hold on
plot([0 15],[0 0],'k');
axis([0 15 -1.8 1.8]);
xlabel('node number')
ylabel('actuator gain')
hold off
```

B.8 Example 9.1

```
% identification of a simple system
%
clear
% stiffness, mass, and  damping matrices:
k1=10;k2=50;k3=50;k4=10;
k=[k1+k2 -k2 0;
  -k2 k2+k3 -k3;
   0 -k3 k3];
m1=1;m2=1;m3=1;
m=[m1 m2 m3];
m=diag(m);
damp=.001*k+.2*m;

% state-space representation:
k=inv(m)*k;
damp=inv(m)*damp;
a=[0*eye(3) eye(3);-k -damp];
eig(a),
b=[0;0;1];
b=inv(m)*b;
b=[0*b;b];
c=[0,0,0,0,0,1];

% discrete-time system:
dt=.1;                  % sampling time
```

```
[ad,bd]=c2d(a,b,dt);
cd=c;

% generating data for system identification:
pp=500;
t=(0:pp-1)*dt;
u=[zeros(2,1);ones(pp-2,1)];
yx=dlsim(ad,bd,cd,0,u);
% data with or without measurement noise:
nn=input('noise yes? (nn=1), or no? (nn=0)')
noise=rand(pp,1)*.01*nn;
mm=mean(noise);
noise=noise-mm;
y=yx+noise;
y_noise=y;
p=32;
q=300;

% output measurement matrix Y:
Y=y(1:q,1)';
u=u';

% input measurement matrix U:
uu=u(1,1:q);
U=uu;
for i=1:p-1
   U=[U;[zeros(1,i) u(1,1:q-i)]];
end

% Markov parameters H:
H=Y*pinv(U);
nn=max(size(H));
y=H;
%y=M(1,2:nn-1);
p=p/2;

% Hankel matrices h1 and h2:
```

```
h1=[];h2=[];
for i=2:p;
i1=i:i+p-2;
i2=i1+1;
h1=[h1;y(1,i1)];
h2=[h2;y(1,i2)];
end;

% identified state-space representation (ao,bo,co):
[v,gg,u]=svd(h1);
g=sqrt(gg);
gam=diag(g);

% reduction:
nn=1:6;
g=g(nn,nn);
u=u(:,nn);
v=v(:,nn);
Q=g*u';
P=v*g;
gi=inv(g);
Pi=gi*v';
Qi=u*gi;
ao=Pi*h2*Qi;
bo=Q(:,1);
co=P(1,:);

figure(1)
% plot Markov parameters:
subplot(211)
stem(H(1:2*p-2),'filled')

% compare responses of the original system (ad,bd,cd)
% and identified system (ao,bo,co):
subplot(212)
t=(0:pp-1)*dt;
u=[zeros(2,1);ones(pp-2,1)];
```

```
y_orig=dlsim(ad,bd,cd,0,u);
y_ident=dlsim(ao,bo,co,0,u);
plot(t,y_orig,t,y_ident,':')

% Hankel singular values
figure(2)
stem(gam,'filled')
```

B.9 Example 10.4.2

```
% dissipative controller design for 2D truss,
% 2 inputs and 2 outputs
clear
%
% mass and stiffness matrices of the 2D truss (see Appendix C):
load c:\truss_2D
%
% damping matrix:
damp=20e-6*k+.3*m;

% inputs:
bo=zeros(16,1);
bo(16,1)=1;
bo(5,2)=1;
bo=inv(m)*bo;

[n1,nu]=size(bo);

ko=inv(m)*k;
dampo=inv(m)*damp;
nd=16;

% state-space representation:
a=[zeros(nd,nd) eye(nd);
    -ko        -dampo];
```

```
b=[zeros(nd,nu);bo];
c=b';

% state-space modal representation:
[va2,ad,bd,cd]=balmod2(a,b,c);
gam1=lyap(ad,bd(:,1)*bd(:,1)');
gam2=lyap(ad,bd(:,2)*bd(:,2)');

gam1=diag(gam1);
gam2=diag(gam2);
G=2*[gam1 gam2];

[zz,ii]=sort(-G);

% set values d_beta:
db=zeros(2*nd,1);
db(ii(1,1),1)=59;
db(ii(2,1),1)=59;
db(ii(1,2),1)=59;
db(ii(2,2),1)=59;

% solve equation (10.23):
k=pinv(G)*db;
% gains:
k=diag(k);

% closed-loop system:
ac=ad-bd*k*cd;
bc=bd*k;
cc=cd;

% open-loop poles:
l=eig(ad);
l=sort(l);

% closed-loop poles:
lc=eig(ac);
```

```
lc=sort(lc);

nl=max(size(l));
l=l(1:2:nl);
lc=lc(1:2:nl);

% relative pole shift:
beta=real(lc)./real(l);

% natural frequencies:
omo=abs(imag(l));

% plot relative shift:
subplot(211)
bar(omo,beta);
xlabel('natural frequency, rad/s')
ylabel('\beta')

% impulse responses:
t=0:.0001:1;
yo=impulse(ad,bd,cd,zeros(2,2),1,t);
yc=impulse(ac,bc,cc,zeros(2,2),1,t);
subplot(212)
plot(t,yo(:,1),'k',t,yc(:,1),'-- k')
axis([0 .5 -20 20])
xlabel('time, s')
ylabel('impulse response')
```

B.10 Example 11.13.1

```
% LQG controller design for a simple system
clear
% stiffness, mass, and damping matrices:
k1=10;k2=3;k3=4;k4=3;
k=[k1+k2 -k2 0;-k2 k2+k3 -k3;0 -k3 k3];
```

```
m1=1;m2=1;m3=1;
m=[m1 m2 m3];
m=diag(m);
damp=.004*k+.001*m;

% natural frequencies:
om2=eig(k,m);
om=sqrt(om2);

% state-space representation:
k=inv(m)*k;
damp=inv(m)*damp;
a=[0*eye(3) eye(3);-k -damp];
eig(a),
b=[0;0;1];
c=[b' 0*b'];
b=inv(m)*b;
b=[0*b;b];

% modal coordinates:
[v,am,bm,cm]=modal2(a,b,c);

% grammians in modal coordinates:
wc=lyap(am,bm*bm');
wo=lyap(am',cm'*cm);

% Hankel singular values:
gam=diag(wc).*diag(wo);
gam=sqrt(gam);
[gam,ig]=sort(-gam);
gam=-gam;

% sorting Hankel singular values in decreasing order:
am=am(ig,ig);
bm=bm(ig,:);
cm=cm(:,ig);
n1=max(size(am));
```

```
% LQG controller:
Q=diag([0.5,0.5,1,1,2.5,2.5]);              % weights
V=Q;                                        % estimator weights
[k,sc,e]=lqr(am,bm,Q,1);                    % CARE
[ke,se,ee]=lqe(am,eye(n1),cm,V,1);          % FARE

% determining LQG singular values, mu:
[abl,bbl,cbl,mu,kcb,keb,Qcb,Qeb,T]=LQG_bal(am,bm,cm,Q,V,1,1)
mu=sort(diag(mu))

% determining approximate mu:
mu_approx=sqrt(diag(sc).*diag(se)),

% closed-loop system:
ac=[am  -bm*k;ke*cm  am-ke*cm-bm*k];
cc=[cm 0*cm];
bc=[bm;bm];

% Bode plots:
figure(2);
w=logspace(-1,1,800);
[ma,pa]=bode(ac,bc,cc,0,1,w);               % closed-loop
[mo,po]=bode(am,bm,cm,0,1,w);               % open-loop
subplot(211)
loglog(w,mo,'k',w,ma(:,1),'-- k');
xlabel('frequency, rad/s')
ylabel('magnitude')
axis([.1 10 1e-3 100])
%
% simulations with the reduced compensator
%
% reduction matrix, Sigma:
sig=gam.*mu;
[ii,jj]=sort(-sig);
sig=-ii;
sig_approx=gam.*mu_approx;
mu=sort(mu);
```

```
mu_approx=sort(mu_approx);

figure(1)
subplot(211)
nn=1:2:6;
stem(om,mu(nn),'k')
hold on
plot(om,mu_approx(nn),'. k')
hold off
xlabel('natural frequency, rad/s')
ylabel('\mu and \mu_a_p_p_r_o_x')
subplot(212)
stem(om,sig_approx(nn),'k')
xlabel('natural frequency, rad/s')
ylabel('\sigma_a_p_p_r_o_x')

% sort states with respect to sigma values:
am=am(jj,jj);
bm=bm(jj,:);
cm=cm(:,jj);
k=k(:,jj);
ke=ke(jj,:);

% reduced-order controller:
nn=[1:4];
acr=[am -bm*k(:,nn);ke(nn,:)*cm am(nn,nn)-ke(nn,:)*cm(:,nn)-
bm(nn,:)*k(:,nn)];
bcr=[bm;bm(nn,:)];
ccr=[cm 0*cm(:,nn)];

% Bode plots of the closed-loop system with the reduced controller:
figure(3);
[mr,pr]=bode(acr,bcr,ccr,0,1,w);
subplot(211)
loglog(w,ma(:,1),'k',w,mr(:,1),'-- k');
xlabel('frequency, rad/s')
ylabel('magnitude'); axis([0.1 10 0.001 10])
```

B.11 Example 12.13.2

```
% H_inf controller for 2D truss
clear
% mass and stiffness matrices of the truss (see Appendix C):
load c:\truss_2D
%
% damping matrix
damp=1*m+.3e-6*k;

% state matrix A:
k=inv(m)*k;
damp=inv(m)*damp;
[n1,n2]=size(k);
A=[0*eye(n1) eye(n1);-k -damp];

% matrices B1 and B2:
scale2=90;              % scaling factor

B2=zeros(n1,1);
B2(6,1)=1;
B2=inv(m)*B2;
B2=[0*B2;B2];

B1=zeros(n1,1);
B1(15,1)=1;
B1=inv(m)*B1;
B1=[0*B1;B1];
B1=[B1 scale2*B2 0*B1];

% matrices C1 and C2:
C2=zeros(1,n1);
C2(1,6)=1;
C2=[0*C2 C2];

C1=zeros(1,n1);
C1(1,14)=1;
```

```
C1=[C1 0*C1];
C1=[C1; scale2*C2;0*C1];

% matrices D12 and D21:
[n1,q]=size(B1);
[n1,r]=size(B2);
[p,n1]=size(C1);
[s,n1]=size(C2);

D12=zeros(p,r);
for i=1:r;D12(p,i)=1;end;
D21=zeros(s,q);
for i=1:s;D21(i,q)=1;end;
%
%
ro=125.1;
% balanced representations:
[ab,bb1,bb2,cb1,cb2,mu,T]=bal_H_inf(A,B1,B2,C1,C2,ro);
mu=diag(mu);
[a1,b1,c1,g12,r1]=balan2(A,B1,C1);
[a2,b2,c2,g22,r2]=balan2(A,B2,C2);
g12=diag(g12);
g22=diag(g22);
sigma1=g22.*mu;
[sigma_2,ir]=sort(-sigma1);
sigma_2=-sigma_2;

% plot mu:
nn=1:2:32;
figure(1)
semilogy(nn,mu(nn),'o');
xlabel('mode number')
ylabel('\mu')

% plot sigma:
figure(2)
semilogy(nn,sigma_2(nn),'o');
```

```
xlabel('mode number')
ylabel('\sigma')
%
[n1,n2]=size(A);
%
% closed-loop system
mu=diag(mu);
kc=-bb2'*mu;
so=inv(eye(n1)-mu*mu/ro/ro);
kf=-so*mu*cb2';

% controller (aco,bco,cco,dco)
[n1,n2]=size(kc);
[n3,n4]=size(kf);

ainf=ab+bb2*kc+kf*cb2+bb1*bb1'*mu/ro/ro;
bco=-kf;
cco=kc;
dco=zeros(n1,n4);

[n1,n2]=size(C1);
[n3,n4]=size(B1);
[n5,n6]=size(C2);
[n7,n8]=size(B2);

dd=[zeros(n1,n4) D12;D21 zeros(n5,n8)];
[acl,bcl,ccl,dcl]=feedback(ab,[bb1 bb2],[cb1;cb2],dd,ainf,-kf,kc,...
0,n4+1,n1+1);

ao=[ab+bb2*kc               -bb2*kc;
    -bb1*bb1'*mu/ro/ro    ab+kf*cb2+bb1*bb1'*mu/ro/ro];
bo=[bb1;bb1+kf*D21];
co=[cb1+D12*kc -D12*kc];

t=0:.0001:.5;
[n1,n2]=size(bo);
[n3,n4]=size(co);
```

```
% closed-loop impulse response:
y=impulse(acl,bcl,ccl,dcl,1,t);

% open-loop impulse response:
yopen=impulse(A,[B1 B2],C1,[zeros(n2,n3) D12],1,t);

% plot impulse responses
figure(3)
plot(t,yopen(:,1),t,y(:,1),':');
xlabel('time, s')
ylabel('open- and closed-loop impulse response')

% Bode plots:
w=logspace(1,4,400);
[mcl,p]=bode(acl,bcl,ccl,dcl,1,w);                    % closed-loop
[m,p]=bode(A,[B1 B2],C1,[zeros(n2,n3) D12],1,w);      % open-loop
figure(4)
loglog(w,m(:,1),w,mcl(:,1),':') ;
xlabel('frequency, rad/s')
ylabel('open- and closed-loop magnitude')

% reduced closed-loop system
nr=10;                % 5 modes, or 10 states
nr=1:nr;
nr=ir(nr,:);
[n1,n2]=size(kc(:,nr));
[n3,n4]=size(kf(nr,:));

% reduced-order controller:
ainf=ab(nr,nr)+bb2(nr,:)*kc(:,nr)+kf(nr,:)*cb2(:,nr)+bb1(nr,:)* …
bb1(nr,:)'*mu(nr,nr)/ro/ro;
bco=-kf(nr,:);
cco=kc(:,nr);
dco=zeros(n1,n4);

[n1,n2]=size(C1);
[n3,n4]=size(B1);
```

```
[n5,n6]=size(C2);
[n7,n8]=size(B2);

dd=[zeros(n1,n4) D12;D21 zeros(n5,n8)];

% closed-loop with reduced-order controller:
[acl,bcl,ccl,dcl]=feedback(ab,[bb1 bb2],[cb1;cb2],dd,ainf, ...
-kf(nr,:),kc(:,nr),0,n4+1,n1+1);

% impulse response of the closed-loop system with reduced-order
% controller:
yr=impulse(acl,bcl,ccl,dcl,1,t);
figure(5)
plot(t,y(:,1),t,yr(:,1),':');
xlabel('time, s')
ylabel('\impulse responses: full- and reduced-order')
```

C
Structural Parameters

This appendix provides the parameters of the 2D truss (shown in Fig. 1.2), the clamped beam (shown in Fig. 1.4), and the Deep Space Network antenna (shown in Figs. 1.5 and 1.6). They allow the reader to check the methods and to exercise her/his own ideas and modifications. No result is final, and no approach is perfect.

C.1 Mass and Stiffness Matrices of the 2D Truss

The mass matrix, M,

$$M = \begin{bmatrix} M_1 & O \\ O & M_1 \end{bmatrix},$$

where

$$M_1 = \begin{bmatrix} 0.41277 & 0 & 0 & 0 & 0 & 0 & 0 & 0 \\ 0 & 0.41277 & 0 & 0 & 0 & 0 & 0 & 0 \\ 0 & 0 & 0.41277 & 0 & 0 & 0 & 0 & 0 \\ 0 & 0 & 0 & 0.41277 & 0 & 0 & 0 & 0 \\ 0 & 0 & 0 & 0 & 0.41277 & 0 & 0 & 0 \\ 0 & 0 & 0 & 0 & 0 & 0.41277 & 0 & 0 \\ 0 & 0 & 0 & 0 & 0 & 0 & 0.23587 & 0 \\ 0 & 0 & 0 & 0 & 0 & 0 & 0 & 0.23587 \end{bmatrix}$$

and

$$O = \begin{bmatrix} 0 & 0 & 0 & 0 & 0 & 0 & 0 & 0 \\ 0 & 0 & 0 & 0 & 0 & 0 & 0 & 0 \\ 0 & 0 & 0 & 0 & 0 & 0 & 0 & 0 \\ 0 & 0 & 0 & 0 & 0 & 0 & 0 & 0 \\ 0 & 0 & 0 & 0 & 0 & 0 & 0 & 0 \\ 0 & 0 & 0 & 0 & 0 & 0 & 0 & 0 \\ 0 & 0 & 0 & 0 & 0 & 0 & 0 & 0 \\ 0 & 0 & 0 & 0 & 0 & 0 & 0 & 0 \end{bmatrix}.$$

The stiffness matrix, K,

$$K = \begin{bmatrix} K_1 & K_2 \\ K_2^T & K_1 \end{bmatrix},$$

where

$$K_1 = 10^6 \times \begin{bmatrix} 3.024 & -1 & 0 & 0 & 0 & 0 & 0 & 0 \\ 0 & 1.909 & 0 & 0 & 0 & 0 & 0 & 0 \\ -1 & 0 & 3.024 & 0 & -1 & 0 & 0 & 0 \\ 0 & 0 & 0 & 1.909 & 0 & 0 & 0 & 0 \\ 0 & 0 & -1 & 0 & 3.024 & 0 & -1 & 0 \\ 0 & 0 & 0 & 0 & 0 & 1.909 & 0 & 0 \\ 0 & 0 & 0 & 0 & -1 & 0 & 1.512 & -0.384 \\ 0 & 0 & 0 & 0 & 0 & 0 & -0.384 & 1.621 \end{bmatrix}$$

and

$$K_2 = 10^6 \times \begin{bmatrix} 0 & 0 & -0.512 & -0.384 & 0 & 0 & 0 & 0 \\ 0 & -1.333 & -0.384 & -0.288 & 0 & 0 & 0 & 0 \\ -0.512 & 0.384 & 0 & 0 & -0.512 & -0.384 & 0 & 0 \\ 0.384 & -0.288 & 0 & -1.333 & -0.384 & -0.288 & 0 & 0 \\ 0 & 0 & -0.512 & 0.384 & 0 & 0 & -0.512 & -0.384 \\ 0 & 0 & 0.384 & -0.288 & 0 & -1.333 & -0.384 & -0.288 \\ 0 & 0 & 0 & 0 & -0.512 & 0.384 & 0 & 0 \\ 0 & 0 & 0 & 0 & 0.384 & -0.288 & 0 & -1.333 \end{bmatrix}.$$

C.2 Mass and Stiffness Matrices of the Clamped Beam Divided into 15 Finite Elements

For $n = 15$ the beam has 42 degrees of freedom (14 nodes, each node has three degrees of freedom: horizontal and vertical displacement, and in-plane rotation). The mass and stiffness matrices are of dimensions 42×42. The mass matrix is a diagonal with the diagonal entries as follows.

The beam mass matrix:

$$M = \begin{bmatrix} M_1 & O & O & O & O & O & O & O & O & O & O & O & O & O \\ O & M_1 & O & O & O & O & O & O & O & O & O & O & O & O \\ O & O & M_1 & O & O & O & O & O & O & O & O & O & O & O \\ O & O & O & M_1 & O & O & O & O & O & O & O & O & O & O \\ O & O & O & O & M_1 & O & O & O & O & O & O & O & O & O \\ O & O & O & O & O & M_1 & O & O & O & O & O & O & O & O \\ O & O & O & O & O & O & M_1 & O & O & O & O & O & O & O \\ O & O & O & O & O & O & O & M_1 & O & O & O & O & O & O \\ O & O & O & O & O & O & O & O & M_1 & O & O & O & O & O \\ O & O & O & O & O & O & O & O & O & M_1 & O & O & O & O \\ O & O & O & O & O & O & O & O & O & O & M_1 & O & O & O \\ O & O & O & O & O & O & O & O & O & O & O & M_1 & O & O \\ O & O & O & O & O & O & O & O & O & O & O & O & M_1 & O \\ O & O & O & O & O & O & O & O & O & O & O & O & O & M_1 \end{bmatrix},$$

where

$$M_1 = 10^{-4} \times \begin{bmatrix} 0.7850 & 0 & 0 \\ 0 & 0.7850 & 0 \\ 0 & 0 & 6.5417 \end{bmatrix},$$

$$O = \begin{bmatrix} 0 & 0 & 0 \\ 0 & 0 & 0 \\ 0 & 0 & 0 \end{bmatrix}.$$

The stiffness matrix is as follows:

$$K = \begin{bmatrix} K_1 & K_2 & O & O & O & O & O & O & O & O & O & O & O & O \\ K_2^T & K_1 & K_2 & O & O & O & O & O & O & O & O & O & O & O \\ O & K_2^T & K_1 & K_2 & O & O & O & O & O & O & O & O & O & O \\ O & O & K_2^T & K_1 & K_2 & O & O & O & O & O & O & O & O & O \\ O & O & O & K_2^T & K_1 & K_2 & O & O & O & O & O & O & O & O \\ O & O & O & O & K_2^T & K_1 & K_2 & O & O & O & O & O & O & O \\ O & O & O & O & O & K_2^T & K_1 & K_2 & O & O & O & O & O & O \\ O & O & O & O & O & O & K_2^T & K_1 & K_2 & O & O & O & O & O \\ O & O & O & O & O & O & O & K_2^T & K_1 & K_2 & O & O & O & O \\ O & O & O & O & O & O & O & O & K_2^T & K_1 & K_2 & O & O & O \\ O & O & O & O & O & O & O & O & O & K_2^T & K_1 & K_2 & O & O \\ O & O & O & O & O & O & O & O & O & O & K_2^T & K_1 & K_2 & O \\ O & O & O & O & O & O & O & O & O & O & O & K_2^T & K_1 & K_2 \\ O & O & O & O & O & O & O & O & O & O & O & O & K_2^T & K_1 \end{bmatrix},$$

where

$$K_1 = 10^5 \times \begin{bmatrix} 4.200 & 0 & 0 \\ 0 & 0.010 & 0 \\ 0 & 0 & 0.336 \end{bmatrix},$$

$$K_2 = 10^5 \times \begin{bmatrix} -2.100 & 0 & 0 \\ 0 & -0.005 & 0.025 \\ 0 & -0.025 & 0.084 \end{bmatrix},$$

and

$$O = \begin{bmatrix} 0 & 0 & 0 \\ 0 & 0 & 0 \\ 0 & 0 & 0 \end{bmatrix}.$$

C.3 State-Space Representation of the Deep Space Network Antenna

The state-space representation (A, B, C) of the Deep Space Network antenna in azimuth axis motion was obtained from the field test data. The following are the state matrices after reduction to 18 states. The state matrix A is in the block-diagonal form

$$A = \begin{bmatrix} A_1 & O & O & O & O & O & O & O & O \\ O & A_2 & O & O & O & O & O & O & O \\ O & O & A_3 & O & O & O & O & O & O \\ O & O & O & A_4 & O & O & O & O & O \\ O & O & O & O & A_5 & O & O & O & O \\ O & O & O & O & O & A_6 & O & O & O \\ O & O & O & O & O & O & A_7 & O & O \\ O & O & O & O & O & O & O & A_8 & O \\ O & O & O & O & O & O & O & O & A_9 \end{bmatrix},$$

where

$$A_1 = \begin{bmatrix} 0 & 0 \\ 0 & -1.104067 \end{bmatrix}, \qquad A_2 = \begin{bmatrix} -0.348280 & 10.099752 \\ 10.099752 & -0.348280 \end{bmatrix},$$

$$A_3 = \begin{bmatrix} -0.645922 & 12.561336 \\ -12.561336 & -0.645922 \end{bmatrix}, \qquad A_4 = \begin{bmatrix} -0.459336 & 13.660350 \\ -13.660350 & -0.459336 \end{bmatrix},$$

$$A_5 = \begin{bmatrix} -0.934874 & 18.937362 \\ -18.937362 & -0.934874 \end{bmatrix}, \qquad A_6 = \begin{bmatrix} -0.580288 & 31.331331 \\ 31.331331 & -0.580288 \end{bmatrix},$$

$$A_7 = \begin{bmatrix} -0.842839 & 36.140547 \\ -36.140547 & -0.842839 \end{bmatrix}, \qquad A_8 = \begin{bmatrix} -0.073544 & 45.862202 \\ -45.862202 & -0.073544 \end{bmatrix},$$

$$A_9 = \begin{bmatrix} -3.569534 & 48.508185 \\ -48.508185 & -3.569534 \end{bmatrix}, \qquad O = \begin{bmatrix} 0 & 0 \\ 0 & 0 \end{bmatrix}.$$

The matrix B,

$$B = \begin{bmatrix} 1.004771 \\ -0.206772 \\ -0.093144 \\ 0.048098 \\ 0.051888 \\ 1.292428 \\ -0.024689 \\ 0.245969 \\ -0.234201 \\ 0.056769 \\ 0.540327 \\ -0.298787 \\ -0.329058 \\ -0.012976 \\ -0.038636 \\ -0.031413 \\ -0.115836 \\ 0.421496 \end{bmatrix},$$

and the matrix C,

$$C = \begin{bmatrix} C_1 & C_2 & C_3 \end{bmatrix},$$

where

$$C_1 = \begin{bmatrix} 1.004771 & -0.204351 & 0.029024 & -0.042791 & -0.322601 & -0.545963 \end{bmatrix},$$

$$C_2 = \begin{bmatrix} -0.098547 & -0.070542 & 0.113774 & 0.030378 & 0.058073 & 0.294883 \end{bmatrix},$$

and

$$C_3 = \begin{bmatrix} 0.110847 & -0.109961 & -0.022496 & -0.009963 & 0.059871 & -0.198378 \end{bmatrix}.$$

References

[1] Aidarous, S.E., Gevers, M.R., and Installe, M.J., "Optimal Sensors' Allocation Strategies for a Class of Stochastic Distributed Systems," *International Journal of Control*, vol. 22, 1975, pp. 197–213.

[2] Anderson, B.D.O., "A System Theory Criterion for Positive Real Matrices," *SIAM Journal of Control*, vol. 5, 1967, pp. 171–182.

[3] Anderson, B.D.O., and Moore, J.B., *Optimal Control*, Prentice Hall, Englewood Cliffs, NJ, 1990.

[4] Athans, M., "On the Design of PID Controllers Using Optimal Linear Regulator Theory," *Automatica*, vol. 7, 1971, pp. 643–647.

[5] Aubrun, J.N., "Theory of the Control of Structures by Low-Authority Controllers," *Journal of Guidance, Control, and Dynamics*, vol. 3, 1980.

[6] Aubrun, J.N., and Margulies, G., "Low-Authority Control Synthesis for Large Space Structures," *NASA Contractor Report 3495*, Contract NAS1-14887, 1982.

[7] Basseville, M., Benveniste, A., Moustakides, G.V., and Rougee, A., "Optimal Sensor Location for Detecting Changes in Dynamical Behavior," *IEEE Transactions on Automatic Control*, vol. AC-32, 1987, pp. 1067–1075.

[8] Bellos, J., and Inman, D.J., "Frequency Response of Nonproportionally Damped, Lumped Parameter, Linear Dynamic Systems," *ASME Journal of Vibration and Acoustics*, vol. 112, 1990, pp. 194–201.

[9] Bendat, J.S., and Piersol, A.G., *Random Data*, Wiley, New York, 1986.

[10] Benhabib, R.J., Iwens, R.P., and Jackson, R.L., "Stability of Large Space Structure Control Systems Using Positivity Concepts," *Journal of Guidance, Control, and Dynamics*, vol. 4, 1981, pp. 487–494.

[11] Bhaskar, A., "Estimates of Errors in the Frequency Response of Non-Classically Damped Systems," *Journal of Sound and Vibration,* vol. 184, 1995, pp. 59–72.

[12] Boyd, S.P., and Barratt, C.H., *Linear Controller Design*, Prentice Hall, Englewood Cliffs, NJ, 1991.

[13] Burl, J.B., *Linear Optimal Control H_2 and H_∞ Methods,* Addison-Wesley, Menlo Park, CA, 1999.

[14] Chandrasekharan, P.C., *Robust Control of Linear Dynamical Systems*, Academic Press, London, 1996.

[15] Chen, T., and Francis, B., *Optimal Sampled-Data Control Systems,* Springer-Verlag, London, 1995.

[16] Chung, R., and Lee, C.W., "Dynamic Reanalysis of Weakly Non-Proportionally Damped Systems," *Journal of Sound and Vibration*, vol. 111, 1986, pp. 37–50.

[17] Clark, R.L., Saunders, W.R. and Gibbs, G.P., *Adaptive Structures, Dynamics and Control,* Wiley, New York, 1998.

[18] Clough, R.W., and Penzien, J., *Dynamics of Structures*, McGraw-Hill, New York, 1975.

[19] Colgren R.D., "Methods for Model Reduction," *AIAA Guidance, Navigation and Control Conference,* Minneapolis, MN, 1988, pp. 777–790.

[20] Collins, Jr., E.G., Haddad, W.M., and Ying, S.S., "Construction of Low-Authority, Nearly Non-Minimal LQG Compensators for Reduced-Order Control Design," *1994 American Control Conference*, Baltimore, MD, 1994.

[21] Cooper, J.E., and Wright, J.R., "To Fit or to Tune? That Is the Question," *Proceedings of the 15th International Modal Analysis Conference*, Orlando, FL, 1997.

[22] Cottin, N., Prells, U., and Natke, H.G. "On the Use of Modal Transformation to Reduce Dynamic Models in Linear System Identification," *IUTAM Symposium on Identification of Mechanical Systems*, Wuppertal, Germany, 1993.

[23] Cronin, L., "Approximation for Determining Harmonically Excited Response of Nonclassically Damped Systems," *ASME Journal of Engineering for Industry*, vol. 98, 1976, pp. 43–47.

[24] DeLorenzo, M.L., "Sensor and Actuator Selection for Large Space Structure Control," *Journal of Guidance, Control and Dynamics*, vol. 13, 1990, pp. 249–257.

[25] Derese, I., and Noldus, E., "Design of Linear Feedback Laws for Bilinear Systems," *International Journal of Control*, vol. 31, 1980, pp. 219–237.

[26] De Silva, C.W., *Vibration, Fundamentals and Practice,* CRC Press, Boca Raton, FL, 2000.

[27] Desoer, C.A., and Vidyasagar, M., *Feedback Systems: Input–Output Properties,* Academic Press, New York, 1975.

[28] Dorato, P., Abdallah, C., and Cerone, V., *Linear Quadratic Control*, Prentice Hall, Englewood Cliffs, NJ, 1995.

[29] Dorf, R.C., *Modern Control Systems*, Addison-Wesley, Reading, MA, 1980.

[30] Doyle, J.C., Francis, B.A., and Tannenbaum, A.R., *Feedback Control Theory*, Macmillan, New York, 1992.

[31] Elliott, S., *Signal Processing for Active Control,* Academic Press, San Diego, CA, 2001.

[32] Enns, D.F., "Model Reduction for Control System Design," Ph.D. Thesis, Department of Aeronautics and Astronautics, Stanford University, Palo Alto, CA, 1984.

[33] Ewins, D.J., *Modal Testing*, 2nd edition, Research Studies Press, Baldock, England, 2000.

[34] Fairman, F.W., *Linear Control Theory, The State Space Approach,* Wiley, Chichester, England, 1998.

[35] Felszeghy, F., "On Uncoupling and Solving the Equations of Motion of Vibrating Linear Discrete Systems," *ASME Journal of Applied Mechanics*, vol. 60, 1993, pp. 456–462.

[36] Francis, B.A., and Wonham, W.M., "The Internal Model Principle of Control Theory," *Automatica*, vol. 12, 1976, pp. 457–465.

[37] Franklin, G.F., Powell, J.D., and Workman, M.L. *Digital Control of Dynamic Systems,* Addison-Wesley, Reading, MA, 1992.

[38] Friswell, M.I., "On the Design of Modal Actuators and Sensors," *Journal of Sound and Vibration,* vol. 241, No. 3, 2001.

[39] Fukata, S., Mohri, A., and Takata, M., "On the Determination of the Optimal Feedback Gains for Multivariable Linear Systems Incorporating Integral Action," *International Journal of Control*, vol. 31, 1980, pp. 1027–1040.

[40] Fuller, C.R., Elliot, S.J., and Nelson, P.A., *Active Control of Vibrations,* Academic Press, London, 1996.

[41] Furuta, K., Sano, A., and Atherton, D., *State Variable Methods in Automatic Control*, Wiley, Chichester, England, 1988.

[42] Gawronski, W., "Antenna Control Systems: From PI to H_∞," *IEEE Antennas and Propagation Magazine*, vol. 43, no. 1, 2001.

[43] Gawronski, W., "Modal Actuators and Sensors," *Journal of Sound and Vibration*, vol. 229, no. 4, 2000.

[44] Gawronski, W., *Dynamics and Control of Structures,* Springer-Verlag, New York, 1998.

[45] Gawronski, W., "Almost-Balanced Structural Dynamics," *Journal of Sound and Vibration*, vol. 202, 1997, pp. 669–687.

[46] Gawronski, W., "Actuator and Sensor Placement for Structural Testing and Control," *Journal of Sound and Vibration*, vol. 208, 1997.

[47] Gawronski, W., *Balanced Control of Flexible Structures,* Springer-Verlag, London, 1996.

[48] Gawronski, W., "Wind Gusts Models Derived from Field Data," *Telecommunication and Data Acquisition Progress Report,* vol. 42-123, 1995, pp. 30–36 (NASA/JPL publication).

[49] Gawronski, W., "Balanced LQG Compensator for Flexible Structures," *Automatica*, vol. 30, 1994, pp. 1555–1564.

[50] Gawronski, W., "Balanced Systems and Structures: Reduction, Assignment, and Perturbations," in: *Control and Dynamics Systems*, ed. C.T. Leondes, vol. 54, Academic Press, San Diego, CA, 1992, pp. 373–415.

[51] Gawronski, W., "Model Reduction for Flexible Structures: Test Data Approach," *Journal of Guidance, Control, and Dynamics*, vol. 14, 1991, pp. 692–694.

[52] Gawronski, W., and Hadaegh, F.Y., "Balanced Input–Output Assignment," *International Journal for Systems Science*, vol. 24, 1993, pp. 1027–1036.

[53] Gawronski, W., and Juang, J.N., "Model Reduction for Flexible Structures," in: *Control and Dynamics Systems*, ed. C.T. Leondes, vol. 36, Academic Press, San Diego, CA, 1990, pp. 143–222.

[54] Gawronski, W., and Lim, K.B., "On Frequency Weighting for the H_∞ and H_2 Control Design of Flexible Structures," *AIAA Structures, Structural Dynamics, and Materials Conference/AIAA Adaptive Structures Forum*, Long Beach, CA, 1998.

[55] Gawronski, W., and Lim, K.B., "Balanced Actuator and Sensor Placement for Flexible Structures," *International Journal of Control*, vol. 65, 1996, pp. 131–145.

[56] Gawronski, W., and Lim, K.B., "Balanced H_∞ and H_2 Controllers," *Proceedings of the 1994 IEEE American Control Conference*, Baltimore, MD, 1994, pp. 1116–1122.

[57] Gawronski, W., and Lim, K.B., "Controllability and Observability of Flexible Structures with Proof-Mass Actuators," *Journal of Guidance, Control, and Dynamics*, vol. 16, 1993, pp. 899–902.

[58] Gawronski, W., and Mellstrom, J.A., "Model Reduction for Systems with Integrators," *Journal of Guidance, Control, and Dynamics*, vol. 15, 1992, pp. 1304–1306.

[59] Gawronski, W., and Mellstrom, J.A., "Control and Dynamics of the Deep Space Network Antennas," in: *Control and Dynamics Systems*, ed. C.T. Leondes, vol. 63, Academic Press, San Diego, CA, 1994, pp. 289–412.

[60] Gawronski W., and Natke, H.G., "Realizations of the Transfer Function Matrix," *International Journal of Systems Science*, vol. 18, 1987, pp. 229–236.

[61] Gawronski, W., and Natke, H.G., "Balanced State Space Representation in the Identification of Dynamic Systems," in: *Application of System Identification in Engineering*, ed. H.G. Natke, Springer-Verlag, New York, 1988.

[62] Gawronski, W., Racho, C., and Mellstrom, J.A., "Application of the LQG and Feedforward Controllers to the Deep Space Network Antennas," *IEEE Transactions on Control System Technology*, vol. 3, 1995, pp. 417–421.

[63] Gawronski, W., and Sawicki, J.T., "Response Errors of Non-Proportionally Lightly Damped Structures," *Journal of Sound and Vibration*, vol. 200, 1997, pp. 543–550.

[64] Gawronski, W., and Williams, T., "Model Reduction for Flexible Space Structures," *Journal of Guidance, Control, and Dynamics*, vol. 14, no. 1, 1991, pp. 68–76.

[65] Genta, G., *Vibration of Structures and Machines,* Springer-Verlag, New York, 1999.

[66] Glover, K., "All Optimal Hankel-Norm Approximations of Linear Multivariable Systems and their L^{∞}-Error Bounds," *International Journal of Control*, vol. 39, 1984, pp. 1115–1193.

[67] Glover, K., and Doyle, J.C., "State-Space Formulae for All Stabilizing Controllers That Satisfy an H_{∞}-Norm Bound and Relations to Risk Sensitivity," *Systems and Control Letters,* vol. 11, 1988, pp. 167–172.

[68] Golub, G.H., and Van Loan, C.F., *Matrix Computations*, The Johns Hopkins University Press, Baltimore, MD, 1989.

[69] Gregory, Jr., C.Z., "Reduction of Large Flexible Spacecraft Models Using Internal Balancing Theory," *Journal of Guidance, Control, and Dynamics*, vol. 7, 1984, pp. 725–732.

[70] Hatch, M.R., *Vibration Simulation Using Matlab and ANSYS,* Chapman and Hall/CRC, Boca Raton, FL, 2000.

[71] Heylen, W., Lammens, S., and Sas, P. *Model Analysis Theory and Testing,* Katholieke Universiteit Leuven, Leuven, Belgium, 1997.

[72] Ho, B.L. and Kalman, R.E., "Effective Construction of Linear State-Variable Models from Input–Output Data," *Regelungstechnik*, vol. 14, 1966, pp. 545–548.

[73] Horn, R.A., and Johnson, C.R., *Matrix Analysis*, Cambridge University Press, Cambridge, England, 1985.

[74] Horta, L.G., Juang, J.-N., and Longman, R.W., Discrete-Time Model Reduction in Limited Frequency Ranges. *Journal of Guidance, Control and Dynamics,* vol. 16, no. 6, 1993, pp. 1125–1130.

[75] Hsu, C.-Y., Lin, C.-C., and Gaul, L., "Vibration and Sound Radiation Controls of Beams Using Layered Modal Sensors and Actuators," *Smart Materials and Structures*, vol. 7, 1998.

[76] Hwang, H., and Ma, F., "On the Approximate Solution of Nonclassically Damped Linear Systems," *ASME Journal of Applied Mechanics*, vol. 60, 1993, pp. 695–701.

[77] Hyland, D.C., "Distributed Parameter Modeling of Flexible Spacecraft: Where's the Beef?" *NASA Workshop on Distributed Parameter Modeling and Control of Flexible Aerospace Systems*, Williamsburg, VA, 1992, NASA Conference Publication 3242, 1994, pp. 3–23.

[78] Hyland, D.C., and Bernstein, D.S., "The Optimal Projection Equations for Model Reduction and the Relationships Among the Methods of Wilson, Skelton, and Moore," *IEEE Transactions on Automatic Control*, vol. AC-30, 1985, pp. 1201–1211.

[79] Inman D.J., *Vibration with Control, Measurements, and Stability*, Prentice Hall, Englewood Cliffs, NJ, 1989.

[80] Johnson, C.D., "Optimal Control of the Linear Regulator with Constant Disturbances," *IEEE Transactions on Automatic Control*, vol. AC-13, 1968, pp. 416–421.

[81] Jonckheere, E.A., "Principal Component Analysis of Flexible Systems—Open Loop Case," *IEEE Transactions on Automatic Control*, vol. AC-27, 1984, pp. 1095–1097.

[82] Jonckheere, E.A., and Silverman, L.M., "A New Set of Invariants for Linear Systems—Application to Reduced-Order Controller Design," *IEEE Transactions on Automatic Control*, vol. AC-28, 1983, pp. 953–964.

[83] Joshi, S.M., *Control of Large Flexible Space Structures*, Springer-Verlag, Berlin, 1989.

[84] Juang, J.-N., *Applied System Identification*, Prentice Hall, Englewood Cliffs, NJ, 1994.

[85] Juang, J.-N., Phan, M., Horta, L.G., and Longman, R.W., "Identification of Observer/Kalman Filter Markov Parameters: Theory and Experiments," *Journal of Guidance, Control and Dynamics,* vol. 16, no. 2, 1993, pp. 320–329.

[86] Juang, J-N., and Rodriguez, G., "Formulations and Applications of Large Structure Actuator and Sensor Placements," *VPI/AIAA Symposium on Dynamics and Control of Large Flexible Spacecraft*, Blacksburg, VA, 1979, pp. 247–262.

[87] Junkins, J.L., and Kim, Y., *Introduction to Dynamics and Control of Flexible Structures,* American Institute of Aeronautics and Astronautics, Washington, DC, 1993.

[88] Kailath, T., *Linear Systems*, Prentice Hall, Englewood Cliffs, NJ, 1980.

[89] Kammer, D., "Sensor Placement for On-Orbit Modal Identification and Correlation of Large Space Structures," *Journal of Guidance, Control and Dynamics*, vol. 14, 1991, pp. 251–259.

[90] Kim, Y., and Junkins, J.L., "Measure of Controllability for Actuator Placement," *Journal of Guidance, Control and Dynamics*, vol. 14, 1991, pp. 895–902.

[91] Kwakernaak, H., and Sivan, R., *Linear Optimal Control Systems*, Wiley, New York, 1972.

[92] Kwon, Y.W., and Bang, H., *The Finite Element Method Using Matlab,* CRC Press, Boca Raton, FL, 1997.

[93] Lee, C.-K., and Moon, F.C., "Modal Sensors/Actuators," *Journal of Applied Mechanics, Transactions ASME*, vol. 57, 1990.

[94] Levine, W.S., and Reichert, R.T., "An Introduction to H_∞ Control Design," *Proceedings of the 29th Conference on Decision and Control*, Honolulu, HI, 1990, pp. 2966–2974.

[95] Li, X.P., and Chang, B.C., "On Convexity of H_∞ Riccati Solutions and its Applications," *IEEE Transactions on Automatic Control*, vol. AC-38, 1993, pp. 963–966.

[96] Lim, K.B., "Method for Optimal Actuator and Sensor Placement for Large Flexible Structures," *Journal of Guidance, Control and Dynamics*, vol. 15, 1992, pp. 49–57.

[97] Lim, K.B., and Gawronski, W., "Actuator and Sensor Placement for Control of Flexible Structures," in: *Control and Dynamics Systems*, ed. C.T. Leondes, vol. 57, Academic Press, San Diego, CA, 1993, pp. 109–152.

[98] Lim, K.B., and Gawronski, W., "Hankel Singular Values of Flexible Structures in Discrete Time," *AIAA Journal of Guidance, Control, and Dynamics*, vol. 19, no. 6, 1996, pp. 1370–1377.

[99] Lim, K.B., "Robust Control of Vibrating Structures," in: *Responsive Systems for Active Vibration Control,* Preumont, A. (Editor), NATO Science Series, Kluwer Academic, Dordrecht, 2002, pp. 133–179.

[100] Lin, C.F., *Advanced Control Systems Design*, Prentice Hall, Englewood Cliffs, NJ, 1994.

[101] Lindberg Jr., R.E., and Longman, R.W., "On the Number and Placement of Actuators for Independent Modal Space Control," *Journal of Guidance, Control and Dynamics*, vol. 7, 1984, pp. 215–223.

[102] Ljung, L., *System Identification,* Prentice Hall, Englewood Cliffs, NJ, 1987.

[103] Longman, R.W., and Alfriend, K.T., "Energy Optimal Degree of Controllability and Observability for Regulator and Maneuver Problems," *Journal of the Astronautical Sciences*, vol. 38, 1990, pp. 87–103.

[104] Maciejowski, J.M., *Multivariable Feedback Design*, Addison-Wesley, Wokingham, England, 1989.

[105] Maghami, P.G., and Joshi, S.M., "Sensor/Actuator Placement for Flexible Space Structures," *IEEE American Control Conference*, San Diego, CA, 1990, pp. 1941–1948.

[106] Maghami, P.G., and Joshi, S.M., "Sensor-Actuator Placement for Flexible Structures with Actuator Dynamics," *AIAA Guidance, Navigation, and Control Conference*, New Orleans, LA, 1991, pp. 46–54.

[107] Maia, N.M.M., and Silva, J.M.M. (Editors), *Theoretical and Experimental Modal Analysis*, Research Studies Press, Taunton, England, 1997.

[108] Meirovitch, L., *Dynamics and Control of Structures*, Wiley, New York, 1990.

[109] Moore, B.C., "Principal Component Analysis in Linear Systems, Controllability, Observability and Model Reduction," *IEEE Transactions on Automatic Control*, vol. 26, 1981, pp. 17–32.

[110] Mustafa, D., and Glover, K., "Controller Reduction by H_∞-Balanced Truncation," *IEEE Transactions on Automatic Control*, vol. 36, 1991, pp. 668–682.

[111] Natke, H.G., *Einführung in Theorie und Praxis der Zeitreihen- und Modalanalyse*, Vieweg, Braunschweig, 1992.

[112] Nicholson, W., "Response Bounds for Nonclassically Damped Mechanical Systems Under Transient Loads," *ASME Journal of Applied Mechanics*, vol. 54, 1987, pp. 430–433.

[113] Opdenacker, P., and Jonckheere, E.A., LQG Balancing and Reduced LQG Compensation of Symmetric Passive Systems," *International Journal of Control*, vol. 41, 1985, pp. 73–109.

[114] Oyadiji, S.O., and Solook, M., "Selective Modal Sensing of Vibrating Beams Using Piezoelectric Beams," *Smart Sensing, Processing, and Instrumentation, Proceedings SPIE,* vol. 3042, 1997.

[115] Panossian, H., Gawronski, W., and Ossman, H., "Balanced Shaker and Sensor Placement for Modal Testing of Large Flexible Structures," *IMAC-XVI*, Santa Barbara, CA, 1998.

[116] Phillips, A.W., and Allemang, R.J., "Single Degree-of-Freedom Modal Parameter Estimation Methods." *Proceedingsof the 14th International Modal Analysis Conference*, Dearborn, MI, 1996.

[117] Popov, V.M., *Hyperstability of Automatic Control Systems*, Springer-Verlag, New York, 1973.

[118] Porter, B., "Optimal Control of Multivariable Linear Systems Incorporating Integral Feedback," *Electronics Letters,* vol. 7, 1971.

[119] Porter, B., and Crossley R., *Modal Control*, Taylor and Francis, London, 1972.

[120] Preumont, A., *Vibration Control of Active Structures*, Kluwer Academic, Dordrecht, 2002.

[121] Preumont, A. (Editor), *Responsive Systems for Active Vibration Control,* NATO Science Series, Kluwer Academic, Dordrecht, 2002.

[122] Shahian, B., and Hassul, M., *Control System Design Using Matlab*, Prentice Hall, Englewood Cliffs, NJ, 1993.

[123] Shahruz, M., and Ma, F., “Approximate Decoupling of the Equations of Motion of Linear Underdamped Systems,” *ASME Journal of Applied Mechanics*, vol. 55, 1988, pp. 716–720.

[124] Skelton, R.E., “Cost Decomposition of Linear Systems with Application to Model Reduction,” *International Journal of Control*, vol. 32, 1980, pp. 1031–1055.

[125] Skelton, R.E., *Dynamic System Control: Linear System Analysis and Synthesis*, Wiley, New York, 1988.

[126] Skelton, R.E., and Hughes, P.C., “Modal Cost Analysis for Linear Matrix Second-Order Systems,” *Journal of Dynamic Systems, Measurements and Control*, vol. 102, 1980, pp. 151–158.

[127] Skelton, R.E., and DeLorenzo, M.L., “Selection of Noisy Actuators and Sensors in Linear Stochastic Systems,” *Journal of Large Scale Systems, Theory and Applications*, vol. 4, 1983, pp. 109–136.

[128] Skelton, R.E., Singh, R., and Ramakrishnan, J., “Component Model Reduction by Component Cost Analysis,” *AIAA Guidance, Navigation and Control Conference*, Minneapolis, MN, 1988, pp. 264–2747.

[129] Skogestad, S., and Postlethwaite, I., *Multivariable Feedback Control,* Wiley, Chichester, England, 1996.

[130] Uwadia, E., and Esfandiari, R.S., “Nonclassically Damped Dynamic Systems: An Iterative Approach,” *ASME Journal of Applied Mechanics*, vol. 57, 1990, pp. 423–433.

[131] Van de Wal, M., and de Jager, B., “A Review of Methods for Input/Output Selection,” *Automatica,* vol. 37, 2001, pp. 487–510.

[132] Voth, C.T., Richards, Jr., K.E., Schmitz, E., Gehling, R.N., and Morgenthaler, D.R., “Integrated Active and Passive Control Design Methodology for the LaRC CSI Evolutionary Model,” *NASA Contractor Report 4580*, Contract NAS1-19371, 1994.

[133] Wen, J.T., “Time Domain and Frequency Domain Conditions for Strict Positive Realness,” *IEEE Transactions on Automatic Control*, vol. AC-33, 1988, pp. 988–992.

[134] Wicks, M.A., and DeCarlo, R.A., “Grammian Assignment of the Lyapunov Equation,” *IEEE Transactions on Automatic Control*, vol. AC-35, 1990, pp. 465–468.

[135] Willems, J.C., “Dissipative Dynamical Systems, Part I: General Theory, Part II: Linear Systems with Quadratic Supply Rates,” *Archive for Rational Mechanics and Analysis,* vol. 45, 1972, pp. 321–351 and pp. 352–392.

[136] Willems, J.C., “Realization of Systems with Internal Passivity and Symmetry Constraints,” *Journal of the Franklin Institute*, vol. 301, 1976, pp. 605–620.

[137] Williams, T., “Closed Form Grammians and Model Reduction for Flexible Space Structures,” *IEEE Transactions on Automatic Control*, vol. AC-35, 1990, pp. 379–382.

[138] Wilson, D.A., "Optimum Solution of Model-Reduction Problem," *IEE Proceedings*, vol. 119, 1970, pp. 1161–1165.

[139] Wilson, D.A., "Model Reduction for Multivariable Systems," *International Journal of Control*, vol. 20, 1974, pp. 57–60.

[140] Wortelboer, P., "Frequency-Weighted Balanced Reduction of Closed-Loop Mechanical Servo Systems: Theory and Tools," Ph.D. Dissertation, Technische Universiteit Delft, Holland, 1994.

[141] Yae, H., and Inman, D.J., "Response Bounds for Linear Underdamped Systems," *ASME Journal of Applied Mechanics,* vol. 54, 1987, pp. 419–423.

[142] Yahagi, T., "Optimal Output Feedback Control in the Presence of Step Disturbances," *International Journal of Control*, vol. 26, 1977, pp. 753–762.

[143] Zhou, K., *Robust and Optimal Control*, Prentice Hall, Englewood Cliffs, NJ, 1995.

[144] Zimmerman, D.C, and Inman, D.J., "On the Nature of the Interaction Between Structures and Proof-Mass Actuators," *Journal of Guidance, Control, and Dynamics*, vol. 13, 1990, pp. 82–88.

Index

N

O

P

R

S

T

W

Mechanical Engineering Series *(continued from page ii)*

E.N. Kuznetsov, **Underconstrained Structural Systems**

P. Ladevèze, **Nonlinear Computational Structural Mechanics: New Approaches and Non-Incremental Methods of Calculation**

A. Lawrence, **Modern Inertial Technology: Navigation, Guidance, and Control, 2nd ed.**

R.A. Layton, **Principles of Analytical System Dynamics**

F.F. Ling, W.M. Lai, D.A. Lucca, **Fundamentals of Surface Mechanics: With Applications, 2nd ed.**

C.V. Madhusudana, **Thermal Contact Conductance**

D.P. Miannay, **Fracture Mechanics**

D.P. Miannay, **Time-Dependent Fracture Mechanics**

D.K. Miu, **Mechatronics: Electromechanics and Contromechanics**

D. Post, B. Han, and P. Ifju, **High Sensitivity Moiré: Experimental Analysis for Mechanics and Materials**

F.P. Rimrott, **Introductory Attitude Dynamics**

S.S. Sadhal, P.S. Ayyaswamy, and J.N. Chung, **Transport Phenomena with Drops and Bubbles**

A.A. Shabana, **Theory of Vibration: An Introduction, 2nd ed.**

A.A. Shabana, **Theory of Vibration: Discrete and Continuous Systems, 2nd ed.**